Experimental Brain Research Series 20

M. R. Klee H. D. Lux E.-J. Speckmann (Eds.)

Physiology, Pharmacology and Development of Epileptogenic Phenomena

With 97 Figures and 23 Tables

Springer-Verlag

Berlin Heidelberg New York
London Paris Tokyo
Hong Kong Barcelona
Budapest

Prof. Dr. med. Manfred Robert Klee
Max-Planck-Institut für Hirnforschung
Abteilung Neurophysiologie
Deutschordenstr. 46, W-6000 Frankfurt 71, FRG

Prof. Dr. med. Hans Dieter Lux
Max-Planck-Institut für Psychiatrie
Abteilung Neurophysiologie
Am Klopferspitz 18a, W-8033 Planegg-Martinsried, FRG

Prof. Dr. med. Erwin-Josef Speckmann
Institut für Physiologie
Universität Münster
Robert-Koch-Str. 27a, W-4400 Münster, FRG

ISBN-13:978-3-642-46734-9   e-ISBN-13:978-3-642-46732-5
DOI: 10.1007/978-3-642-46732-5

25/3140 543210 – Printed on acid-free paper

PREFACE

In July of 1989 the 31. International Congress of Physiological Sciences was held in Helsinki. This event in Finland gave us a chance to organize and maintain a satellite symposium on problems of epilepsy from July 4th to 8th, 1989, in Frankfurt-Main.

In order to reduce the number of satellite symposia outside of Finland, the president of the IUPS asked different national committees to organize joint meetings. Previously, Dr. Mareš and Dr. Rokyta had planned to organize a symposium on developmental aspects of epilepsy in Prague, while a meeting in Frankfurt had been planned as an update of a former satellite symposium on physiology and pharmacology of epileptogenic phenomena, held here in 1980. However we were able to combine our efforts by organizing one satellite symposium in Frankfurt entitled: 'PHYSIOLOGY, PHARMACOLOGY and DEVELOPMENT of EPILEPTOGENIC PHENOMENA'. Scientists are generally disappointed about the slow progress in knowledge of their field. We found that by comparing the outcome of the two symposia in 1980 and 1989 a remarkable progress in different, some completely new aspects of the field of experimental epileptogenesis could be realized. Also the new interest in the development of the brain contributed enormously to our understanding of the mechanisms underlying epileptogenic phenomena. We would like to thank all of our colleagues who accepted our invitation to present their recent data and for sending us their manuscripts.

The organizing committee would like to thank the German EEG society for its generous support of this meeting. We also wish to thank the Czechoslovak Physiological Society and the Max-Planck-Society for their financial help, as well as the following companies: Bayer AG, Wuppertal; Boehringer Ingelheim International GmbH, Ingelheim; CIBA-GEIGY GmbH, Wehr; Hoffmann-La Roche AG, Grenzach-Wyhlen; Knoll AG, Ludwigshafen; LABAZ GmbH, München; Merz & Co. GMBH, Frankfurt-M.; nbn Medizin-Elektronik GmbH, Herrsching, and Schering AG, Berlin. We want to thank the Springer-Verlag, Heidelberg, for publishing this book, especially Dr. T. Thiekötter and Mrs. S. Benko who co-ordinated the publication. Finally we are greatly indebted to Mrs. M. Duesmann for her supervision of the book, and to Dr. P. Germroth, Frankfurt-M., Dr. P. Boerrigter, Münster and Dr. P. Küster, München for their technical assistance.

Manfred R. Klee   H. Dieter Lux
Pavel Mareš   Richard Rokyta
Erwin-J. Speckmann

ACKNOWLEDGEMENT

We want to thank the Springer-Verlag, Heidelberg, for publishing this book, especially Dr. T. Thiekötter and Mrs. S. Benko who co-ordinated the publication. Finally we are greatly indebted to Mrs. M. Duesmann for her supervision of the book, and to Dr. P. Germroth, Frankfurt-M., Dr. P. Boerrigter, Münster and Dr. P. Küster, München for their technical assistance.

The Editors:

Manfred R. Klee   H. Dieter Lux

Erwin-J. Speckmann

CONTENTS

XI

LIST OF CONTRIBUTORS *

Abbott N.J. 35 [1]
Abdul-Ghani M. 99
Aitken P.G. 103
Albrecht D. 17
Allgeier H. 205
Altrup U. 61, 65
Artola A. 81
Baimbridge K.G. 75, 111
Baker R.E. 233
Batini C. 49
Belan P.V. 45
Bentivoglio M. 131
Bernášková K. 187
Berretta S. 49
Bigalke H. 217
Bingmann D. 95, 233
Binscheck T. 217
Bossu J.L. 53
Brailowsky S. 107
Braun M.S. 81
Brdička R. 123
ten Bruggencate G. 115
Cavalheiro E.A. 131
Chagnac-Amitai Y. 11
Champagnat J. 107
Connors B.W. 11
Corner M.A. 175
Coulter D.A. 39, 201
Cusimano G. 159
Czuczwar S.J. 255
Dupont J.L. 53
Erdélyi L. 27
Fagg G.E. 205
Fagni L. 53
Fehér O. 31
Feltz A. 53
Feltz P. 57
de Feo M.R. 159
Gasior M. 255
Grantyn R. 3
Haas H.L. 251
Hablitz J.J. 155
Hamann M. 57
Heckendorn R. 205
Heinemann U. 17

Holmes G.L. 143, 147, 151
Hoyer J. 119, 241
Huguenard J.R. 39, 201
Jaffe D.B. 103
Jeker A. 205
Jensen M.S. 71, 85
Jones R.S.G. 127
Kado R.T. 49
Kajiwara K. 259
Kato N. 81
Klancnik J.M. 111
Klebs K. 205
Klee M.R. 119, 223
Kleinrok Z. 255
Koch R. 217
Köhr G. 17
Konnerth A. 71
Kostyuk P.G. 45
Krishtal O.A. 89
Kubík V. 187
Kunz M. 123
Lambert J.D.C. 127
Lee W.L. 155
Lehmenkühler A. 23, 61, 65
Lindner K. 245
Löscher W. 191
Lücke A. 61
Luhmann H.J. 139
Lux H.D. 3
MacDonald J.F. 75
Madeja M. 65
Mareš J. 123, 163
Mareš P. 143, 147, 151, 167, 171, 179, 183
Marešová D. 167, 171
Marešová S. 167, 187
McLean M.J. 211
Mecarelli O. 159
Meiri H. 99
Menini C. 107
Misgeld U. 3, 7
Mody I. 75
Morain P. 49
Moshé S.L. 143, 147, 151
Nadler J.V. 103
Naquet R. 107

XIV

I BIOPHYSICAL MECHANISMS
OF EPILEPTOGENIC ACTIVITY

Do Proton-Activated Transient Sodium Currents Contribute to Paroxysmal Depolarization?

H. D. Lux, R. Grantyn and U. Misgeld

Abteilung Neurophysiologie, Max-Planck-Institut für Psychiatrie, Am Klopferspitz 18a, 8033 Planegg-Martinsried, FRG

Introduction

Neurons are known to be very sensitive to changes in pH levels. Alkaline shifts in the neuronal environment, as observed during stimulation-induced repetitive activity (Kraig et al. 1983), tend to produce an increase in neuronal excitability. Acidic shifts could be the consequence of an excess of strongly dissociated anions by the release of metabolically generated lactate (Kraig et al. 1983) or by the loss of cations (Na^+, Ca^{2+}) due to activated neuronal inward currents.

The consequence of a rapid decrease in extracellular pH is in marked contrast to the effect of slow acidification, which depresses neuronal activity. Experiments with dissociated cell cultures (Krishtal and Pidoplichko 1980; Konnerth et al. 1987; Grantyn and Lux 1988) showed that neurons react to a rapid increase in $[H^+]_0$ by strong depolarization (Fig. 1) due to a transient current that is predominantly carried by Na^+ ions. This Na current in response to a step change in $[H^+]_0$, termed $I_{Na(H)}$, is unaffected by tetrodotoxin but is blocked by inorganic Ca^{2+} channel blockers. In chick dorsal root ganglion (DRG) neurons, voltage-activated Ca^{2+} currents have been shown to be abolished during $I_{Na(H)}$ (Konnerth et al. 1987; Davies et al. 1988). $I_{Na(H)}$ is nearly fully activated at pH 6 under usual conditions of the external medium.

The effects of step changes in pH, as shown in Fig. 1, were recorded in hypothalamic neurons maintained in culture for 3 - 6 weeks. At this time, the neurons developed extensive synaptic connections and discharged spontaneously in a burstlike manner. The bursts are driven by synaptic activity when inhibition is blocked by picrotoxin (20 μM). Persistent step pH changes from 7.8 to 7.0 result in transient depolarizations resembling paroxysmal depolarization shifts (see Fig. 1), an

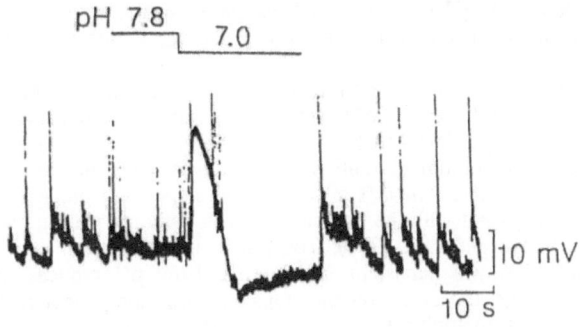

Fig. 1. Rapid transient depolarization of a cultured hypothalamic neuron of rat during a step change in pH. Subsequent hyperpolarizing depression of activity. During the pH change toward 7.8 and 7.0 (bars) by rapid superfusion, the external Ca concentration was set to 0.3 mM maintaining 4 mM $[Mg]_0$ to mimic the drop in $[Ca]_0$ usually observed during intense epileptic interictal paroxysm (see Heinemann et al. 1977). After 2 weeks invitro these neuronal cultures form a synaptic network and generate spontaneous burstlike activity (Misgeld and Swandulla 1989)

effect which is modified by external Ca concentration. With 1 mM external $[Ca^{2+}]$, the depolarization induced by the pH change amounts to 5 - 10 mV from membrane potentials of -80 mV.

The depolarization is accompanied by strong synaptic activity that triggers spikes. Decreasing the external $[Ca^{2+}]$ to 0.3 mM reduces the strength and frequency of the synaptically driven bursts, but synaptic activity is still evident. In this case, the step change in pH causes a transient depolarization of up to 60 mV. It initially triggers spikes, but subsequent inactivation often occurs, as may be concluded from the disappearance of the spiking activity. Spontaneous recovery, with a half-life of about 5, takes place at a maintained pH of 7.0. Voltage clamp recordings show currents similar to the proton-activated Na current described for other peripheral and central vertebrate neurons.

$I_{Na(H)}$ represents a property of many types of neurons and appears early in development

Our studies on $I_{Na(H)}$ were carried out on mature (Konnerth et al. 1987; Davies et al. 1988) and developing (Gottmann et al. 1989) DRG neurons and on a variety of cells from the mammalian central nervous system, including neurons from the rat retina tectum, hippocampus, neocortex, and oligodendrocytes (see Sontheimer et al. 1989). $I_{Na(H)}$ can be elicited in any of these neurons but not in rat cardiac myocytes and skeletal muscle fibers (Davies et al. 1988 and unpublished). The $I_{Na(H)}$ of central neurons decays more slowly and also has a significantly higher sensitivity to the blocking action of divalent cations than that of DRG neurons while the pH dependency of the activation and inactivation processes appears to be similar. In developing neurons, $I_{Na(H)}$ channels appear to be expressed ahead of Ca and Na channels (Gottmann et al. 1989; Grantyn et al. 1989) and ahead of channels activated by amino acids (Grantyn et al. 1989).

The activation of $I_{Na(H)}$ was found to be facilitated by slow alkalization and decreased levels of extracellular $[Ca^{2+}]_0$ and $[Mg^{2+}]_0$. In central neurons, Ca^{2+} appears to act as an $I_{Na(H)}$ channel blocker, the half-blocking concentration of Ca^{2+} being 1.25 mM, as observed by Grantyn et al. on tectal neurons. In the later cells and in hypothalamic neurons, $I_{Na(H)}$ was always markedly increased by reducing $[Ca^{2+}]_0$ below physiological levels. Reduction in $[Mg^{2+}]_0$ are similarly effective.

Rapid application of low Ca^{2+} solutions or NMDA activates a current similar to $I_{Na(H)}$

In view of the double effect of $[Ca^{2+}]$ on $I_{Na(H)}$, it could be expected that the largest current amplitudes would be elicited by combining a step change from to low $[Ca^{2+}]_0$ with a stepwise increase in $[H^+]_0$. This is, indeed, the case. Moreover, a sudden transition from high to low $[Ca^{2+}]_0$ in the absence of pH changes may activate a current similar to $I_{Na(H)}$ (Hablitz et al. 1986; Fig. 2C).

In the search for other factors that may activate currents similar to $I_{Na(H)}$ in the absence of larger shifts in $[H^+]_0$ and $[Ca^{2+}]_0$, we tested a number of transmitter substances, including excitatory amino acids. It was found that N-methyl-D-aspartate (NMDA) at a concentration of 100 - 200 µM induced transient sodium currents which decayed with the same time constant as $I_{Na(H)}$ (Fig. 2B). In the presence of 1.5 mM $[Ca^{2+}]_0$ a voltage-dependent block was seen with the persistent component of I_{NMDA} but not with the transient NMDA-activated current, which thus behaved similar to that of $I_{Na(H)}$ (Grantyn and Lux 1988). The transient I_{NMDA} is not evoked at low pH, also in analogy to I_{NMDA}, which shows persistent inactivation. Low pH reduces, to some extent, all ionic conductances. However, the effect on the transient $I_{Na(H)}$ component is far more pronounced than that on other currents, including the persistent NMDA-activated current (Fig. 2D). Finally, it was shown in rat tectal neurons that during $I_{Na(H)}$ NMDA failed to elicit a transient sodium current, a finding which could be explained by convergent actions on proton- and NMDA-receptor sites at the transduction system (Grantyn and Lux 1988). By contrast, full summation was displayed by the persistent NMDA-activated current and $I_{Na(H)}$.

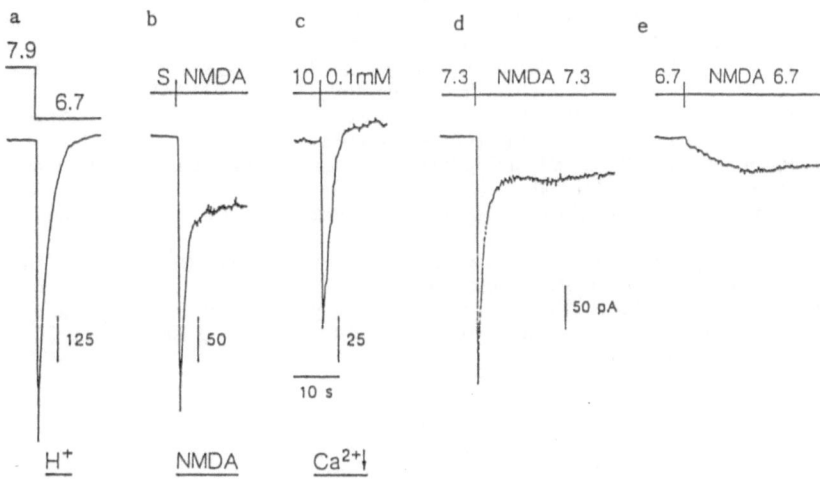

Fig. 2A-E. Configuration of transient sodium inward currents in rat mesencephalic neurons, as induced by a rapid pH decrease (**a**), application of 200 M NMDA (**b**) after standard saline *(S)* perfusion, and rapid decrease in extracellular $[Ca^{2+}]_0$ (**c**). Voltage-clamp recording from a cultured neuron of the rat superior collicullus. **d,e** Influence of pH on NMDA-activated currents in a rat tectal neuron. Note the complete inactivation of the transient NMDA-induced component at constant low pH (**e**)

Conclusion

The transient depolarizing effects of step changes in proton concentrations may bear some resemblance to paroxysmal depolarizations in ictal episodes. Particularly noteworthy is that epileptogenic conditions such as low $[Ca^{2+}]_0$ and extracellular alkaline shifts also favor the production of the proton-activated transient Na current. The increase of a glutamate-induced transient current component at high pH in the physiological range and its interaction with the proton-induced Na current may deserve attention in respect of the strong depolarization and excitability changes during synaptically generated paroxysmal events. A detailed discussion would be speculative, however. A release-related increase in proton concentration appears possible, but information on pH changes in situ within the time domain of synaptic and even paroxysmal events is not yet available.

Acknowledgements. The authors dedicate this paper to Professor Manfred Klee on the occasion of his 60th birthday.

References

Davies NW, Lux HD, Morad M (1988) Site and mechanism of activation of proton-induced sodium current in chick dorsal root ganglion neurones. J Physiol (Lond) 400: 159-187

Gottmann K, Dietzel ID, Lux HD, Ruedel C (1989) Proton-induced Na+ current develops prior to voltage-dependent Na+ and Ca2+ currents in neuronal precursor cells from chick dorsal root ganglion. Neurosci Lett 99: 90-94

Grantyn R, Lux HD (1988) Similarity and mutual exclusion of NMDA- and proton-activated transient Na+ currents in rat tectal neurons. Neurosci Lett 89: 198-203

Grantyn R, Perouansky M, Lux HD (1988) Proton-gated sodium rents in mammalian central neurons. Abst Neurosci Soc 14: 597

Grantyn R, Perouansky M, Rodriguez-Tebar A, Lux HD (1989) Expression of depolarizing voltage- and transmitter-activated currents in neuronal precursor cells from the rat brain is preceded by a proton activated sodium current. Dev Brain Res 49: 150-155

Hablitz JJ, Heinemann U, Lux HD (1986) Step reductions in extracellular Ca^{2+} activate a transient inward current in chick dorsal root ganglion cells. Biophys J 50: 753-757

Heinemann U, Lux HD, Gutnick MJ (1977) Extracellular free calcium and potassium during paroxysmal activity in the cerebral cortex of the cat. Exp Brain Res 27: 237-243

Konnerth A, Lux HD, Morad M (1987) Proton-induced transformation of calcium channel in chick dorsal root ganglion cells. J Physiol (Lond) 386: 603-633

Kraig RP, Ferreira-Filho CR, Nicholson C (1983) Alkaline and acid transients in cerebellar microenvironment. J Neurophysiol 49: 831-850

Krishtal OA, Pidoplichko VI (1980) A receptor for protons in the nerve cell membrane. Neuroscience 5: 2325-2327

Misgeld U, Swandulla D (1989) Quisqualate-receptor mediated rhythmic bursting of rat hypothalamic neurons in dissociated cell culture. Neurosci Lett 98: 291-296

Sontheimer H, Perouansky M, Hoppe D, Lux HD, Grantyn R, Kettenmann H (1989) Glial cells of the oligodendrocyte lineage express proton-activated Na$^+$ channels. J Neurosci Res 24: 496-500

Synaptic Synchronized Burst Activity
in a Hypothalamic Network

U. Misgeld* and D. Swandulla

Max-Planck-Institute Psychiatry, Department Neurophysiology, Am Klopferspitz 18A, 8033 Planegg-Martinsried, FRG
* Present address: I. Physiologisches Institut Universität Heidelberg, Im Neuenheimer Feld 326, 6900 Heidelberg, FRG

Introduction

Periodic events during which neurons fire a burst of action potentials in synchrony and then become silent can be observed after the blockade of GABA receptors by penicillin, bicuculline, or picrotoxin in a variety of mammalian CNS tissues. This type of activity has provoked considerable interest as a model for neuronal epileptic behavior, in particular for interictal spiking. However, synchronization of phasic activity in a neuronal network might also be required for it to play its functional role. For example, endocrine cells of the hypothalamus, for which phasic activity is characteristic, vary their degree of synchronization according to the changing needs for hormone secretion (Poulain and Wakerley 1982). We have succeeded (Misgeld and Swandulla 1989; Swandulla and Misgeld 1990) in growing embryonic rat hypothalamic cells in long-term dissociated culture in which the cells display burst discharge activity after blockade of GABA-ergic inhibition. The preparation allowed us to study a synaptic mechanism involved in the synchronization of burst activity.

Methods

Ventral hypothalamic tissue from the 14- to 15 day-old embryos as mechanically dissociated without prior enzymatic treatment. Cells were plated on a background layer of glial cells. The whole cell and cell-attached configurations (Hamill et al. 1981) were used to record membrane potentials and transmitter-activated currents under current clamp or voltage clamp, respectively, from cells 7 - 90 days in culture (DIC). Recordings were performed with a single electrode clamp amplifier (NPI, Tamm, FRG). Drugs dissolved in external solution were applied either by replacing the standard solution by a drug-containing solution in the culture dish or by superfusing the cell from a wide bore (tip diameter = 150 μm) 3 - 5 barrelled pipette (Carbone and Lux 1987) positioned 100 - 200 μm from the cell. The extracellular solution contained (in mM): 145 NaCl, 5 KCl, 5 $CaCL_2$, 2 $MgCl_2$, 15 - 30 glucose, 10 HEPES (pH 7.3). Recording pipettes were filled with (in mM): 10 NaCl, 100 KCl, 5 EGTA, 0.25 $CaCl_2$, 10 glucose, 10 HEPES (pH 7.3). Experiments were carried out at room temperature (20 - 22°C).

Results

Hypothalamic cells displayed two patterns of spontaneous synaptic activity. The first was characterized by large and prolonged depolarizations which occurred rhythmically and were superimposed by action potential discharges. The second pattern consisted of brief discrete depolarizations which persisted in TTX and in low calcium. Synaptic activity was comprised of GABAergic and excitatory synaptic potentials, since part of the activity was blocked by picrotoxin, whereas the remaining activity was blocked by the excitatory amino acid antagonists kynurenic acid (0.2 - 1 mM) and CNQX (10 μM) (Fig. 1A). GABA (5 - 10 μM) or muscimol (0.5 - 5 μM) induced inward currents in all cells tested. In contrast, cells 7 - 14 days in culture (DIC) either did

8

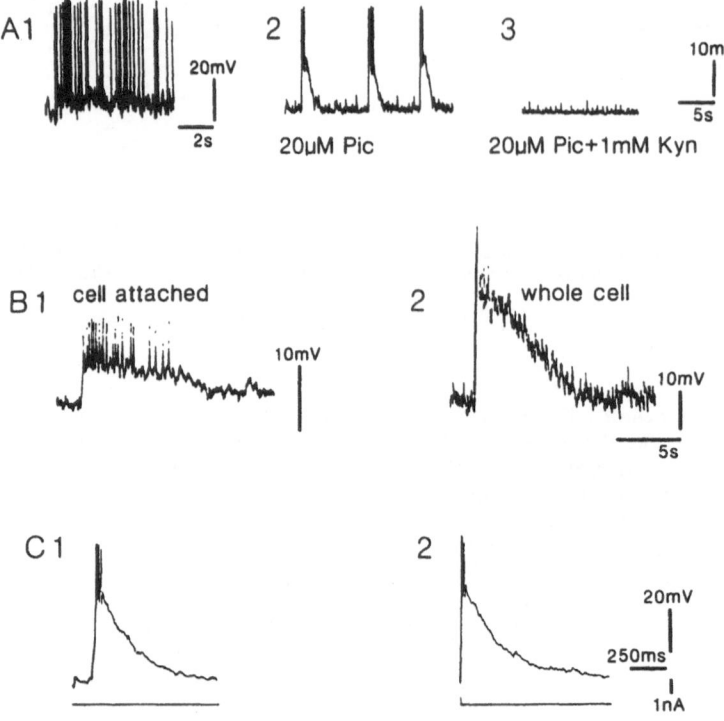

Fig. 1. A*1* Spontaneous activity in a culture dish 21 DIC; *2* Burst activity in the same culture dish after blockade of GABA$_A$ receptors. *3* Activity in the same culture dish after additional blockade of amino acids excitation. **B** Spontaneous burst activity in the presence of picrotoxin in a cell 41 DIC. **C** Spontaneous burst (*1*) and burst elicited by a 10 ms current pulse (*2*) in a single cell

not respond or responded only weakly to glutamate (100 μM), whereas cells > 21 DIC invariably responded to glutamate (100 μM) or quisqualate (0.1 - 10 μM). Quisqualate-induced current amplitudes steadily increased with increasing agonist concentration (0.1 - 0.5 μM). At higher concentrations (1 - 10 μM), quisqualate elicited biphasic currents (see also Tang et al. 1989) consisting of a fast and a sustained component, the latter decreasing with increasing concentrations. During the fast component, intense spiking was triggered in the current clamp.

In the presence of GABA antagonists picrotoxin (20 μM) or bicuculline (20 μM), the majority (61%) of neurons ≥ 21 DIC (n = 268) burst (Fig. 1). In normal solution, only a fraction of neurons burst, but the occurrence of bursts gradually increased with respect to time in culture: from 23% of cells 7 DIC (n = 16), 19% of cells 14 DIC (n = 51), and 23% of cells 21 DIC (n = 44) to 31% > 21 DIC (n = 95). The burst frequencies ranged from 0.07 - 0.32 Hz (mean frequency 0.18 Hz, n = 26). Hyperpolarizing current injections through the recording pipette, which prevented action potential firing, revealed the phasic occurrence of large excitatory postsynaptic potentials (EPSPs), the frequency of which was unaltered by hyper- or depolarizing current injections. In a very few (about 2%) cells > 21 DIC, all burst activity disappeared when the cell was hyperpolarized, suggesting that these cells were producing endogenous bursts. Eventually, two types of burst activity with different characteristics could be recorded simultaneously in a single cell (Fig. 2A). Whereas one disappeared with hyperpolarization of the membrane, the other one persisted (Fig. 2B). Under voltage clamp, EPSCs corresponding only to the latter type could be observed; slow current waves which could have generated the first type were not seen.

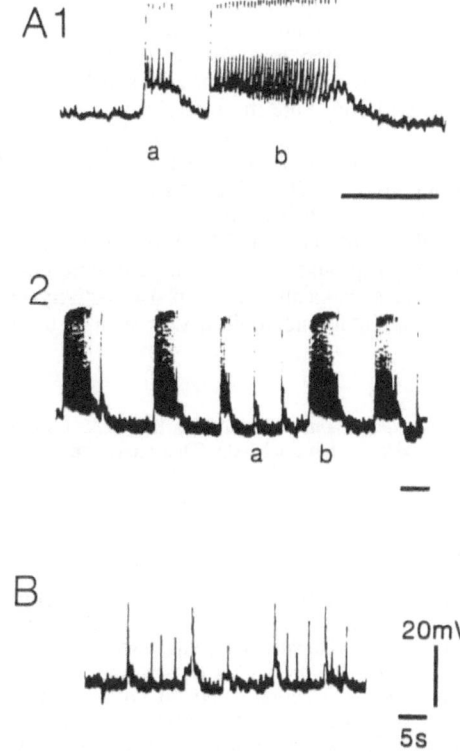

Fig. 2. A Two types (*a, b*) of spontaneous burst activity recorded at -50 mV. B At -64 mV only the *a* type of burst activity is seen

In a group of cells, injection of brief depolarizing current pulses (10 ms) elicited bursts which outlasted the activating pulse. The appearance of these bursts could not be distinguished from spontaneously occurring synaptically triggered bursts in the same cells (Fig. 1C). Under voltage clamp, short depolarizations (10 ms) to 0 mV elicited an inward current which decayed slowly. The time course of this current matched that of the depolarizing wave underlying driven and spontaneous bursts.

Burst activity was readily blocked by TTX (0.3 µM), the calcium antagonist cadmium (10 µM) and reduced calcium (1 mM). Kynurenic acid (0.2 - 1 mM), CNQX (10 µM) and low quisqualate concentrations (0.1 - 0.5 µM) blocked bursts reversibly; however APV and mecamylamine failed to do so. Reducing the number of cell to cell connections by isolating a small cell cluster from the surrounding culture abolished burst activity in this particular cell island, although the individual cells could be driven to discharge by current injection.

Discussion and Conclusion

In organotypic hypothalamic postnatal culture phasic activity is transmitted synaptically from pacemaker to follower cells after blockade of inhibition (Gähwiler and Dreifuss 1979). As shown here, a similar activity pattern develops in dissociated culture, and an excitatory amino acid, acting through a quisqualate receptor, transmits rhythmic activity. No evidence was found for the involvement of NMDA receptors, which seem to contribute to burst generation in organotypic hippocampal cultures (McBain et al. 1988). Endogenous pacemaker cells, present in organotypic cultures, were only rarely encountered in dissociated culture.

Synaptic activity developed progressively, and > 21 DIC cells were under an intense synaptic barrage. At this time, cells vigorously responded to quiqualate and glutamate. The observed increase in responsiveness to excitatory agonists was indicative of the progressive development of synaptic connections: The occurrence of synchronous bursts triggered from excitatory synaptic potentials increased with time in culture.

In conclusion, syncnronized burst activity is observed in a network formed by dissociated cells containing a small number of pacemaker cells connected through excitatory synapses to cells which can be driven to burst by current injection. The activity spreads through the entire network under conditions of reduced GABAergic inhibition. The strength of synaptic coupling might be due to collateral sprouting of axons in long term cultures (Frotscher and Gähwiler 1988). Indeed, we could observe synaptically driven phasic activity in the presence of picrotoxin in cultures derived from very different embryonic tissues such as the forebrain and the midbrain (unpublished observation).

Acknowledgements. The authors thank A. Lewen and H. Tyrlas for excellent technical assistance and E. Schroeder for editorial help. This study was supported by the Deutsche Forschungsgemeinschaft.

References

Carbone E, Lux HD (1987) Kinetics and selectively of a low-voltage-activated calcium current in chick and rat sensory neurons. J Physiol 386: 547-570

Frotscher M, Gähwiler BH (1988) Synaptic organization of intracellularly stained CA3 pyramidal neurons in slice cultures of rat hippocampus. Neuroscience 24: 541-551

Gähwiler BH, Dreifuss JJ (1979) Phasically firing neurons in long-term cultures of the rat hypothalamic supraoptic area: Pacemaker and follower cells. Brain Res 177: 95-103

Hamill PP, Marty A, Neher E, Sakmann B, Sigworth FJ (1981) Improved patch-clamp techniques for high resolution current recording from cells and cell-free membrane patches. Pflügers Arch 391: 85-100

McBain CJ, Boden O, Hill RG (1988) The kainate/quisqualate receptor antagonist, CNQX, blocks the fast component of spontaneous epileptiform activity in organotypic cultures of rat hippocampus. Neurosci Lett 93: 341-345

Misgeld U, Swandulla D (1989) Quisqualate-receptor mediated rhythmic bursting of rat hypothalamic neurons in dissociated cell culture. Neurosci Lett 98: 291-296

Poulain DA, Wakerley JB (1982) Electrophysiology of hypothalamic magnocellular neurons secreting oxytocin and vasopressin. Neuroscience 7: 773-808

Swandulla D, Misgeld U (1990) Synaptic mechanisms for fast synchronization of rhythmic bursting in a network of rat hypothalamic neurons grown in culture. J Neurophysiol (in press)

Tang CM, Dichter M, Morad M (1989) Quisqualate activates a rapidly inactivation high conductance ionic channel in hippocampal neurons. Science 243: 1474-1477

Synaptic Inhibition, Intrinsically Bursting Neurons and Synchronization in Neocortex

B. W. Connors[1] and Y. Chagnac-Amitai[2]

[1] Section of Neurobiology, Division of Biology and Medicine, Brown University, Providence, RI 02912 USA
[2] Department of Physiology, Faculty of Health Sciences, Ben-Gurion University of the Negev, Beer-Sheva, Israel

Introduction

Changes in the strength of synaptic inhibition have profound effects on the functions of the neocortex. Increasing its strength with drugs contributes to sedation and anesthesia (Olsen and Venter 1987), while decreasing its strength alters the sensory transformations of cortical circuits (Dykes et al. 1984; Sillito 1986), facilitates long-term plasticity of excitatory synapses (Artola and Singer 1987), and leads to the highly synchronized discharges of epileptic seizures (Dichter and Ayala 1987). The strongest system of inhibition in neocortex uses the neurotransmitter γ-aminobutyric acid (GABA) acting on postsynaptic $GABA_A$ receptors, which open chloride channels (Connors et al. 1988). GABA-utilizing synapses and $GABA_A$ receptors may be under various forms of regulatory control (Hendry and Jones 1986; Sigel and Baur 1988; Stelzer et al. 1988). The striking sensitivity of the cortex to the level of $GABA_A$-mediated inhibition, and the potential for regulation of inhibition, suggest that inhibitory efficacy may be locally or globally modulated during normal behavior.

We have studied the consequences of small reductions in $GABA_A$-mediated inhibition on the activity of isolated neocortical circuitry (Chagnac-Amitai and Connors 1989a,b), and the main results are reviewed here. The primary questions addressed were: How sensitive is the neocortex to disinhibition? What is the character of epileptiform activity in the presence of strong inhibition? What is the neuronal mechanism for the highly synchronized activity that occurs when inhibition is reduced?

Sensitivity of Neocortex to Disinhibition

Many studies have examined the effects of reducing inhibition in neocortex, but almost none have quantified the disinhibition necessary for a particular effect. The recording and stimulating arrangement is shown schematically in Fig. 1A. The $GABA_A$ antagonist bicuculline or bicuculline methiodide was added to the bathing medium to reduce the efficacy of $GABA_A$-mediated inhibition. The initial concentration was 0.1 μM, and this was raised in 0.1 μM increments.

Neocortex proved to be exquisitely sensitive to suppression of its $GABA_A$ function. In the control state evoked activity was sharply confined to a vertical strip of cortex no more than about 800 μm lateral to the stimulus site (Fig. 1B). However, at drug concentrations between 0.3 and 0.5 μM, the horizontal limits of evoked activity approximately doubled; at higher concentrations evoked epileptiform activity appeared.

The actions of bicuculline and its analogues have been very well studied. In vertebrate nervous systems they are competitive antagonists of the $GABA_A$ receptor, and the best estimates are that 1 μM concentrations reduce $GABA_A$-generated chloride currents or IPSPs by about 10% - 20% (e.g. Alger and Nicoll 1982; Gallagher et al. 1978; Yakushiji et al. 1987). Thus, even a small (<20%) depression of IPSPs can have a profound effect on cortical function.

A

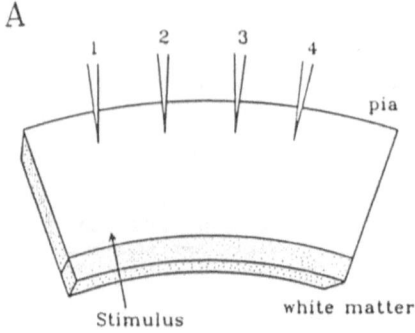

B

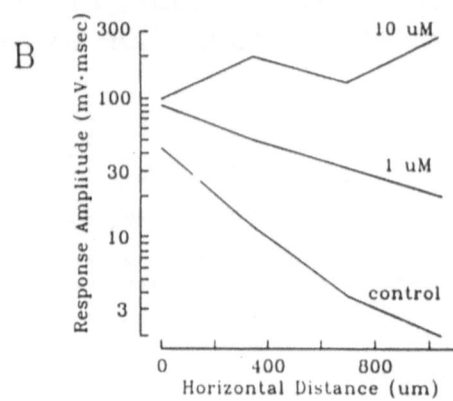

C

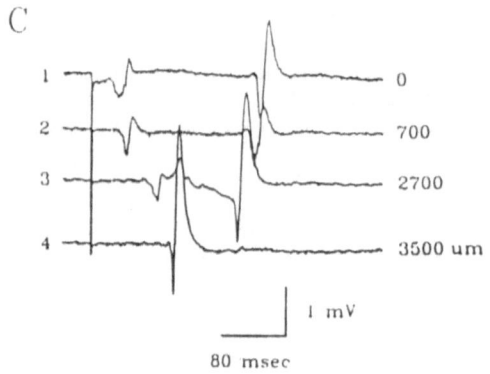

D

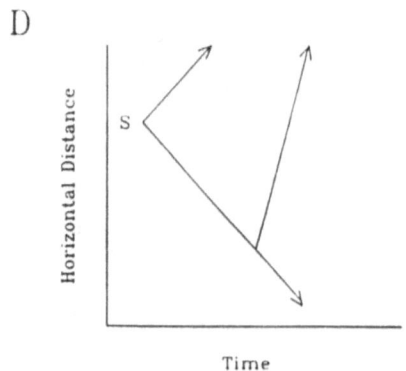

Fig. 1A-D. Low doses of GABA antagonists greatly increase the horizontal spread of activity in neocortex. **A** Experimental arrangement. Coronal slices of rat SI were stimulated in layer VI, and field potentials were simultaneously measured from four sites in layers II/III. **B** Under control conditions field potentials were largest at the site vertical to the stimulus (0 μm), and declined to baseline 800 μm horizontally. Addition of 1 μM bicuculline methiodide greatly expanded the horizontal spread from the same strength stimulus, and 10 μm caused epileptiform discharge that spread without decrement. **C** Epileptiform activity in the presence of 0.4 μM bicuculline. Single shock (vertical artifact) evoked a horizontally spreading primary event and a secondary event that was reflected back. Horizontal positions of each recording site marked to the right. **D** Schematized flow of activity recorded in C. S, Stimulus site. (Data in **B** from RD Chervin and BW Connors, unpublished; data in **C** and **D** from Chagnac-Amitai and Connors 1989a)

Epileptiform Activity in the Presence of Inhibition

When GABA$_A$ inhibition is virtually abolished, slices of mature neocortex generate interictal spikelike events (Gutnick et al. 1982; Connors 1984; Chervin et al. 1988). These consist of large discharges of synchronized synaptic excitation in all neurons of an isolated slice. The discharge latency is very variable, and horizontal propagation is unidirectional and reliable. However, it is likely that substantial inhibitory function is retained in most clinically epileptic cortex. To investigate the effect of inhibition, we established the antagonist concentrations that were threshold for epileptiform activity: 0.4 - 0.7 μM for bicuculline and 0.7 - 1.0 μM for bicuculline methiodide.

Threshold-dose epileptiform activity was quite different from that in high doses of antagonist. The field potentials were small and variable in amplitude (both from trial to trial and as they propagated across the cortex), they sometimes failed to propagate past specific sites in cortex, and they were often reflected back into the area of cortex from which they were initiated. Some of these characteristics are illustrated by the example shown in Fig. 1C,D. Apparently $GABA_A$-mediated inhibition plays a major role in sculpting the form and movement of epileptiform events.

Synchronized EPSPs and IPSPs

Neurons of the neocortex synchronize their activity under a variety of behavioral and pathological conditions. While input from the thalamus may impose some forms of synchrony upon the cortex (Steriade and Llinás 1988), other forms are clearly dependent upon neocortical circuitry alone. In isolated slices of cortex this is necessarily the case. Neurons of the neocortex are extremely diverse, and it would be valuable to know how different neuronal types contribute to synchronized behavior. Fortunately, there are interesting correlates between intrinsic neuronal physiology and morphology. Previous work has identified three major classes of intrinsic firing patterns in neurons of the neocortex (reviewed in Connors and Gutnick, 1990): *Regular-spiking* (RS) neurons are the majority; these have strongly adapting firing frequencies, and pyramidal cell morphology with somata in layers II to VI. *Intrinsically bursting* (IB) neurons are also pyramidal cells, but their somata are found only in layers IV and V; they generate a cluster of spikes (the burst) at threshold. RS and IB cells thus constitute the excitatory cells of the cortex. *Fast-spiking* (FS) neurons display very brief action potentials, high firing rates, and little or no adaptation of frequency during long stimuli; FS cells have been identified as GABAergic interneurons (McCormick et al. 1985; Huettner and Baughman 1988).

In low $GABA_A$ antagonist concentrations, synchronized population activity was signaled by large extracellular field potentials (see Fig. 1C), while single neuron activity was simultaneously measured with intracellular recordings. The three physiological classes of cortical neuron displayed distinctive patterns of synchronized synaptic and spike activity. In phase with each synchronous population event RS cells were primarily inhibited, while both IB cells and FS cells were strongly excited (Fig. 2). Inhibition of the RS cells was accomplished by synchronous barrages of both $GABA_A$- and $GABA_B$-mediated IPSPs (Connors et al. 1988), while excitation of all cells was driven by EPSPs. Thus, IB cells were the only excitatory cortical neurons generating substantial synchronized output, while the strong output of the FS cells presumably generated synchronous inhibition onto the RS cell majority.

Conclusions

Relatively small (<20%) changes in the efficacy of inhibition yield large changes in the horizontal spread of excitation. Propagating, synchronized epileptiform activity can occur even when the large majority of cortical inhibition is intact. Different types of neocortical neurons make different contributions to the synchronized activity: IB cells of the middle layers are the only neurons with excitatory output that are strongly activated; the majority of cortical pyramidal cells are synchronously inhibited. We suggest that a slight reduction in $GABA_A$ inhibition allows an interconnected network of IB cells to dominate cortical activity, and that other neurons are in turn synchronized via excitatory and inhibitory synaptic interactions. This mechanism may have significance for both epilepsy and certain forms of normal synchronous rhythms.

Acknowledgements. This work was supported by the Klingenstein Fund, and by grants NS-01271 and NS-25983 from the NIH.

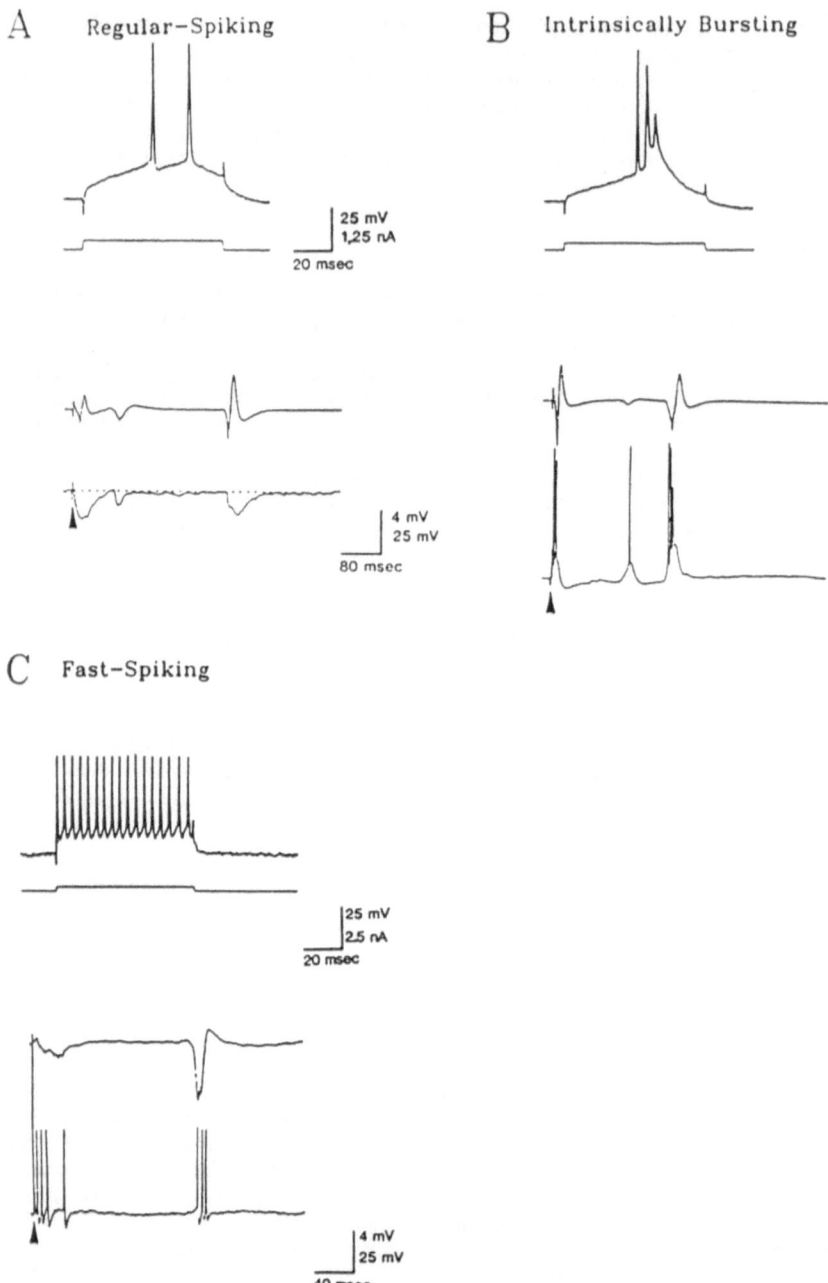

Fig. 2A-C. Correlation between synchronized synaptic activity and intrinsic membrane properties of individual neurons. *Top traces*, response of each neuron to an intracellular current pulse; *bottom traces*, simultaneous responses of the extracellular field potential and intracellular voltage. **A** Regular-spiking neurons showed IPSPs synchronized with population events. **B** Intrinsically bursting neurons, in contrast, showed strong, synchronized excitation. Neurons in **A** and **B** were both in layer V and were recorded sequentially from the same slice. **C** Fast-spiking neurons (presumed inhibitory interneurons) were also phasically excited. Bicuculline methiodide 0.9 µM. (Data from Chagnac-Amitai and Connors 1989b)

15

References

Alger BE, Nicoll RA (1982) Pharmacological evidence for two kinds of GABA receptor on rat hippocampal pyramidal cells studied in vitro. J Physiol (Lond) 328: 125-141
Artola A, Singer W (1987) Long-term potentiation and NMDA receptors in rat visual cortex. Nature 330: 649-652
Chervin RD, Pierce PA, Connors BW (1988) Periodicity and directionality in the propagation of epileptiform discharges across neocortex. J Neurophysiol 60: 1695-1713
Chagnac-Amitai Y, Connors BW (1989a) Horizontal spread of synchronized activity in neocortex, and its control by GABA-mediated inhibition. J Neurophysiol 61: 747-757
Chagnac-Amitai Y, Connors BW (1989b) Synchronized excitation and inhibition driven by intrinsically bursting neurons in neocortex. J Neurophysiol 62: 1149-1162
Connors BW (1984) Initiation of synchronized neuronal bursting in neocortex. Nature 310: 685-687
Connors BW, Gutnick MJ (1990) Intrinsic firing patterns of diverse neocortical neurons. Trends Neurosci 13: 99-104
Connors BW, Malenka RC, Silva LR (1988) Two inhibitory postsynaptic potentials, and GABA$_A$ and GABA$_B$ receptor-mediated responses in neocortex of rat and cat. J Physiol (Lond) 406: 443-468
Dichter MA, Ayala GF (1987) Cellular mechanisms of epilepsy: A status report. Science 237: 157-164
Dykes RW, Landry P, Metherate R, Hicks TP (1984) The functional role of GABA in cat primary somatosensory cortex: Shaping the receptive field of cortical neurons. J Neurophysiol 52: 1066-1093
Gallagher JP, Higashi, H, Nishi S (1978) Characterization and ionic basis of GABA-induced depolarizations recorded in vitro from cat primary afferent neurones. J Physiol (Lond) 275: 263-282
Gutnick MJ, Connors BW, Prince DA (1982) Mechanisms of neocortical epileptogenesis in vitro. J Neurophysiol 48: 1321-1335
Hendry SHC, Jones EG (1986) Reduction in number of immunostained GABAergic neurons in deprived-eye dominance columns of monkey area 17. Nature 320: 750-753
Huettner JE, Baughman RW (1988) The pharmacology of synapses formed by identified corticocollicular neurons in primary culture of rat visual cortex. J Neurosci 8: 160-175
McCormick DA, Connors BW, Lighthall JW, Prince DA (1985) Comparative electrophysiology of pyramidal and sparsely spiny neurons of the neocortex. J Neurophysiol 54: 782-806
Olsen RW, Venter JC (1987) Benzodiazepine/GABA receptors and chloride channels: Structural and functional properties. Liss, New York
Sigel E, Baur R (1988) Activation of protein kinase C differentially modulates neuronal Na$^+$, Ca^{2+}, and γ-aminobutyrate type A channels. Proc Natl Acad Sci USA 85: 6192-6196
Sillito AM (1986) Functional considerations of the operation of GABAergic inhibitory processes in the visual cortex. In Cerebral cortex. Vol. 2. Functional properties of cortical cells, Jones EG, Peters A, (eds) Plenum, New York, pp. 91-118
Stelzer A, Kay AR, Wong RKS (1988) GABA$_A$-receptor function in hippocampal cells is maintained by phosphorylation factors. Science 242: 339-341
Steriade M, Llinás RR (1988) The functional states of the thalamus and the associated neuronal interplay. Physiol Rev 68: 649-742
Yakushiji T, Tokutomi N, Akaike N, Carpenter DO (1987) Antagonists of GABA responses, studied using internally perfused frog dorsal root ganglion neurons. Neurosci 22: 1123-1133

Nonsynaptic Spread of Epileptiform Activity in Rat Hippocampal Slices

U. Heinemann, D. Albrecht, G. Köhr, G. Rausche, J. Stabel and T. Wisskirchen

Institut für Neurophysiologie, Zentrum für Physiologie und Pathophysiologie, Universität zu Köln, Robert-Koch-Str. 39, 5000 Köln 41, FRG

Introduction and Methods

It is well documented that lowering extracellular Ca^{2+} to levels below 0.3 mM induces seizure like events (SLE) characterized by 20 to 30 mV sustained depolarizations of hippocampal pyramidal cells (HPC) (Jefferys and Haas 1982; Konnerth et al. 1984). Extracellularly, these events are associated with up to 8 mV slow negative field potential (fp) shifts superimposed by prolonged bursts of population spikes. Typically the population spikes start some 500 to 800 ms after the onset of the negative events.

SLEs, apart from the $[Ca^{2+}]_o$, also depend on the $[K^+]_o$ (Konnerth et al. 1986; Yaari et al. 1986). They can only be induced when $[K^+]_o$ is above 4.5 mM. Interestingly, these events persist and even spread along the pyramidal cell (SP) layer in the absence of evocable synaptic transmission, typically migrating from area CA1a towards area CA3. Extracellular K^+ accumulation is probably involved in mediating the spread of SLEs, since increases in $[K^+]_o$ precede onset of SLEs at secondarily recruited sites. Indeed, direct stimulation of the SP induces only spreading events when $[K^+]_o$ is locally elevated above 5 mM. Moreover, artificial injection of K^+ can induce a spreading event. Finally, measures which block the sodium potassium pump such as application of ouabain can also induce SLEs.

The critical level for inducing an epileptiform event is about 1 mM above the already elevated baseline of 5 mM. With such a small $[K^+]_o$ elevation, membrane depolarization is typically only a few mV (Konnerth et al. 1986). This demands an amplifying mechanism which should be K^+ dependent and support the seizure spread. Here, we have investigated the hypothesis that the role of glial cells in K^+ homeostasis provides such an amplifying mechanism.

Glial cells are involved in K^+ homeostasis in three ways. First they are able to take up K^+ and Cl^- by one or more cotransport mechanisms (Bourke and Nelson 1972). Second, the glial Na-K-ATPase is, like the neuronal ATPase, very easily activated by extracellular K^+ accumulation and thus provides K^+ uptake into glia in exchange for Na^+. Third, glial cells form a more or less widespread syncytium which permits spatial K^+ buffering by which K^+ is redistributed from sites of maximal neuronal activity to remote inactive places (Orkand et al. 1966; Kettenmann et al. 1982). Spatial K^+ buffering depends primarily on the fact that the membrane potential of glial cells is almost exclusively determined by the transmembrane K^+ concentration gradient (Kettenmann et al. 1983). At sites where K^+ accumulates very strongly, glial cells should depolarize. The induced depolarization would spread through the glial syncytium. The charge transfer would lead to a slightly less strong depolarization of glial cells at sites of maximal K^+ accumulation and to an overly large depolarization of glial cells at remote sites. This would provide a driving force for K^+ uptake into glia at sites of maximal extracellular K^+ accumulation and for K^+ release from glia at remote sites. Associated with the K^+ accumulation are negative fp changes at the site of maximal K^+ accumulation and positive fps at sites of K^+ release (Dietzel et al. 1980, 1982; Somjen 1973). Shrinkage of the extracellular space at sites of maximal K^+ accumulation and widening of the extracellular space at remote sites (Dietzel et al. 1980, 1982) may also be associated with such a

18

mechanism. We assumed that spatial K⁺ buffering could be the amplifying mechanism for the generation of the SLEs. This would be the case if a large part of the slow negative fps in the hippocampal SP layer are due to spatial K⁺ buffering. The negative fps in hippocampal SP would then add to the transmembrane potential of nerve cells and provide an additional depolarization.

Results and Discussion

In order to obtain evidence for this hypothesis, we measured the laminar profile of extracellular slow fps in relation to rises in $[K^+]_o$ (Heinemann et al. 1983; Albrecht et al. to be published). Maximal elevations in $[K^+]_o$ were found in SP of area CA1, both during ortho- and antidromic stimulation in normal medium. Blocking synaptic transmission resulted in a large decrease in $[K^+]_o$ elevations in SP and in a smaller reduction of antidromically evoked increases in $[K^+]_o$. Whereas the site of maximal K⁺ accumulation shifted under conditions of blocked synaptic transmission from SP into stratum radiatum (SR), the site of maximal K⁺ accumulation remained in SP during antidromic stimulation. Elevations in $[K^+]_o$ associated with SLEs were also maximal in SP (up to 5 mM). These rises in $[K^+]_o$ were accompanied by negative fps in SP of up to 8 mV and by positive fps in SR and stratum moleculare (SM). Similar large negative fps also accompanied rises in $[K^+]_o$ induced by antidromic stimulation in normal as well as in low Ca^{2+} medium in SP whereas the slow fps were positive in SR and SM.

If spatial buffering contributed to the rise in $[K^+]_o$ preceding SLEs, it is expected that this would occur in all directions. We therefore investigated whether elevations in $[K^+]_o$ were also seen in dentate gyrus, which is not innervated by area CA1 (Albrecht et al. 1988). Simultaneous recordings in the dentate gyrus revealed, indeed, that positive fps occurred synchronously with the slow negative fps in SP of area CA1. These positive slow fps in dentate gyrus were even accompanied by rises in $[K^+]_o$ in the leaflet of dentate gyrus neighboring area CA1 (Fig. 1). We suggest that

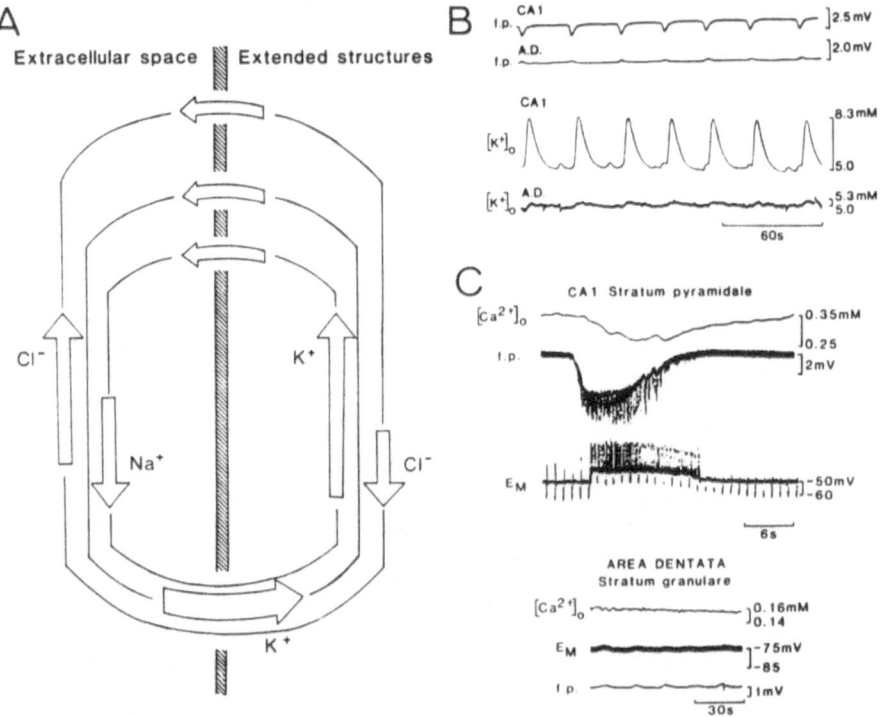

Fig. 1. A schematic drawing of the spatial K⁺ buffering mechanism through glial cells. For details, see text. B Simultaneous recording of changes in $[K^+]_o$ and fp in area CA1 and dentate gyrus. C Intracellular recordings during low Ca^{2+} seizures para to nearby extracellular recordings of $[Ca^{2+}]_o$ and fp

these rises in $[K^+]_o$ result from spatial K^+ buffering across the fissure. This conclusion is based on two findings: First, whereas pyramidal cells in area CA1 displayed pronounced depolarizations during SLEs, dentate gyrus granule cells usually remained silent. Second, the rise and decay times of $[K^+]_o$ elevations in SM and sometimes even in stratum granulare of dentate gyrus were too fast to be accounted for by diffusion alone (Cordingly and Somjen 1978).

These findings also support the idea that a good part of the slow negative fps in SP of area CA1 are due to spatial K^+ buffering through glial cells. However, transfer of actively generated depolarizations into dendrites would cause release of K^+ in SR/SM associated with positive fps, whereas the inward currents underlying depolarizations of SP somata would account for their negative fps. This would imply that dendrites behave passively during SLEs. In order to test this hypothesis, we measured the decreases in $[Ca^{2+}]_o$ and $[Na^+]_o$ along the dendritic tree of SP (Fig. 2).

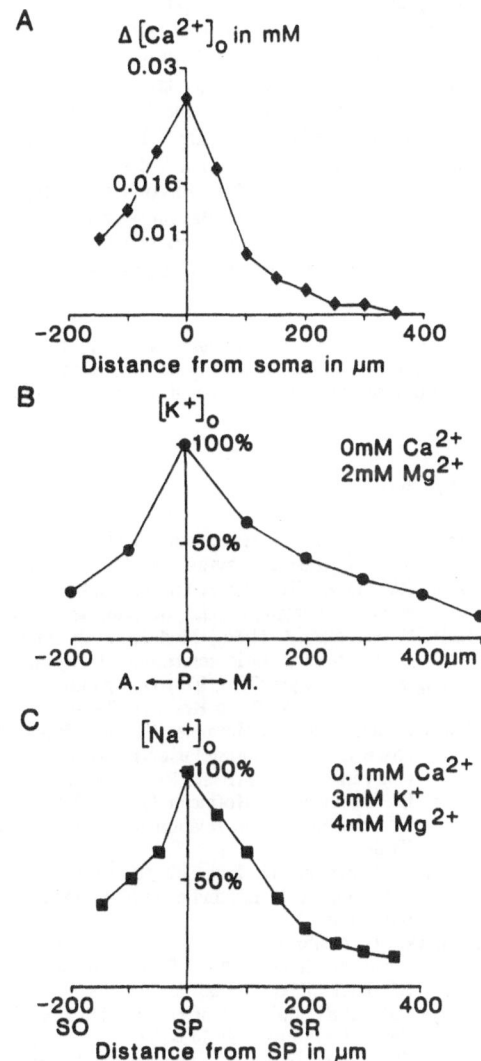

Fig. 2. Laminar profiles of changes in Ca^{2+}, K^+ and Na^+ during low Ca^{2+} seizures. Note that the rise in $[K^+]_o$ continues across the fissure; this is not the case for changes in $[Na]_o$ and $[Ca_2^+]_o$

We found that decreases in $[Ca^{2+}]_o$ and in $[Na^+]_o$ accompanied SLEs. The largest decreases were observed in SP, but notable decreases of $[Na^+]_o$ and $[Ca^{2+}]_o$ were still seen at distances of 400 µm above the SP. This would indicate that the dendrites of hippocampal SP actively participate in epileptogenesis. This conclusion is also supported by findings with local administration of adenosine. Various groups had shown that adenosine is a powerful anticonvulsant that acts through A1 receptors. Local administration of adenosine showed that the site, where adenosine most easily suppresses SLEs is about 150 µm from SP in SR (Lee et al. 1984). Active participation of dendrites with pronounced inward currents would also generate negative fps in SR. Hence, a positive fp appearing in SR and SM cannot result from passive depolarization from apical dendrites and must be ascribed to spatial K^+ buffering.

If this were so, transmembrane currents underlying slow negative fps should be very large. Therefore, we determined the sum of transmembrane currents in SP by performing a current source density analysis of the slow fps in area CA1 during low Ca^{2+} seizures and antidromic stimulation in the three spatial dimensions (see also Dietzel et al. 1989). The voltage gradients amounted to values of up to 40 mV/mm. This value is far above that needed for cell excitation in dentate gyrus granule cells (Jefferys 1981) and in cerebellar Purkinje cells (Chan and Nicholson 1986). Assuming a volume fraction for the SP layer of 7%, this would imply transmembrane currents of the order of 0.3 mA/mm³ tissue (Heinemann et al. 1983). Based on reasonable estimates of the number of SP cells in such a volume, one arrives at the conclusion that the individual cellular currents are too large for the observed depolarizations in SP. Hence, a good part of the transmembrane currents in the hippocampal SP layer are probably due to spatial K^+ buffering currents. These would contribute to SLEs by adding to the transmembrane potential of SP. In addition, the spatial redistribution of K^+ would support the seizure spread.

Acknowledgements. This research was supported by DFG grant He 1128/2-4 and by the SFB 200, Teilprojekt C 8. We gratefully acknowledge the technical assistance of M. Groenenwald and G. Heske in performing the experiments and in preparing the manuscript.

References

Albrecht D, Rausche G, Heinemann U (1989) Reflections of low calcium epileptiform activity from area CA1 into dentate gyrus in the rat hippocampal slice, Brain Res 480: 393-396

Bourke RS, Nelson KM (1972) Further studies on the K^+-dependent swelling of primate cerebral cortex in vivo: the enzymatic basis of chloride. J Neurochem 19: 663-685

Chan CY, Nicholson C (1986) Modulation by applied electrical fields of Purkinje and Stellate cell activity in the isolated turtle cerebellum. J Physiol (Lond) 371: 89-114

Cordingley GE, Somjen GG (1978) Dissipation of locally accumulated extracellular potassium in the cat cerebral cortex. Brain Res 151: 291-306

Dietzel I, Heinemann U, Hofmeier G, Lux HD (1980) Transient changes in the size of the extracellular space in the sensorimotor cortex of cats in relation to stimulus induced changes in potassium concentration. Exp Brain Res 40: 432-439

Dietzel I, Heinemann U, Hofmeier G, Lux HD (1982) Stimulus induced changes in extracellular Na^+ and Cl^- concentration in relation to changes in the size of the extracellular space. Exp Brain Res 46: 73-84

Dietzel I, Heinemann U, Lux HD (1989) Relations between slow extracellular potential changes, glial potassium buffering, and electrolyte and cellular volume changes during neuronal hyperactivity in cat brain. Glia 2: 25-44

Heinemann U, Neuhaus S, Neuhaus I, Dietzel I (1983) Aspects of K^+ regulation in normal and gliotic brain tissue. In: Cerebral blood flow, metabolism and epilepsy. Baldy-Moulinier MJ, Ingvar DH, Meldrum BS (eds) Libbey, Paris, pp 271-278

Jefferys JGR (1981) Influence of electric fields on the excitability of granule cells in guinea-pig hippocampal slices. J Physiol 319: 143-152

Jefferys JGR, Haas HL (1982) Synchronized bursting of CA1 hippocampal pyramidal cells in the absence of synaptic Transmission. Nature 300: 448-450

Kettenmann H, Orkand RK, Schachner M (1982) Coupling among identified cells in nervous system cultures. J Neurosci 3: 506-516

Kettenmann H, Sonnhof U, Schachner M (1983) Exclusive K$^+$ dependence of the membrane potential in cultured oligodendrocytes. J Neurosci 3: 500-505

Konnerth A, Heinemann U, Yaari Y (1984) Slow transmission of neural activity in hippocampal area CA1 in absence of active chemical synapses. Nature 307: 69-71

Konnerth A, Heinemann U, Yaari Y (1986) Nonsynaptic epileptogenesis in the mammalian hippocampus in vitro I. Development of seizure like activity in low extracellular calcium. J Neurophysiol 56: 409-423

Lee KS, Schubert P, Heinemann U (1984) The anticonvulsive action of adenosine: a postsynaptic, dendritic action by a possible endogenous anticonvulsant. Brain Res 321: 160-164

Orkand RK, Nicholls JG, Kuffler SW (1966) Effect of nerve impulses on the membrane potential of glial cells in the central nervous system of amphibia. J Neurophysiol 29: 788-806

Somjen GG (1973) Electrogenesis of sustained potentials. Prog Neurobiol 1: 199-237

Yaari Y, Konnerth A, Heinemann U (1986) Nonsynaptic epileptogenesis of the mammalian hippocampus in vitro. II. Role of extracellular potassium. J Neurophysiol 56: 424-438

Interictal Discharges: Changes in Size of Extracellular Space in Relation to Changes in Extracellular K+ and Na+ Concentration

A. Lehmenkühler and A. Richter

Institut für Physiologie, Universität Münster, Robert-Koch-Straße 27a, 4400 Münster, FRG

Introduction

Epileptic seizure activity is associated with changes in extracellular ion concentrations in the brain (for reviews see Caspers et al. 1987; Lux et al. 1986). Using iontophoresis of extracellular probe ions in the sensorimotor cortex Dietzel and Heinemann (1986) found decreases of up to 30% in the extracellular space (ECS) during seizure. They proposed that shrinkage of the ECS is a consequence of the removal of extracellular K^+ by glial cells (Dietzel and Heinemann 1986). The processes, however, responsible for the release of K^+ into the ECS are not well understood. In the present experiments, a fast decrease in the size of the ECS during interictal discharges within a penicillin-induced epileptic focus is described. This observation led to the hypothesis that seizure-related rises in $[K^+]_o$ are not only due to neuronal depolarization, but also due to a swelling-induced exit of K^+ (see Falke and Misler 1989; Sackin 1989; Sigurdson and Morris 1989).

Methods

The experiments were carried out on rats weighing between 350 and 400 g. The rats were anesthetized with urethane (750 mg/kg injected i.p.), paralyzed with suxamethonium chloride (20 mg/kg/h) and artificially ventilated with oxygen. The motor cortex was superfused with artificial cerebrospinal fluid containing penicillin (25 mmol/l) in exchange for an equimolar amount of chloride. This procedure induced an epileptic focus involving all cortical layers (Lehmenkühler and Richter 1988). For measurements of fluctuations of the size of the ECS, tetramethylammonium ion (TMA^+) was used as an extracellular marker and added to the superfusate (Lehmenkühler et al. 1985; Nicholson and Rice 1988). The experiments started when a standing gradient of extracellular TMA^+ concentration ($[TMA^+]_o$) in the cortex was achieved. Recordings of field potentials, $[K^+]_o$, $[Na^+]_o$, $[TMA^+]_o$ were made with triple-barreled ion-selective microelectrodes (Lehmenkühler and Richter 1989).

Results and Discussion

ECS and $[K^+]_o$ Profile
Changes in $[TMA^+]_o$ were restricted to cortical depths between 1000 and 2000 μm (Fig. 1B). In these layers, increases in $[TMA^+]_o$, most pronounced at a depth of 1500 μm, were observed. They showed a fast rise time ranging between 90 and 130 ms and represented a rapid decrease in the extracellular volume of up to 8%. (Fig.2).
In comparison with K^+ signals simultaneously recorded at the same depths (Fig. 1B), the TMA^+ signal reached its maximum during the first half of the ascending phase of the K^+ signal. $[TMA^+]_o$ had already returned to baseline shortly after the maximum increase in $[K^+]_o$. Maximal increases in $[K^+]_o$ and $[TMA^+]_o$ were not located at the same depth (Fig. 1B). Furthermore, the slopes of the ascending and descending phases of the K^+ signals were not related to the amplitude of the TMA^+ signals (Fig. 1B).

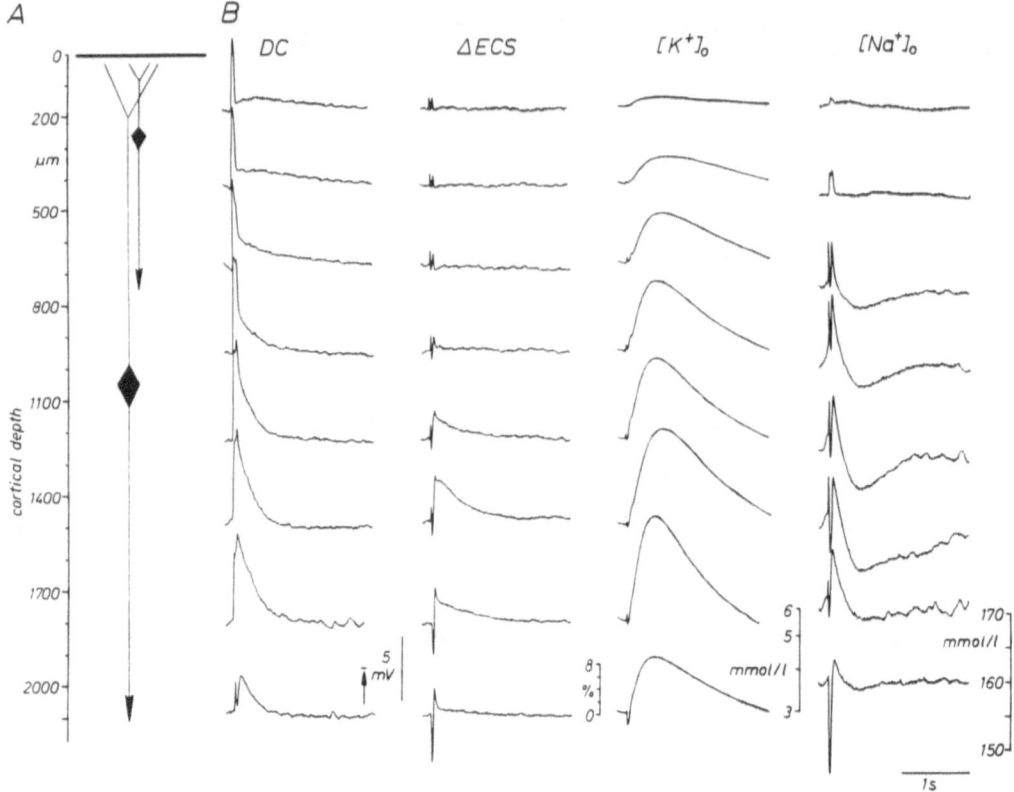

Fig. 1A,B. Records from an acute experimental penicillin focus in the motor cortex of a rat. **A** Schematic drawing of the cortical laminae. **B** Registrations of penicillin-induced DC field potentials, of changes in the size of the extracellular space (*ECS*), of extracellular K^+ concentration ($[K^+]_o$) and of extracellular Na^+ concentration ($[Na^+]_o$). The recording sites are related to the scheme in **A**. Position of calibration scales refers to the bottom traces

The question arose as to whether the relatively lower response time of the K^+-selective electrode, based on valinomycin, could be responsible for this observation. Therefore, in control experiments $[K^+]_o$ profiles were also recorded with K^+-selective microelectrodes containing the classical Corning exchanger. In such experiments the superfusate did not contain TMA^+. The time course of the K^+ signals were found to be independent of the type of electrodes used. Thus, the longer response time of the K^+-selective electrode based on valinomycin does not explain the different time courses of the $[TMA^+]_o$- and $[K^+]_o$ recordings.

$[Na^+]_o$ Profile
The behavior of $[Na^+]_o$ during interictal epileptic discharges was studied, since it has often been assumed that an influx of sodium and chloride may cause swelling of neurons during excitation processes.
The laminar distribution of changes in $[Na^+]_o$ is displayed in Fig. 1B. The Na^+ signal is composed of different components. The first component is an artifact due to incomplete subtraction of the bioelectrical DC field potential from the reference barrel. The second component represents an increase and the third one a decrease in $[Na^+]_o$ (Lehmenkühler et al. 1982). The decrease in $Na[^+]_o$ was observed in cortical depths of 600 to 2000 μm, whereas significant changes in

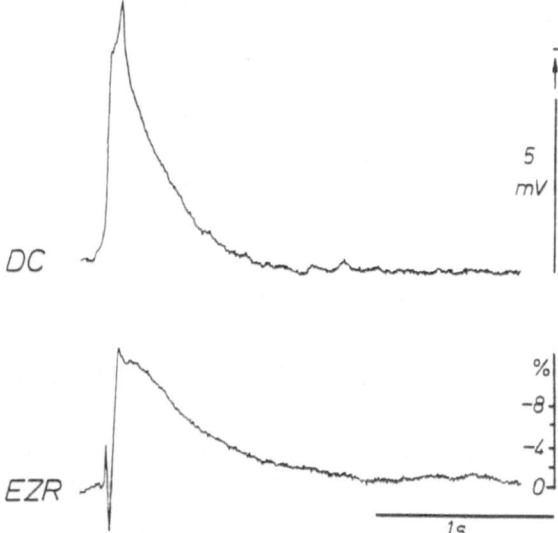

Fig.2. Simultaneous recordings of the field potential (*DC*) and of the relative change in the size of the extracellular space (*EZR*) during interictal seizure activity in the rat motor cortex. Recording depth below pial surface: 1500 μm. (Reprinted with permission Lehmenkühler 1988)

$[TMA^+]_0$ occurred only in depths of ca. 1000 to 2000 μm. The decline in $[Na^+]_0$ was about three times greater in amplitude than the concomitant increases in $[K^+]_0$ and reached its maximum of 8 mM at a depth of 1500 μm and at a time when the shrinkage of the ECS had recovered almost completely.

Hypothetical Mechanisms Underlying the Shrinkage of ECS
These observations might be interpreted in the following way: The transmembrane currents that occur during an interictal spike cause a primary increase in the number of intracellular particles and thus an increase in osmotic pressure in the intracellular space. This leads to swelling of the cells and consequently to a stretch of the membranes. The increase in membrane tension increases the open probability of stretch-activated K^+ channels, as described previously (Falke and Misler 1989; Sackin 1989; Sigurdson and Morris,1989). A subsequent exit of K^+ down its electrochemical gradient would re-decrease cell volume by a concomitant exit of an anion, probably bicarbonate (Lehmenkühler and Richter 1989), and by osmotic water movement into the ECS. Recovery of the ECS from shrinkage is, however, disturbed by K^+-carrying spatial buffering currents and by water movement across glial membranes (Dietzel and Heinemann 1986). Since an influx of Na^+ is still taking place when the ECS has already recovered from shrinkage, an influx of sodium, together with chloride, is probably not the primary cause of cell swelling.

References

Caspers H, Speckmann E-J, Lehmenkühler A (1987) DC potentials of the cerebral cortex. Seizure activity and changes in gas pressures. Rev Physiol Biochem Pharmacol 106: 127-178
Dietzel I, Heinemann U (1986) Dynamic variations of the brain cell microenvironment in relation to neuronal hyperactivity. Ann NY Acad Sci 481: 72-84
Falke LC, Misler S (1989) Activity of ion channels during volume regulation by clonal N1E115 neuroblastoma cells. Proc Natl Acad Sci USA 86: 3919-3923

Lehmenkühler A (1988) Änderungen des Mikromilieus von Nervenzellen in der Hirnrinde bei epilep tischen Anfällen - Experimentelle Beobachtungen. EEG-Labor 10: 145-161

Lehmenkühler A, Caspers H, Kersting U (1985) Relations between DC potentials, extracellular ion activities and extracellular volume fraction in the cerebral cortex with changes in pCO_2. In: Kessler M, Harrison DK, Höper J (eds) Ion measurements in physiology and medicine. Springer, Berlin Heidelberg New York, pp 199-205

Lehmenkühler A, Richter A (1988) Fokale epileptische Aktivität: Zeitliche Beziehungen zwischen Änderungen des extrazellulären Volumens und der extrazellulären K^+- und Na^+ Konzentration. In: Speckmann E-J, Palm DG (eds) Epilepsie 87. Einhorn, Reinbek, pp 298-302

Lehmenkühler A, Richter A (1989) Cellular water uptake in relation to extracellular K^+ and Na^+ concentration shifts during focal epileptic discharges. In: Höper J, Kessler M, Harrison DK (eds) Theory and application of ion-selective electrodes in biology and medicine. Springer, Berlin Heidelberg New York, in press

Lehmenkühler A, Zidek W, Caspers H (1982) Changes of extracellular Na^+ and Cl^- activity in the brain cortex during seizure discharges. In: Klee MR, Lux HD, Speckmann E-J (eds) Physiology and pharmacology of epileptogenic phenomena. Raven, New York, pp 37-45

Lux HD, Heinemann U, Dietzel I (1986) Ionic changes and alterations in the size of the extra cellular space during epileptic activity. In: Delgado-Escueta AV, Ward AA Jr, Woodbury DM, Porter RJ (eds) Basic mechanisms of the epilepsies, molecular and cellular approaches. Raven, New York, pp 619-639

Nicholson C, Rice ME (1988) Use of ion-selective microelectrodes and voltammetric microsensors to study brain cell microenvironment. In: Boulton AA, Baker GB, Walz W (eds) Neuromethods; the neuronal microenvironment. Humana, Clifton, pp 247-361

Sackin H (1989) A stretch-activated K^+ channel sensitive to cell volume. Proc Natl Acad Sci USA 86: 1731-1735

Sigurdson WJ, Morris CE (1989) Stretch-activated ion channels in growth cones of snail neurons. J Neurosci 9: 2801-2808

Convulsant Hydroxybenzenes Differentially Modulate A-Currents in Molluscan Neurons

L. Erdélyi

Department of Comparative Physiology, Jozsef Attila University, H-6701 Szeged, P.O.Box 533, Hungary

Introduction

Convulsants induce complex electrophysiological effects in neuronal membranes. Their action includes suppression of the A-currents (Erdélyi and Such 1988; Ito and Maeno 1986; Williamson and Crill 1976), which in turn leads to a decrease in the interspike interval and may evoke abnormal neuronal discharges (Connor and Stevens 1971; Neher 1971). At the same time, synaptic efficacy may also be modified as a consequence of modulation of A-currents (Daut 1973). Some phenol derivatives are known to be convulsants in both vertebrates and invertebrates (for a review, see (Kaila 1982). For example, catechol was recently reported to be a potent inhibitor of A-currents in frog sensory neurons (Ito and Maeno 1986) and *Helix* ganglion cells (Erdélyi and Such 1988).

Materials and Methods

Experiments were performed on identified and some unidentified neurons of the land snail, *Helix pomatia L*. Left and right parietal neurons (LPa and RPa 1, 2, 3), right parietal bursting cells and surrounding pacemaker cells were used most often. The double pulse method was applied to study the activation and inactivation of the A-currents. Detailed methods concerning physiological preparation, electrophysiological recording and solutions were published previously (Erdélyi 1987). Evoked postsynaptic potentials (PSPs) were recorded after stimulation of one of the main nerves (anal, intestinal, left and right parietal) by the use of teflon-coated silver wire pairs (10 V, 0.1 ms). Compounds were dissolved in physiological solution and administered in perfusate at room temperature (22° - 25°C).

Results

4-Cl-phenol depolarized the neuronal membrane and moderately prolonged the action potential duration (APD) (Fig. 1). The action of 4-Cl-phenol on the APD was a time-and dose-dependent event, as can be seen in Fig. 1 A. A convulsant-like action of 4-Cl-phenol is shown in Fig. 1B. 4-Cl-phenol reversibly depolarized this beating pacemaker neuron and induced spike doublets, triplets and sometimes paroxysmal depolarizing shifts 5 - 6 min after application. Excitatory synaptic bombardment of the neuron increased during 4-Cl-phenol exposure, which served as a trigger for burst-like transformation of the spike generation. The direct membrane effect of 4-Cl-phenol was highly dependent on the synaptic input organization of the neuron studied. When the presynaptic influence of 4-Cl-phenol facilitated inhibitory transmitter liberation, an anticonvulsant-like influence of the compound predominated. This is the situation with an RPa 1 burster, as can be seen in Fig. 1C during a typical experiment.

The direct membrane effect of 4-Cl-phenol under voltage clamp is closely related to the suppression of the fast potassium current, with much less influence of the compound on other voltage-gated

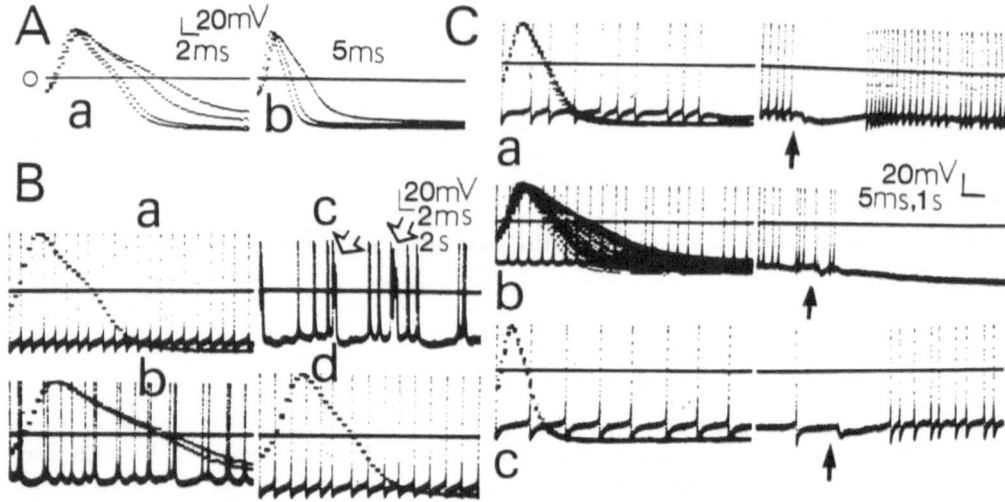

Fig. 1. A Gradual prolongation of the APD induced by 1.6 (a) and 3.2 (b) mM 4-Cl-phenol; 0, 2, 4, and 6 min after application in a and 0, 1, and 3 min after application in b. B 4-Cl-phenol (1.6 mM) depolarized the neuronal membrane, prolonged the APD, and induced abnormal neuronal discharges. a control; b and c, 5 and 6 min after 4-Cl-phenol application; d recovery after 25 min washing. C 4-Cl-phenol potentiates the stimulus-evoked inhibitory postsynaptic response in an RPa 1 burster. a control activity; b activity in 4-Cl-phenol (1 mM, 1 min); c partial recovery after 27 min washing. *Open arrows*: paroxysmal depolarizing shifts; *Closed arrows*: single pulse stimulation of the left parietal nerve

membrane ionic currents. Figure 2A shows the effects on A-currents in an LPa 1 neuron. It can be seen that 4-Cl-phenol decreased the amplitude and increased the rate of activation and inactivation of A-currents. The voltage-dependence of the steady state activation and inactivation of A-currents was not significantly influenced, however (not shown). The action of 4-Cl-phenol on A-currents was dose-dependent. Figure 2B shows the dose-response relationship derived from three different experiments ($K_d = 0.7$ +/- 0.2 mM; mean +/- SD). A semilogarithmic plot of the relative amplitude of A-currents after peak vs time is shown in the inset. It is clear that the single exponential character of A-current decay did not change during 4-Cl-phenol exposure, but the time constant of decay decreased. Phenol acted similarly to 4-Cl-phenol but proved less potent ($K_d = 5$ mM). The benzene ring was necessary for the effectivity of the molecule, as cyclohexanol (10 mM, 15 min) was ineffective with respect to A-currents. It seems that a hydroxy-substituted benzene ring is necessary for the observed effects. 4-OH-phenol (hydroquinone) and with decreasing potency 3-OH-phenol (resorcinol) and 2-OH-phenol (catechol) decreased the amplitude and the rate of activation and inactivation of A-currents. The corresponding thiophenols were as active as the phenol derivatives.

Discussion

4-Cl-phenol induced dual effects on the neurons studied. First, it suppressed A-currents in a dose-dependent way ($K_d = 0.7$ mM) and moderately prolonged the APD. Convulsant phenols have been reported to produce an inhibition of voltage-dependent potassium conductance and a dose-dependent prolongation of the APD in the medial giant axon of the crayfish (Kaila and Saarikoski 1980), which reflects the differences and similarities of the actions found in two neuronal preparations and species. Nevertheless, similar to other convulsants (Connor and Stevens 1971), 4-Cl-phenol may have the ability to modify potassium conductance in a nonspecific way. This may be the root of its presynaptic mode of action in various preparations (Kaila 1982). Second, potentiated transmit-

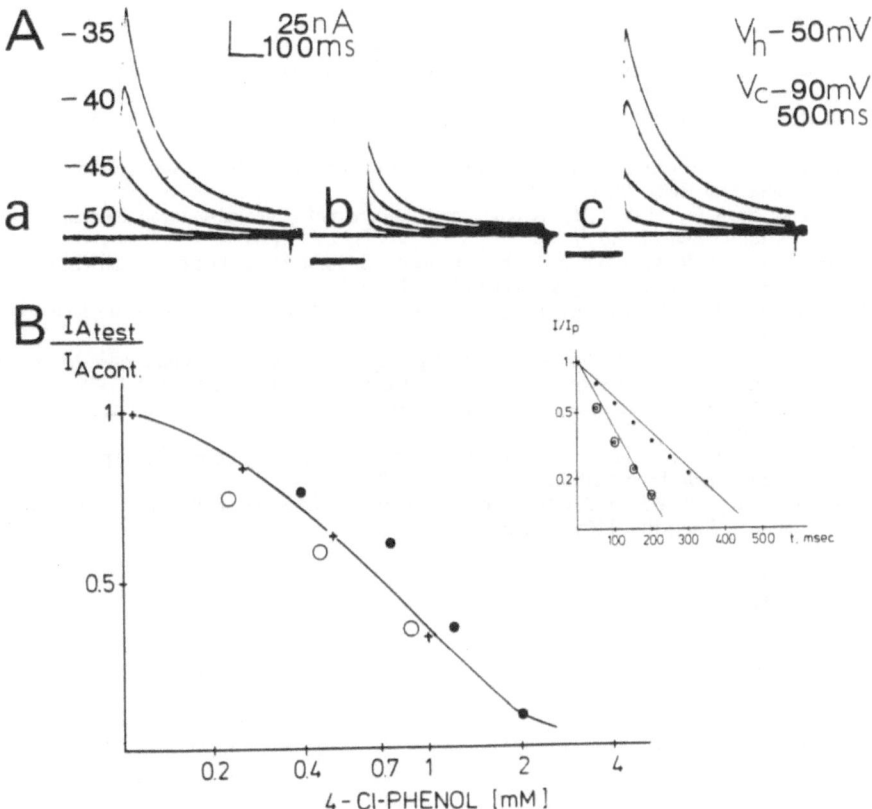

Fig. 2. A 4-Cl-phenol (1.2 mM, 3 min) reversibly suppressed A-currents in an LPa 1 neuron. **a** control A-currents; **b** A-currents in the test solution; **c** recovery after 10 min washing. **B** Semilogarithmic plot of relative amplitude of A-currents vs 4-Cl-phenol dose in mM. The symbols relate to three different estimations (K_d = 0.7 mM). The inset shows a semilogarithmic plot of A-current amplitude after the peak versus time under control circumstances (*closed circle*) or in the presence of 0.8 mM 4-Cl phenol (*open circle*)

ter liberation can increase or decrease the neuronal excitability, depending on the transmitters involved and the excitatory or inhibitory input organization of the neurons, as demonstrated in two types of *Helix* neurons. Convulsant-like and anticonvulsant-like electrophysiological actions of 4-Cl-phenol could be recorded in different neurons.

The structure-activity analysis showed that the aromatic ring is a very important part of the molecule since cyclohexanol was ineffective in suppressing A-currents. In contrast to cyclohexanol, hexanol has been reported to be the most effective of the alkanols studied on A-currents in three identified neurons of Aplysia (Treitsman and Wilson 1987). Our experiments showed that, in ring systems substituted with -OH or -SH groups, aromaticity of the ring was necessary for activity, and further substituents could differentially modulate the amplitude and kinetics of A-currents. It is suggested that the modulatory actions of convulsant phenol derivatives on potassium currents and on transmitter liberation processes form the basis of the mechanisms of their convulsion-inducing ability.

References

Connor JA, Stevens CF (1971) Prediction of repetitive firing behavior from voltage clamp data on an isolated neurone soma. J Physiol (Lond) 213: 31-53

Daut J (1973) Modulation of the excitatory synaptic response by fast transient K^+ currents in snail neurons. Nature New Biol 246: 193-196

Erdélyi L (1987) Effects of extracellular Ca and Ca-channel blockers on A-currents in snail brain neurons. Acta Biol Hung 38: 299-314

Erdélyi L, Such Gy (1988) The A-type potassium current: catechol-induced blockage in snail neurons. Neurosci Lett 92: 46-51

Ito I, Maeno T (1986) Catechol: a potent and specific inhibitor of the fast potassium channel in frog primary efferent neurones. J Physiol (Lond) 373: 115-127

Kaila K (1982) Cellular neurophysiological effects of phenol derivatives. Comp Biochem Physiol 73C: 231-241

Kaila K, Saarikoski J (1980) Inhibition of voltage-dependent potassium conductance by convulsant phenols in the medial giant axon of the crayfish. Comp Biochem Physiol 65C: 17-24

Neher E (1971) Two fast transient current components during voltage clamp on snail neurons. Gen Physiol 58: 36-53

Treitsman SN, Wilson A (1987) Alkanol effects on early potassium currents in Aplysia neurons depend on chain length. Proc Natl Acad Sci USA 84: 9299-9303

Williamson TL, Crill WE (1976) The effects of pentylenetetrazol on molluscan neurones. II. Voltage clamp studies. Brain Res 116: 231-249

Analysis of the Slow Inward Current Induced by Pentylenetetrazol

A. Papp and O. Fehér

Department of Comparative Physiology, Jozsef Attila University, H-6701 Szeged, POB. 533, Hungary

Introduction

It is generally accepted that paroxysmal depolarization shifts (PDSs) are the most important phenomena in chemically induced (and, possibly, naturally occurring) epileptic activity (Ducreux and Gola 1975; Matsumoto and Ajmone-Marsan 1969; Speckmann and Caspers 1973). Generation of PDSs is most probably explained by assuming that chemical convulsants, such as pentylenetetrazol (PTZ), induce an abnormal, slow inward current (SIC). Membrane current phenomena, induced by PTZ, have been found and described by several authors (Ducreux and Gola 1975; Walden et al. 1988). Characteristic changes in intracellular free calcium and calcium distribution following administration of PTZ are also well-documented (Sugaya et al. 1987). Previously, we described the effects of PTZ on normal membrane parameters and currents in central neurons of *Helix pomatia* (Fehér et al. 1987) and some properties of the SIC with respect to ionic dependence and sensitivity to blocking agents (Papp et al. 1987). The experiments detailed below focussed on the effect of organic Ca antagonists and antiepileptic drugs on the SIC and on possible parallels between effects of various treatments on SIC and fast inward current (FIC).

Materials and Methods

Details of the applied methods have been given elsewhere (Papp et al. 1987). In short, the connective tissue sheath was peeled off the subesophageal ganglion group and the latter was fixed in a Sylgard-plated chamber continually superfused with Ringer's solution (in mM): NaCl 80; KCl 4; $CaCl_2$ 7; $MgCl_2$ 5; Tris-Cl 5. The pH was 7.4, at room temperature. The drugs applied were: PTZ 50 mM, tetraethylammonium-Br (TEA) 30 mM, tetrodotoxin (TTX) 10 µM, phenobarbital 5 mM, diphenylhydantoin (DPH) 1 mM, verapamil 1 mM, and diltiazem 1 mM.

Voltage clamping was performed using a single electrode device. Neurons were impaled with a microelectrode of 2 - 7 MΩ resistance, filled with 1:1 M KCl : K citrate. Slow ramp as well as square pulses were used as commands. Ramps were used for directly obtaining current - voltage characteristics on an X-Y plotter. Membrane currents evoked by square pulses were visualized on a storage scope screen and photographed.

Results

A great majority of *Helix* neurons were sensitive to PTZ. On application of 50 mM PTZ the I-V curves underwent characteristic changes. At 15 - 25 mV depolarization a negative slope region appeared, corresponding to a slowly inactivating inward current. Simultaneous administration of 30 mM TEA depressed the slow K currents and the SIC could be seen free of interferences.

The SIC was found to be sensitive to various blocking agents. Although it proved to be resistant to 10 µM TTX (Papp et al. 1987), it was readily depressed by Ca channel blocking cations like

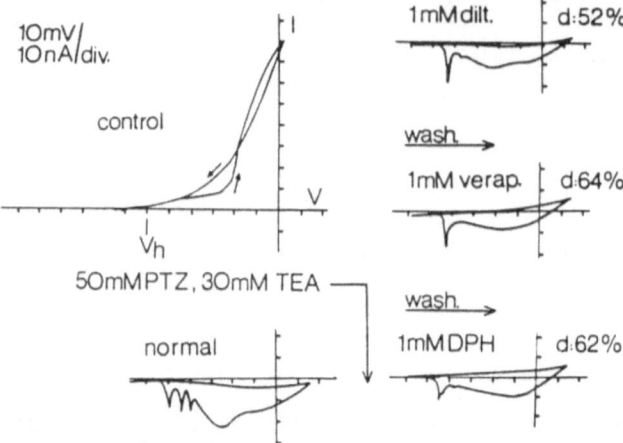

Fig. 1. Current-voltage characteristics of neuron RPa 77, obtained with ramp commands ranging from -60 to +20 mV with 20 mV/s steepness. The SIC induced by PTZ+TEA (*normal*) was reduced by 1 mM diltiazem, verapamil, and diphenylhydantoin

Ni^{2+}, Co^{2+}, Mn^{2+} (8 - 15 mM). Thus, supported by findings of others (Walden et al. 1988), it was plausible to test the effects of organic Ca antagonists on the SIC induced by PTZ. Fig. 1 displays I-V curves of the neuron RPa77 (numbered after Kerkut et al. 1975). On administration of PTZ+TEA the I-V curve showed changes indicating negative slope resistance (first and second curves, vertically). The maximum of the SIC was at about -20 mV.

Treatment with 1 mM diltiazem greatly reduced the SIC. The intensity of depression (d) is given as a percent, calculated from the areas enclosed by the voltage axis and the upstroke portion of the I-V curve. A similar effect was obtained with 1 mM verapamil.

Effects of several antiepileptic drugs were then examined on the SIC and compared to those of Ca antagonists. Using the same neuron, a substantial depression of the SIC was obtained with 1 mM DPH. On another neuron, 5 mM phenobarbital, also a known antiepileptic, was applied and led to the same depression as DPH. It could be seen that different drugs, known either as Ca antagonists or antiepileptics, exert the same effect on the SIC. They do not shift the maximum point of SIC on the voltage axis and the depression caused by them is stronger with higher depolarizations.

In a series of experiments the pharmacological responsiveness and ionic dependence of the SIC and the FIC (underlying spikes) were compared. Fig. 2 shows current records (A) and I-V plots (B) based on those (see legend). Both currents proved sensitive to Ni^{2+}, verapamil and Na removal. In Na-free Ringer, the FIC completely disappeared and the SIC was reduced strongly, but not fully. Changes in the currents using Ca-channel blockers also showed some differences between the SIC and the FIC. Ni^{2+} (8 mM) caused a total block of the SIC whereas a small fraction of the FIC persisted. Reduction of both currents by verapamil was unequivocal. In other experiments both currents proved resistant to 10 μM TTX. Records of the SIC in Fig. 2A suggest that currents remaining after Ni^{2+} and verapamil treatment are kinetically different. When blocked by Ni^{2+}, the SIC became more constant in time and its initial, steep phase disappeared. This unmasked a rapidly inactivating outward current with a relatively high activation threshold. Under influence of verapamil, however, the initial phase was unchanged, but the plateau inactivation increased. These kinetic features need further investigation.

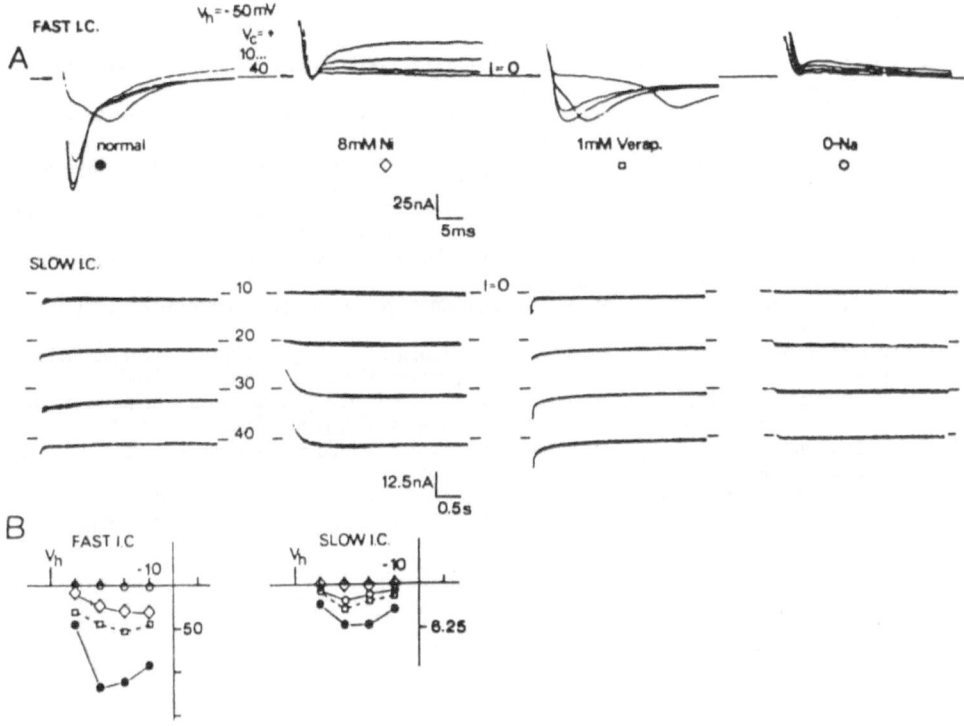

Fig. 2. A Records of FIC (*upper line*) and SIC (*lower lines*) in the presence of 50 mM PTZ and 30 mM TEA. Zero current level, holding and command potentials and calibration are given in the insets. Effects of Na removal and two different Ca channel blockers are shown. **B** Current-voltage curves plotted on the basis of currents in part A. For FIC, the peak current and for SIC, the current level at 1 s after pulse onset, was plotted against the actual membrane potential

Discussion

PDS has been considered as a general mechanism by which chemical - and probably other - convulsants exert their action (Ducreux and Gola 1975; Matsumoto and Ajmone-Marsan 1969; Speckmann and Caspers 1973). In invertebrates PDS is probably a non-synaptic event (Faugier-Grimaud 1974; Speckmann and Caspers 1973) in which voltage-gated currents play a central role.

Effects of various channel blockers on SIC, the current underlying PDSs have been investigated for a long time (Witte et al. 1987). In accordance with several authors we found that inorganic and organic Ca channel blockers can greatly reduce the SIC. We also compared the effects of the blocking agents on the SIC and the natural FIC. Based on the similarities discovered, one can propose that channels transmitting SIC are a subtype of the channels of the FIC, modified by PTZ. This channel subtype in *Helix* is relatively slow and permeable mostly to Ca. On administration of PTZ, these channels become slowed down, predominantly in their inactivation process, and become permeable for Na (and other cations). The strong and long-lasting inward current of the PDS would thus be carried mainly by Na ions through channels which are still sensitive to Ca channel blockers.

Two drugs known in mammals as antiepileptics were found also to be effective in *Helix* on PTZ-induced convulsive activity. Their effect was similar to that of organic Ca antagonists, except that they left the FIC intact. Thus, the mode and/or site of their action must be different.

34

References

Ducreux C, Gola M (1975) Ondes paroxysmales induites par le metrazol (PTZ) sur les neurones d'Helix p.: Modele fonctionnel. Pflügers Arch 361: 43-53
Faugier-Grimaud S (1974) Extrasynaptic mechanisms of cardiazol-induced epileptiform activity of invertebrate neurons. Brain Res 69: 354-360
Fehér O, Erdélyi L, Papp A (1987) The effect of pentylenetetrazol on the metacerebral neuron of Helix pomatia. Gen Physiol Biophys 7: 505-516
Kerkut GA, Lambert JDC, Gayton RJ, Loker JE, Walker RJ (1975) Mapping of nerve cells in the suboesophageal ganglia of Helix aspersa. Comp Biochem Phisiol 50A: 1-26
Matsumoto H, Ajmone-Marsan C (1969) Cortical cellular phenomena in experimental epilepsy: Interictal manifestations. Exp Neurol 9: 286-304
Papp A, Fehér O, Erdélyi L (1987) The ionic mechanism of pentylenetetrazol convulsions. Acta Biol Hung 38: 349-362
Speckmann EJ, Caspers H (1973) Paroxysmal depolarization and changes in action potentials induced by pentylenetetrazol in isolated neurons of Helix pomatia. Epilepsia (Amst) 14: 397-408
Sugaya E, Furuichi H, Takagi T, Kijiwara K, Komatsuhara J (1987) Intracellular calcium concentration during pentylenetetrazol-induced bursting activity in snail neurons. Brain Res 416: 183-186
Walden J, Speckmann EJ, Witte OW (1988) Membrane currents induced by pentylenetetrazol in identified neurons of Helix pomatia. Brain Res 473: 294-305
Witte OW, Walden J, Speckmann EJ (1987) Antiepileptic effects of calcium antagonists in animal experiments. In: The Epileptic Focus. Wieser HG, Speckmann EJ, Engel J (eds) Libbey, London, pp 194-207

Potassium Homeostasis in the Extracellular Space Surrounding Squid Axons during Repetitive Activity

Y. Pichon [1] and N. J. Abbott [2]

[1] Département de Biophysique, Laboratoire de NBCM du CNRS, F-91198 Gif sur Yvette CEDEX, France
[2] Department of Physiology, King's College, Strand, London WC2R 2LS, UK

Introduction

It is generally believed that nerve activity in the giant axons of the squid is associated with a substantial accumulation of potassium ions around the axons. This belief relies on a small number of experiments, most often on damaged animals. We have decided to reexamine this phenomenom first in situ, using the small squid species: *Alloteuthis subulata*, then in vitro on both *Alloteuthis subulata* and *Loligo forbesi*. The main findings are summarized in this paper.

Methods

Fresh specimens of *Alloteuthis* or of *Loligo* were caught in the vicinity of the Marine Laboratories (Plymouth for the former, Roscoff for the latter) and kept alive in circulating sea water. The animals used in the present experiments were selected for their good physiological state. They were decapitated and, for in vitro experiments, the two larger giant axons dissected from the mantle. The membrane potential of the giant axons was generally recorded using high impedance capillary microelectrodes filled with 2.5 M KCl. The axon was stimulated electrically to produce bursts of action potentials. The frequency of these bursts was adjusted to avoid sodium inactivation. The resting potential as well as the action potential (spike and afterpotentials) were monitored continuously on an oscilloscope and a chart-pen recorder and recorded on tape for off-line analysis. They were simultaneously digitized on 12 bits using a 'Data Translation DT2801-A' data acquisition board and stored on the hard disk of a PC compatible microcomputer. In other experiments, the giant axons of *Loligo forbesi* were voltage-clamped using the conventional voltage-clamp system (Pichon 1981) and changes in the extraaxonal potassium concentrations analyzed from the tail currents. In these experiments a 'Hewlett Packard 9000' series based data acquisition system (HP DASYS) was used to stimulate the axon and collect the data. Fast automatic analysis of the digitized data was performed using a program named analysis (Pichon and Pichon 1989) and the resulting data fitted using 'Enzfitter' (Elsevier Biosoft).

Results

In Situ Recordings
Impalement of the axons in situ proved relatively easy. The mean resting potentials and spike amplitudes were not significantly different from those recorded in vitro; however, the negative afterpotential remained constant during repetitive stimulation, indicating the absence of potassium accumulation.

In Vitro Recordings
In the early studies of the isolated *Alloteuthis* axon, the afterhyperpolarization was found to decrease almost continuously during the experiment. Repetitive stimulation under these conditions was associated with a sharp exponential-like decay in the afterhyperpolarization due to a large

A   B

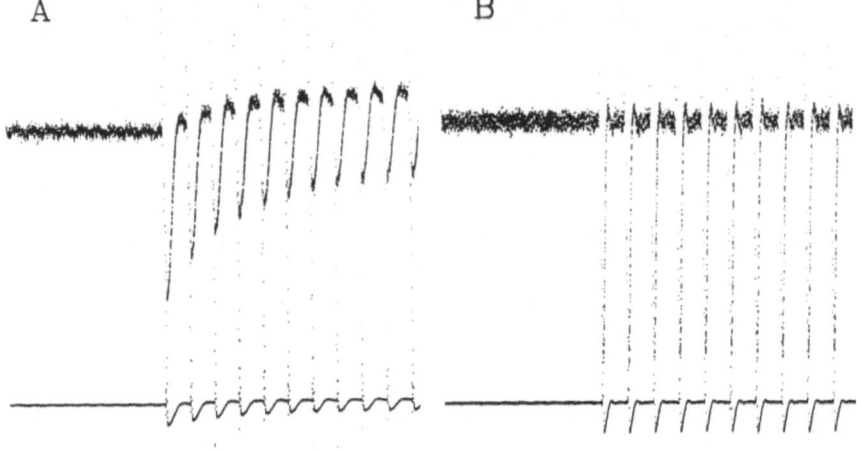

Fig. 1A,B. Time course of the positive afterpotential decay in a strongly accumulating (A) and a non-accumulating (B) isolated giant axon of *Alloteuthis*. Horizontal bar: 100 ms; Vertical bar: 6.25 mV (upper traces) or 50 mV (lower traces)

potassium accumulation in the extracellular space (Fig.1A). Part of this effect was due to the high temperature of the sea water used during the dissection. The critical temperature was found to lie around 12°C. In later studies, the preparation was maintained at a cooler temperature, and the preparations were stable and exhibited very little accumulation (Fig.1B), even during extended stimulation periods (several tens of seconds).

Quantitative Analysis
To investigate the mechanisms involved in potassium homeostasis, we analyzed quantitatively the effects of short and long bursts of action potentials on the afterhyperpolarization. Various treatments, including changes in the osmotic concentration of the saline, changes in temperature, changes in ionic concentrations or application of specific blockers, i.e. ouabain (blocker of the ATP driven Na/K pump) or bumetamide (blocker of the Na/K,Cl cotransport system), were used to try to dissociate the different components of the regulatory mechanisms.

The time course of the drop in the afterhyperpolarization amplitude during a long (5-10 s) stimulation period could not be fitted, as proposed earlier, with a single exponential. Two to three exponential functions were usually needed to fit the curve with good accuracy (Fig.2). The fast time constant was of the order of 100 ms, the medium time constant of the order of 3 s and the slow time constant of 10 s more. The respective size of each component was found to change following various treatments.

The correspondence between these components, the geometry of the system as revealed by recent EM studies and the physiological processes, remains speculative. Whereas the faster component (which was the only one used in previous studies) is most probably associated with passive diffusion of ions in the extracellular space and the extensive glial tubular system, the two other components might be related to the activity of the transport systems.

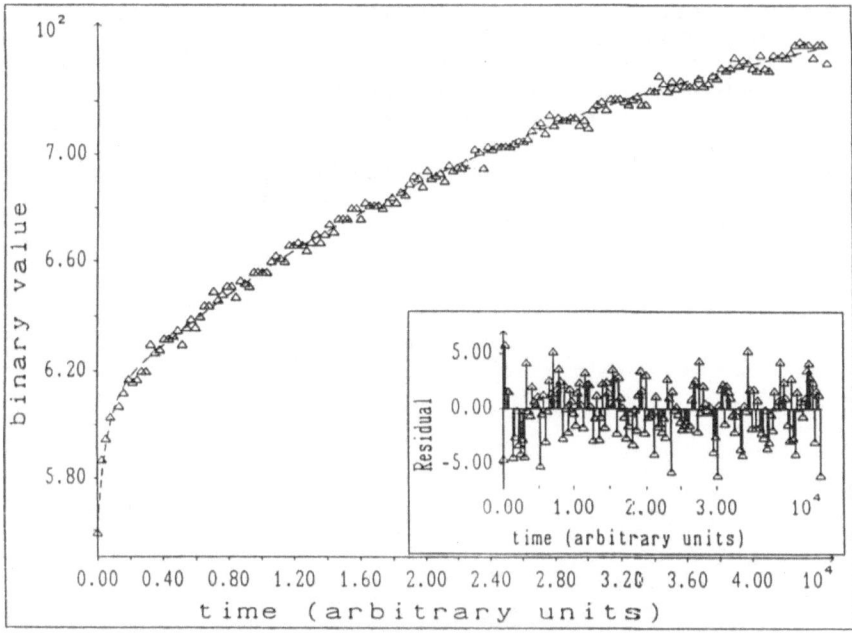

Fig. 2. Curve fit of the afterpotential decay with 3 exponentials. Non-linear regression of the peak afterhyperpolarization during a 4 s stimulation at 30 Hz. The fit yields a fast time constant of 61.6 ms, a medium time constant of 2.9 s and a slow time constant of 3.6 s. The time unit is 100 μs, the amplitude unit 50 μV (i.e. the total change in potential is 7.5 mV and the residual usually smaller than 250 μV)

Conclusion

It is now fairly obvious that potassium ions do not accumulate in significant concentrations around the axons during repetitive activity (a conclusion also reached from voltage-clamp experiments of giants axons from freshly caught *Loligo forbesi*, see Pichon et al. 1987); however, the mechanisms which are responsible for potassium homeostasis remain unknown. A ouabain sensitive Na/K pump in the axonal membrane seems involved. On the other hand, there are indications that the Schwann cells which surround the giant axon are involved. Their first (and may be most important) role would be to provide the space for free diffusion of ions around the axons. They may also act as spacial buffers, although this is unlikely, or insure long term regulation through their indirect responses to axonal stimulation (see Villegas 1984; Evans et al. 1986 and Lieberman et al. 1989). Simultaneous monitoring of membrane potential changes in axon and associated Schwann cells (Pichon et al. 1990) are likely to be of great help to elucidate the respective roles of axons and Schwann cells in potassium homeostasis.

References

Evans PD, Reale V, Villegas J (1986) Peptidergic modulation of the membrane potential of the Schwann cells of the squid giant nerve fibre. J Physiol (London) 379: 61-82

Lieberman EM, Abbott NJ, Hassan S (1989) Evidence that glutamate mediates axon-to-Schwann cell signaling in the squid. Glia 2: 94-102

Pichon C, Pichon Y (1989) A program for automatic analysis of action potentials in squid giant axons. J Physiol (London) 415: 4P

Pichon Y (1981). Pharmacological characterization of ionic channels in unmyelinated axons.
J Physiol (Paris) 77:1119-1127

Pichon Y, Abbott NJ, Lieberman EM, Larmet Y (1987) Potassium homeostasis in the nervous system of cephalopods and crustacea. J Physiol (Paris) 82:346-356

Pichon Y, Abbott NJ, Brown ER (1990) Long-term recordings in Schwann cells surrounding squid axons reveal multiple resting states. Biophys J 57:532a

Villegas J (1984) Axon-Schwann cell relationship. Curr. Topics Membranes Transport 22:547-571

Properties of the Low-Threshold Calcium Current in Thalamic Relay Neurons: Recovery from Inactivation in Relation to the Control of Repetitive Burst Generation

J. R. Huguenard, D. A. Coulter and D. A. Prince

Department of Neurology and Neurological Sciences Room M016 Stanford University Medical Center Stanford, California 94305, USA

Introduction

Relay neurons (RNs), the principle cells of the thalamus, exhibit two different firing patterns in response to excitation. One pattern, burst firing, occurs with more hyperpolarized resting membrane potentials and is associated with certain behavioral states including slow wave sleep (Steriade and Llinás 1988) and some models of petit mal epilepsy (Gloor and Fariello 1988). On the other hand, during normal wakeful states and REM sleep, when resting membrane potential in RNs tends to be more depolarized, the same stimulation will produce regular repetitive firing Steriade and Llináa 1988). Burst discharge in RNs is associated with a Ca^{2+}-dependent low-threshold spike (LTS, Jahnsen and Llinás 1984), and we have previously characterized the essential voltage-dependent properties of the prominent transient (T type) Ca^{2+} current which underlies this event (Coulter et al. 1989). However, one relatively unexplored area is the process of recovery from inactivation, which in part determines the rate at which Ca^{2+} dependent bursts can be repetitively generated. The LTS is thought to be involved in the generation of repetitive spike-wave (SW) discharges that typify absence (petit mal) epilepsy, and thus recovery from inactivation of T current is an important factor that regulates this type of rhythmic firing behavior.

The recovery process for T current has been studied in different neuronal systems and there are some major differences. We have shown that in rat somatosensory RNs a large portion of current recovers with a time constant (τ) of a few hundred milliseconds at 24°C, and that the process is highly temperature dependent, with a value of Q_{10} greater than 2 (Coulter et al. 1989). In chick dorsal root ganglion cells a somewhat slower process of recovery has been shown ($\tau = 0.4\text{-}1.4$ s; Carbone and Lux 1987) and in rat cranial neurons recovery is governed by two separate processes with time constants of hundreds of milliseconds and tens of seconds (Bossu and Feltz 1986). Recently there have been three additional reports (Crunelli et al. 1989; Hernández-Cruz and Pape 1989; Suzuki and Rogawski 1989) on T current properties in RNs of various species. In two of these studies (Crunelli et al. 1989; Suzuki and Rogawski 1989) $\tau_{recovery}$ is on the order of a few hundred milliseconds, while the other (Hernández-Cruz and Pape 1989) demonstrates that recovery requires up to 10 seconds to be complete. Therefore in the experiments reported here we examined long term recovery from inactivation of T current in somatosensory RNs in order to test for the existence of an ultraslow recovery process.

Methods

Standard dissociation and recording procedures were used (Coulter et al. 1989). In addition an ATP regeneration system (Forscher and Oxford 1985) was included in the intracellular perfusion system since it was found that this dramatically improved the quality and longevity of whole cell recordings. Recovery was studied by holding the voltage continuously at a potential in the middle of the activation range for T current (-45 mV), but outside of the range for activation of other Ca^{2+} currents. The membrane potential was then briefly stepped to hyperpolarized potentials and returned to -45 mV, and we measured the asymptotic rise in T current amplitude with increasing duration of hyperpolarization.

This protocol is different than the paired pulses used in a previous study (Coulter et al. 1989), and, because of the prolonged period spent in the inactivated state, would tend to enhance the presence of any slow inactivation mechanism (Bossu and Feltz 1986).

Results

The process of recovery from inactivation, as measured by this protocol, is shown in Fig. 1. There is an initial rapid phase of recovery which brings the T current amplitude to within 70% of maximum levels within 350 - 650 ms. This is followed by a slower recovery phase which lasts for a few seconds. Due to alleviation of steady-state inactivation, as more hyperpolarized recovery potentials are used, the maximum current becomes larger. At the same time the recovery process becomes faster. The arrowheads indicate apparent time constants of recovery which become progressively

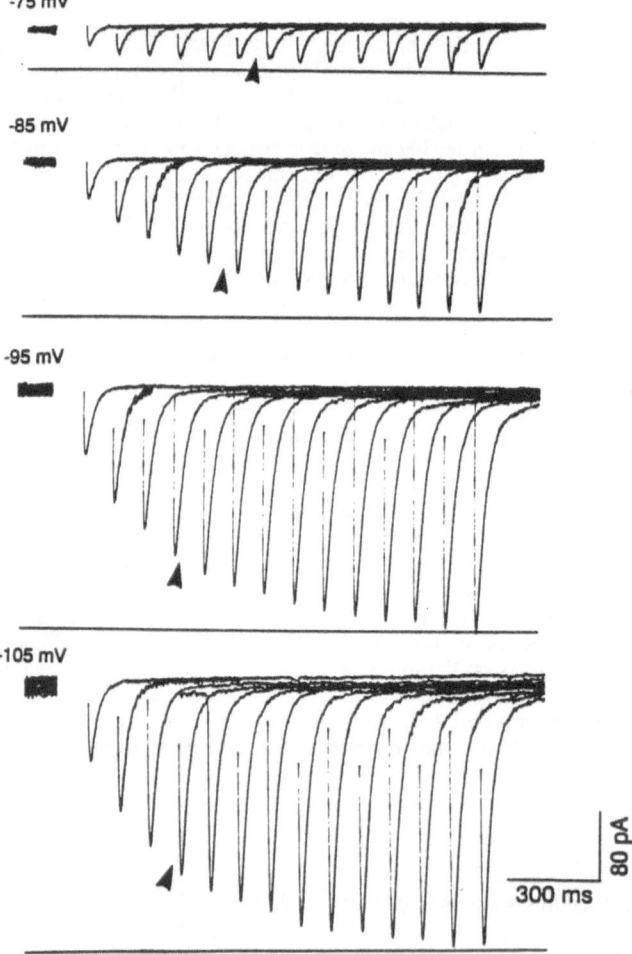

Fig. 1. Voltage dependence of recovery from inactivation. This relay neuron was voltage-clamped at a holding potential of -45 mV, hyperpolarized to the indicated potentials for periods of 100 to 1400 ms, and then returned to -45 mV to elicit T current. The *line* below each set of traces is the maximal T current obtained with 3 s hyperpolarizations, and the currents during the hyperpolarizing pulses have been blanked for clarity. *Arrowheads* indicate apparent time constants, i.e., those times which correspond to recovery of 63% of maximal current. The time constants at -75, -85, -95 and -105 mV were 660, 550, 420, and 380 ms, respectively

shorter as the recovery potentials are varied between -75 and -105 mV. The lines below each set of current traces indicate the maximum activatable current. Note that recovery is nearly complete within 1.4 s at all potentials, i.e., there is little evidence for contribution of a slow component with a time constant of several seconds.

Best fitted curves were generated for the recovery process as shown in Fig. 2A.
Fractional inactivated current was plotted on a semilog graph (inset), and was non-linear indicating a complicated decay process with at least two time constants. The best fitted curve with two components is shown as the solid line in both linear and semilog plots. For comparison, a single exponential curve, which clearly does not fit the data, is shown as the dashed line in the linear plot.

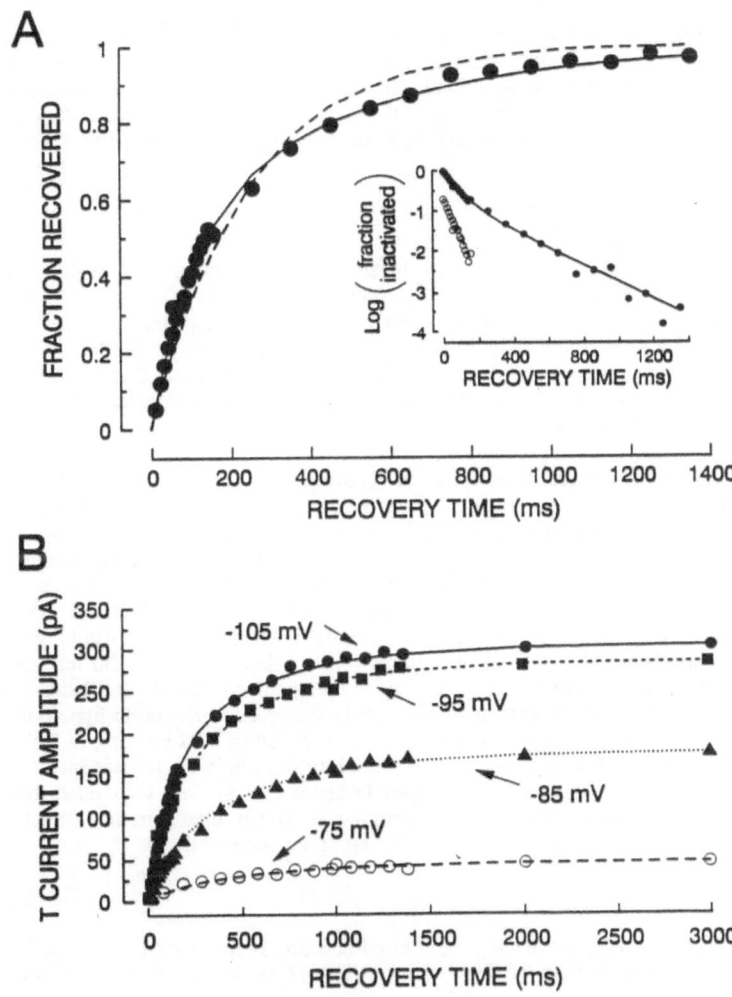

Fig. 2. Kinetics of T current recovery. A Protocols similar to those used in Fig. 1 were used to examine recovery between10 and 1400 ms at - 105 mV. The fraction of recovered current amplitude is shown ( ), as well as curves fitted to one (*dashed line*) or two (*solid line*) exponentials. A single exponential did not describe the data well; however, the double exponential decay provided a good fit. The *inset* shows time-dependent removal of the inactivation on a semilog plot. *Closed symbols* ( ) and *solid line* show the double exponential decay, while *open symbols* ( ) are the isolated fast component ($\tau = 100$ ms). B Recovery for periods up to 3 s at potentials between-75 and -105 are fitted with double exponential curves as in part A. The fitting parameters for the curves are given in Table 1

Table 1. Fitting parameters for Fig. 2B

Recovery potential (mV)	τ_{fast} (ms)	τ_{slow} (ms)	% Fast	Max. current (pA)
-75	85	570	20	-46
-85	100	530	26	-175
-95	96	466	35	-284
-105	101	483	50	-304

In Fig. 2B, are shown the best fitted curves for the data in Fig. 1, and the fitting parameters are given in Table 1. Although the time constants were not markedly voltage dependent, the percentage of fast recovery showed a systematic increase with stronger hyperpolarization. Thus fast recovery is progressively more prominent with increasingly hyperpolarized recovery potentials, and this, in large part, accounts for the voltage-dependent change in apparent time constant.

Discussion

In conclusion, recovery from inactivation of T current in RNs is well described by a double exponential process with time constants near 100 and 500 ms at 24° C. As far as the differences in recovery rate obtained in different preparations, it appears that when a large depolarizing command is used (Bossu and Feltz 1986; Hernández-Cruz and Pape 1989), a very slow component of recovery is evident. One possible explanation for such findings is that these stronger depolarizing commands activate multiple Ca^{2+} conductances, and that the time constant of recovery in this case represents a composite of properties of several currents.

Given the voltage dependence of recovery, one can then predict time-dependent fractional recovery of activatable T current at a given hyperpolarized (recovery) potential. If we also take into account the temperature dependence of this process (Coulter et al. 1989), $\tau_{recovery}$ will be approximately threefold faster at physiological temperatures than in the experiments reported here. During LTS bursting, which occurs during slow wave sleep (Steriade and Llinás 1988) and presumably during spike-wave activity (Gloor and Fariello 1988), the depolarization that accompanies each burst will essentially inactivate all of the available T channels. We would predict that 30% - 40% of the channels would return to activatable form during typical 70-150 ms post-burst hyperpolarization to - 80 − -85 mV, which occurs during sleep spindles (Steriade and Llinás 1988). This may help explain why LTSs are not generated after every consecutive rhythmic hyperpolarization during spindles. To the extent that spike-wave activity is also related to LTS bursting in RNs, the longer interburst interval (200-300 ms) during spike-waves would more completely reverse inactivation (>50%) and allow LTS related bursts to occur more regularly at the rate of 3/s. Of course this is also highly dependent on interburst hyperpolarization potential, and as yet there are not direct data available regarding the membrane potential oscillations during spike-wave activity.

Acknowledgements We are grateful to Edward Brooks for expert technical assistance. This work was supported by National Institutes of Health Grants NS06477 and NS12151 and the Morris and Pimley Research Funds.

References

Bossu J - L, Feltz A (1986) Inactivation of the low-threshold transient calcium current in rat sensory neurons: evidence for a dual process. J Physiol 376: 341-357

Carbone E, Lux H D (1987) Kinetics and selectivity of a low-voltage-activated calcium current in chick and rat sensory neurones. J Physiol 386: 547-570

Coulter D A, Huguenard J R, Prince D A (1989) Calcium current in rat thalamocortical relay neurones: kinetic properties of the transient, low-threshold current. J Physiol 414: 587-604

Crunelli V, Lightowler S, Pollard C E (1989) A T-type Ca^{2+} current underlies low-threshold Ca^{2+} potentials in cells of the cat and rat lateral geniculate nucleus. J. Physiol. 413: 543-561

Forscher P, Oxford G S (1985) Modulation of calcium channels by norepinephrine in internally dialyzed avian sensory neurons. J Gen Physiol 85: 743-763

Gloor P, Fariello R G (1988) Generalized epilepsy: some of its cellular mechanisms differ from those of focal epilepsy. TINS 11: 63-68

Hernández-Cruz A, Pape H - C (1989) Identification of two calcium currents in acutely dissociated neurons form the rat lateral geniculate nucleus. J Neurophysiol 61: 1270-1283

Jahnsen H, Llinás R (1984) Ionic basis for the electroresponsiveness and oscillatory properties of guinea-pig thalamic neurones *in vitro*. J Physiol 349: 227-247

Steriade M, Llinás R R (1988) The functional states of the thalamus and the associated neuronal interplay. Physiol Rev 68: 649-742

Suzuki S, Rogawski M A (1989) T-type calcium channels mediate the transition between tonic and phasic firing in thalamic neurons. Proc Natl Acad Sci USA 86: 7228-7233

Free Calcium Transients and Oscillations in Nerve Cells

P. G. Kostyuk, P. V. Belan and A. V. Tepikin

Bogomoletz Institute of Physiology, Kiev, USSR

Introduction

Changes in cytoplasmic Ca^{2+} concentration in isolated *Helix pomatia* neurons were studied with the aid of fura-2 fluorescence, conventional current and potential clamp methods, and pressure and iontophoretic injection.

Results and Discussion

Xanthines, such as caffeine, theophylline, and isobutylmethylxanthine (IBMX) are well known as liberators of stored Ca^{2+} from the sarcoplasmic reticulum in muscle fibers (Fabiato 1985; Palade 1987a,b) and from intracellular stores in nerve cell soma (Kostyuk et al. 1987; Lipscombe et al. 1988). In our experiments Ca^{2+} release occurred after reaching a threshold of $[Ca^{2+}]_i$ of 0.3 - 0.5 μM. Minimal concentrations of caffeine, theophylline, and IBMX necessary for the release were, respectively, 2.0, 5.0, and 0.5 mM.

Thymol is another substance which, even at low concentrations, induces Ca^{2+} release from intracellular stores (Hisayama and Takayanagi 1986; Palade 1987a,b). In our experiments a calcium transient similar to the caffeine-induced one could be elicited by thymol at a concentration of 400 μM.

Second messengers, such as inositolphosphate (IP3) and cyclic adenosine monophosphate (cAMP), also participate in the regulation of $[CA^{2+}]_i$. Injection of IP3 caused an increase in $[Ca^{2+}]_i$ that remained after replacement of the normal medium by a Ca^{2+}-free medium (Fig. 1a). Our data on *Helix* neurons resemble those obtained for *Aplysia* cells (Fink et al. 1988); however, in our case, the increase in calcium was somewhat smaller.

The injection of guanosine triphosphate (GTP) did not significantly change $[Ca^{2+}]_i$, but injection of Gpp/NH/p (a nonhydrolyzable analogue of GTP) led to a strong increase in $[Ca^{2+}]_i$. The rise in Ca^{2+} was unchanged in Ca^{2+}-free medium (Fig. 1b). Thus we conclude that Gpp/NH/p and IP3 induce Ca^{2+} release from intracellular stores.

The injection of cAMP into neurons also induced an increase in $[Ca^{2+}]_i$. Elimination of Ca^{2+} from the medium had different effects on different cells. In some neurons this procedure resulted in a decrease in Ca^{2+} transients evoked by injection of cAMP. By contrast, in other cells superfusion with Ca^{2+}-free medium did not change the amplitude of Ca^{2+} transients evoked by cAMP injections (Belan et al. 1989). It seems quite plausible to suggest that there are two sources of cAMP-induced Ca^{2+} increases. The first is Ca^{2+} flow through appropiate channels of the plasma membrane; the second is a release of Ca^{2+} from intracellular stores.

It is known from experiments on muscle fibers that caffeine-induced Ca^{2+} release results in exhaustion of intracellular pools so that, usually, repetitive Ca^{2+} transients cannot occur (benham and Bolton 1986); however, in nerve cells such repititive release is quite possible (Kostyuk et al. 1987). In some snail neurons spontaneous oscillations of $[Ca^{2+}]_i$ occurred in the presence of xanthines (Fig. 2). These observations are in agreement with recent findings in both amphibian and mammalian neurons (Kuba and Nishi 1976; Lipscombe et al. 1988). Such oscillations may play a key role in triggering caffeine-induced epileptic activity. We failed to regularly induce such oscilla-

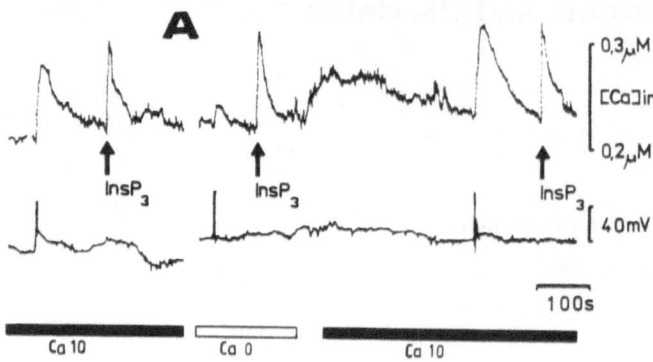

A

InsP₃ InsP₃ InsP₃

$0,3\mu M$

$[Ca]in$

$0,2\mu M$

$40mV$

100s

Ca 10   Ca 0   Ca 10

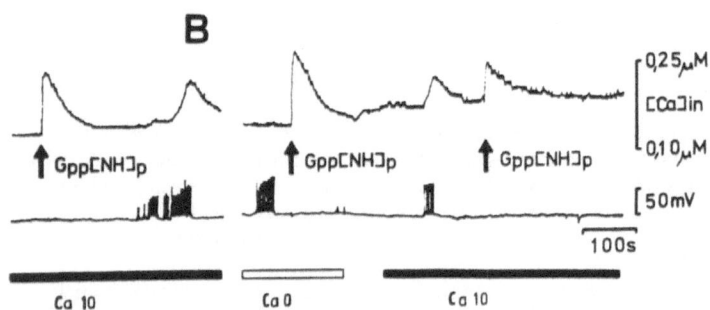

B

GppCNH]p   GppCNH]p   GppCNH]p

$0,25\mu M$

$[Ca]in$

$0,10\mu M$

$50mV$

100s

Ca 10   Ca 0   Ca 10

Fig. 1A,B. Effect of intracellular IP3 and Gpp/NH/p injection on $[Ca^{2+}]$ levels. $[Ca^{2+}]_i$ changes are shown on upper traces, membrane potentials on lower ones. Injections are shown by arrows. **A** Effect of extracellular Ca^{2+} removal on $[Ca^{2+}]_i$ transients induced by membrane depolarization and IP3 injections. **B** Effect of extracellular Ca^{2+} removal on $[Ca^{2+}]_i$ transients induced by membrane depolarizations and Gpp/NH/p injections

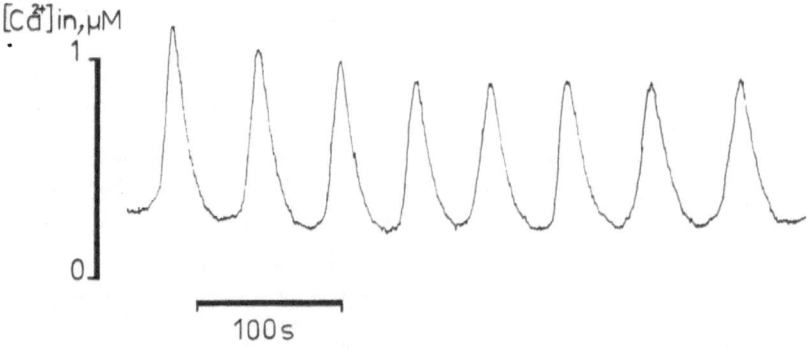

$[Ca^{2+}]in, \mu M$

1

0

100s

Fig. 2. Caffeine-induced oscillation of $[Ca^{2+}]_i$ in *Helix* neurons. Caffeine concentration in the extracellular solution = 5m*M*

tions in neurons by injection of IP3 or cAMP. This may be explained by their rapid inactivation inside the cell prior to the oscillation period (the period of caffeine-induced oscillation was in the range of several tens of seconds). Further experiments with nonmetabolizable analogues of intracellular messengers are necessary to solve this problem.

References

Belan PV, Mironov SL, Osipenko ON, Tepikin AV (1989) The effect of iontophoretic cAMP injection on the changes in intracellular calcium concentration and transmembrane currents in snail neurons. Neurophysiology (Kiev) 21: 396-403

Benham CD, Bolton TB (1986) Spontaneous transient outward currents in single visceral and vascular smooth muscle cells of the rabbit. J Physiol (Lond) 381: 385-406

Fabiato A (1985) Time and calcium dependence of activation and inactivation of calcium-induced release of from the sarcoplasmic reticulum of skinned canine cardiac Purkinje cell.
J Gen Physiol 85: 247-290

Fink LA, Connor JA, Kaczmarek LK (1988) Inositol triphosphate releases intracellularly stored calcium and modulates ion channels in molluscan neurons. J Neurosci 8: 2544-2555, Natl Acad Sci USA 83: 942-946

Hisayama T, Takayanagi I (1986) Some properties and mechanisms of thymol-induced release of calcium from the calcium-store in guinea pig taenia-caecum. J Pharmacol 40: 69-82

Kostyuk PG, Tepikin AV, Belan PV, Mironov SL (1987) Mechanisms of changes in Ca ion concentration in the cytoplasm of snail neurons with participation of intracellular Ca depot.
Biol Membranes (Moscow) 4: 932-936

Kuba K, Nishi S (1976) Rhytmic hyperpolarization and depolarization of sympathetic ganglion cells induced by caffeine. J Neurophysiol 39: 547-563

Lipscombe D, Madison DV, Poenie M, Reuter H, Tsien RW, Tsien RY (1988) Imaging of cytosolic Ca^{2+} transients arising from Ca^{2+} stores and Ca^{2+} channels in sympathetic neurons.
Neuron 1: 355-365

Palade P (1987a) Drug-induced Ca^{2+} release from isolated sarcoplasmic reticulum. I. Use of Pyrophosphate to study caffeine-induced Ca^{2+} release. J Biol Chem 262: 6135-6141

Palade P (1987b) Drug-induced Ca^{2+} release from isolated sarcoplasmic reticulum. II. Release involving a Ca^{2+}-induced Ca^{2+} release channel. J Biol Chem 262: 6142-6148

Intracellular Calcium Regulation and Neuronal Modulation

R. T. Kado[1], P. Morain[1], S. Berretta[2] and C. Batini[2]

[1] Laboratoire de Neurobiologie cellulaire et moléculaire, CNRS, Gif sur Yvette, 91198 France
[2] Laboratoire de Physiologie de la Motricité URA385, CNRS CHU Pitié Salpetrière, 91 Blvd. de l'Hôpital, Paris, 75013 France

Introduction

Intracellular free calcium has been identified with activation of membrane channels, cytoplasmatic processes, and maintenance of cellular integrity. It is now well established that the cell normally has a very low free calcium (Ca^{2+}) level in its cytoplasmatic space. The low Ca^{2+} level is thought to be maintained by a sodium-calcium exchanger and an ATP-driven calcium pump in the plasma membrane which can move the calcium out of the cell as well as by sequestration into cytoplasmatic stores and various specific calcium binding proteins. Cytoplasmatic Ca^{2+} levels are increased by two mechanisms.
(1) Entry across the membrane, through both voltage and agonist activated channels, or
(2) released from binding sites within the cytoplasm or from intracellular Ca^{2+} stores, probably in the endoplasmatic reticulum (Exton 1988 and see chapter on calcium channels, calcium channel blockers and nonspecific currents, this volume). While a great deal has been learned about how Ca^{2+} enters the neuron, not much is known about the effects of increases in cytoplasmatic Ca^{2+}. It is known that excessive Ca^{2+} can destroy the cell (Siesjö 1988), but what happens to Ca^{2+} dependent processes at Ca^{2+} levels higher than needed but not high enough to kill the cell?

At this symposium, R. Llinás showed a video in which both the spike in a neuron and a Fura-II image were recorded. The timing of the spike and the rise in Ca^{2+} seemed to show very little difference. The entering calcium depolarized the membrane and encountered a Fura-II molecule at least as easily as a cytoplasmatic Ca^{2+} buffering site. Given that the quantity of injected Fura-II did not exceed the neuronal buffer concentration, we can conclude from this video as well as from other such data presented at this symposium that the cytoplasmatic Ca^{2+} buffer, in fact, has a very limited capacity. It is able to keep cytoplasmatic Ca^{2+} near the inner membrane surface at a low level only if the quantity of entering or released Ca^{2+} is very small.

If the cytoplasmatic buffering capacity is limited, it could be expected that conditions leading to Ca^{2+} increases beyond the buffering and elimination capacities would leave the cytoplasm with a Ca^{2+} overload. Experimental conditions, such as cell damage during preparation or electrode impalement and forced channel opening by prolonged depolarizations or applied agonists, lead to a cytoplasmatic Ca^{2+} overload and aberrant membrane potential behavior.

We have been examining the differences in properties between normal and neurotoxically deafferented Purkinje cell (PC) (Desclin and Escubi, 1974; Batini et al. 1985) in slice preparations from rat cerebella. Whereas most intracellular recordings from PC in both preparations showed normal simple spike (sodium dependent, SS) discharges, some showed firing of SS and dendritic spikes (calcium dependent, DS) grouped into bursts. PC not showing bursting when impaled could made to fire in bursts by depolarizing currents and the bursts could be suppressed by hyperpolarizing currents (Kado et al. 1989). The induced bursts were indistinguishable from those obtained by impalement in some PC. We interpret this result to mean that the prolonged depolarization caused a large Ca^{2+} influx and that subsequent DS allowed more Ca^{2+} to enter, thus sustaining an overload of the buffer and elimination mechanisms. Thus the burst firing is a consequence of the Ca^{2+} overload.

Experimental Brain Research Series 20
© Springer-Verlag Berlin · Heidelberg 1991

This interpretation seems to gain support from the results of another kind of experiment. Burst firing as also obtained in PC showing no prior bursting when the external sodium content was reduced. The onset was faster with zero sodium in the external medium than when it was reduced by one half. We interpret these results to mean that there is probably a sodium-calcium exchanger in the PC membrane which, in the absence of external sodium, allows calcium to enter instead of leave the PC. This was not due to voltage dependent Ca^{2+} channels since, in the absence of external sodium, the membrane potentials were hyperpolarized. The hyperpolarization could be due not only to a resting sodium conductance but also to Ca^{2+} activated potassium channels. Here again, a cytoplasmatic Ca^{2+} overload probably occurred.

Since the entering Ca^{2+} increases $[Ca^{2+}]_i$ by several orders of magnitude (Ross 1989) and most of this increase is at the inner membrane surface, it could be expected that at least some surface negative charge screening by the Ca^{2+} would occur. Since the screening occurs at the inner surface, its effect would be to reduce the polarization of the membrane and its channels. We propose that this mechanism, the mirror image of that occurring at the outer surface (Hille 1984), may explain the channel openings producing the DS even for hyperpolarized intracellular potentials.

Whereas burst firing could be induced by depolarizing currents in both normal PC and those deprived of their climbing fibers, the deafferent PC required a greater depolarization to initiate bursts. They also had more stable resting potentials on impalement than did the normals. This difference seems to be related to an inward rectification which is present in the deafferented PC (Morain et al. 1989; Kado et al. 1989). A similar rectifying property has been shown to be present in PC of neonatal rats (Kapoor et al. 1988). Such a rectification would tend to clamp the membrane potential and present a low input resistance thus stabilizing the membrane potential at a hyperpolarized level. It has also been shown that growing processes of neurons show a higher resting Ca^{2+} level than other parts of the same cell (Connor 1986; Cohan et al. 1987). These results indicate that regional differences in the Ca^{2+} buffering capacity are maintained by the cytoplasm according to local needs.

The differences in electrical behavior between the normal and deafferented PC may be due to a restarting of growth processes in the dendrites of the PC deprived of their climbing fiber inputs (Sotelo et al. 1975). Reactivation of the growth processes in the fully differentiated PC may involve a reversion of the growing parts of the PC to an earlier stage of development. Electrically, these localized membrane modifications manifest themselves as modifications in the current/voltage relations recorded by the impaling electrode and in the behavior of the entire PC.

The arguments presented above are intended to support the idea that neuronal Ca^{2+} homeostasis can be easily disrupted. Experimental or diseased conditions which modify a neurons ability to deal with cytoplasmic Ca^{2+} can produce aberrant behavior of the membrane potential through a cytoplasmatic Ca^{2+} overload.

References

Batini C, Billard JM, Daniel M (1985) Long term modification of cerebellar inhibition after inferior olive degeneration. Exp Brain Res 59: 404-409
Cohan CS, Connor JA, Kater SB (1987) Electrically and chemically mediated increases in intracellular calcium in neuronal growth cones. J Neurosci 7: 3588-3599
Connor JA (1986) Digital imaging of free calcium changes and of spatial gradients in growing processes in single, mammalian central nervous system cells. Proc Natl Acad Sci (USA) 83: 6179-6183
Exton JH (1988) Mechanisms of action of calcium mobilizing agonists: some variations on a young theme. FASEB J 2: 2670-2676
Desclin JC, Escubi J (1974) Effects of 3-Acetylpyridine on the central nervous system of the rat as demonstrated by silver methods. Brain Res 77: 349-364
Hille B (1984) Ionic Channels of Excitable Membranes. Sinauer Associates Inc Sunderland MA
Kado RT, Batini C, Morain P (1989) Cerebellar Purkinje cells: membrane property changes on partial deafferentation. J Physiol (Paris) 83: 172-180

Kapoor R, Jaeger CB, Llinás R (1988) Electrophysiology of the mammalian cerebellar cortex in organ culture. Neuroscience 26: 493-507

Morain P, Kado RT, Batini C (1989) Electrophysiological effect of partial deafferentation in the rat cerebellar Purkinje cell. Exp Brain Res Series 17: 231-234

Ross WN (1989) Changes in intracellular calcium during neuron activity. Ann Rev Physiol 51: 491-506

Sjesjö BK (1988) Historical overview: Calcium, Ischemia and death of brain cells. In: Van Houtte Paoletti and Govoni (eds) Calcium Antagonists, Ann NY Acad Sci 522: 638-661

Sotelo C, Hillman DE, Zamora AJ, Llinás R (1975) Climbing fiber deafferentation: its action on Purkinje cell dendritic spines. Brain Res 98: 574-581

Dendritic Ca^{2+} Channels of Rat Purkinje Cells Maintained in Culture

J. L. Bossu, J. L. Dupont, L. Fagni and A. Feltz

Laboratoire d'Etude des Régulations Physiologiques, associé à l'Université Louis Pasteur, Centre National de la Recherche Scientifique 23 rue Becquerel, F-67087 Strasbourg, France

Introduction

Purkinje cells (PC) of the cerebellar cortex provide a good model for investigating the role of calcium entry in neurons of the CNS. Intracellular recordings and microfluorometric imaging from PC in cerebellar slices have shown that the main site of Ca entry in these cells is localized mainly, if not exclusively, to the dendritic trunk (Llinas and Sugimori 1980; Ross and Werman 1987; Tank et al. 1988). Ca entry is induced by at least two voltage-dependent Ca conductances with a distinct dendritic location and function (Llinas and Sugimori 1980). One entry site is localized mainly in the tertiary branches of the dendritic trunk and is a plateau-generating Ca conductance which regulates the discharge frequency of PC. The second mode of Ca entry is by a large spike-generating Ca conductance and arises in the primary and secondary dendritic tree. More recently a third type of voltage-dependent Ca conductance has been characterized in PC (Crepel and Penit-Soria 1986; Kapoor et al. 1988). The latter is elicited at a low threshold when the PC are hyperpolarized enough to overcome its inactivation; the inhibitory action of noradrenaline on PC firing may be explained by a reduction of this current (Crepel et al. 1986). The present study focuses on investigating the dendritic Ca conductances of rat PC with the improved recording conditions obtained by using cultured cells.

Methods

Culture procedures and identification of PC have been previously described (Bossu et al. 1989). PC were identified by their morphology (Fig. 1A), recognized as early as 5 days after plating. Dendritic Ca currents and channels were recorded using whole cell recording (WCR) and cell-attached configurations. For total Ca current recordings, the bath solution contained (in mM): choline Cl 130, TEA 7.5, CaCl$_2$ 10, no MgCl$_2$, Hepes/Tris 5 (pH 7.4), glucose 10 and the internal solution CsCl 120, TEA 20, MgCl$_2$ 2, EGTA/CsOH 11, CaCl$_2$ 1, (pCa 8), Hepes/CsOH 10 (pH 7.2).

Ca channels were recorded using an isotonic BaCl$_2$ solution in the pipette and an isotonic KCl solution in the bath to zero the cell membrane potential. The bath contained (in mM): K gluconate 140, EGTA/KOH 10, MgCl$_2$ 2, Hepes/KOH 20 (pH 7.4) and the pipette : BaCl$_2$ 110, Hepes/Tris 10 (pH 7.4). BAY K 8644 (10^{-5} M) was added to the pipette solution. Data stored on a videotape were analyzed on a LSI 11/73. Traces were digitized at 2 and 10 kHz for total current and channel recordings, respectively, and filtered at one-fourth the sampling frequency with a -3 dB/octave 8 pole Bessel filter.

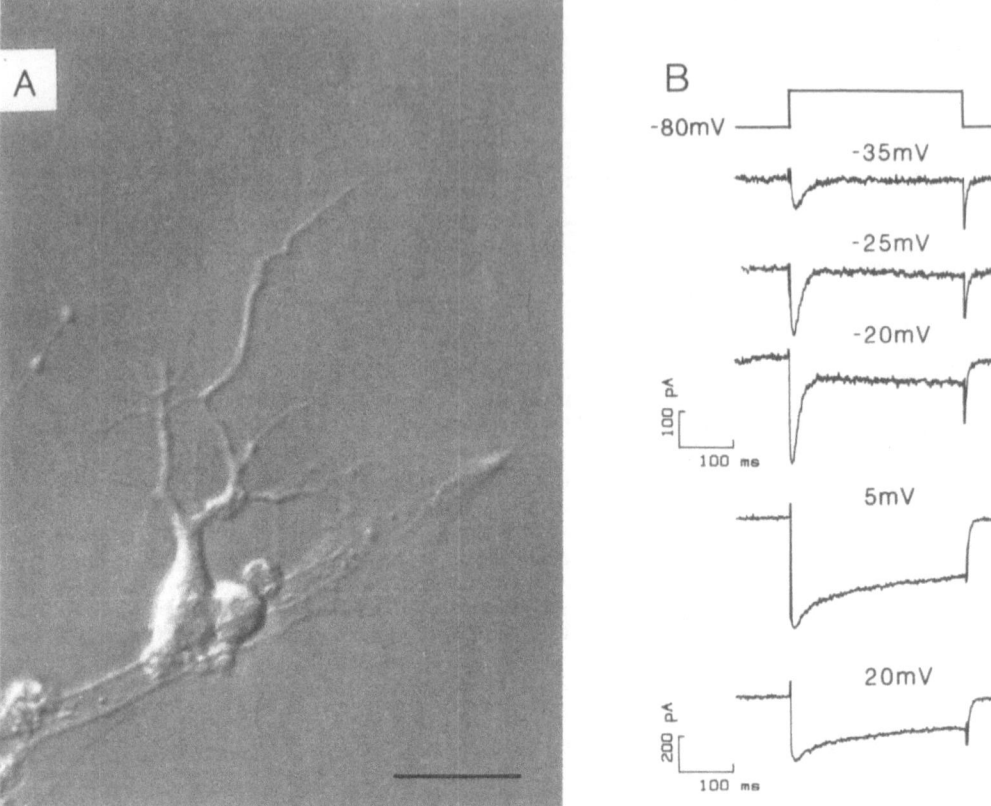

Fig. 1A,B. Identification of PC in culture and its dendritic Ca current. **A** Purkinje-like cell identified among dissociated rat cerebellar cells by a large dendritic-like process arising from a pear-shape soma (2 week old culture). Unit bar: 20 μm. **B** Ca current traces obtained from the dendrite of a PC in WCR configuration

Results

Current Recordings. The inward current that could be recorded from a dendrite PC under conditions devised to isolate Ca currents are illustrated in Fig.1B. In most cells, a voltage jump above -35 mV from a holding potential of -80 mV elicited a transient inward current. The latter reached a maximal amplitude (about 250 pA) at -20 mV. Typically, the rise and decay phases of this inward current were more rapid at more depolarized voltages. Above -20 mV a more sustained inward current appeared (Fig.1B) and depolarizations beyond -20 mV elicited a large inward current (about 700 pA at 0 mV) with a complex decay kinetic.

Channel Recordings. In the presence of BAY K 8644 in the pipette solution two distinct types of voltage-activated channels could be recorded (Fig. 2A). One type of channel was activated at low threshold, at -55 mV when the potential was held at -80 mV (Fig.2A). Elementary current amplitude was about -0.5 pA at -20 mV, corresponding to a slope conductance of 10 pS (Fig.2A, open circles on the plot). At -20 mV, larger events (-1.8 pA) were elicited (Fig. 2A). By contrast to the former channel, this one is not inactivated by holding the patch at -40 mV and further it has long

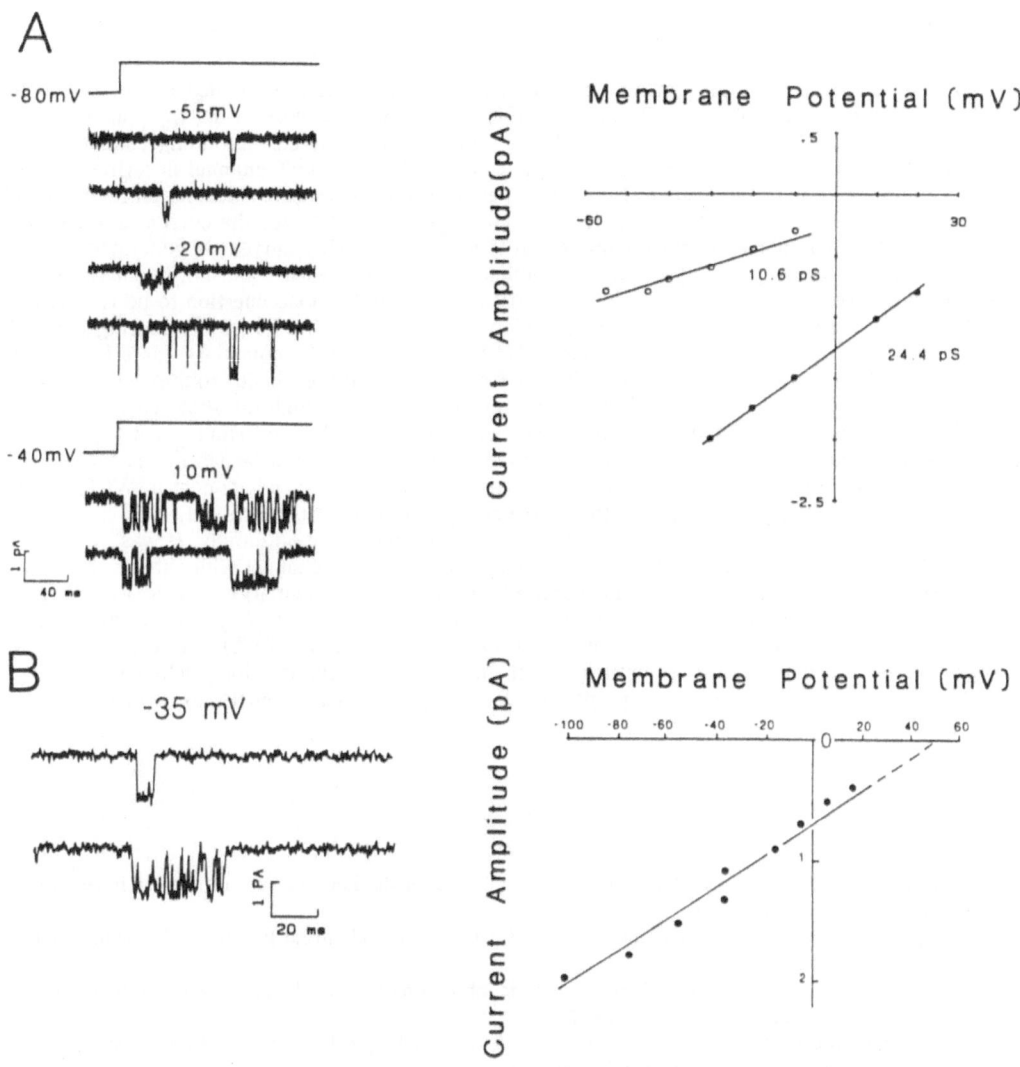

Fig. 2A,B. Three types of channel are seen on the dendritic membrane of PC in cell-attached patch recordings. A Left, Current traces obtained in the presence of BAY K 8644 during depolarizing steps (protocol on upper panels) applied either from a holding potential of -80 mV or -40 mV. Note, at -20 mV, two distinct unitary events, and that from a holding potential of -40 mV, only the openings of large amplitude are elicited. Right, I/V curves for these two voltage-dependent channels. B Left, at -35 mV a spontaneously active channel. Right, Corresponding I/V curve. Linear fit yielded a slope conductance of 16 pS

open times. Its slope conductance was 24 pS (Fig.2A, black circles on the plot). In the absence of BAY K 8644 this channel also appeared but at more depolarized potentials with rare brief openings, whereas the 10 pS channel was still elicited with identical characteristics. In addition to these two voltage-activated channels, we have also recorded a spontaneously active channel (Fig. 2B). This channel is open at the resting membrane potential (-80 mV) and has an amplitude of about -1 pA at -35 mV. The estimated slope conductance for this channel was about 15 pS (plot in Fig. 2B). The reversal potential of this current was found to be around +50 mV (dashed line) suggesting a similar selectivity of the 15 and 24 pS channels for barium ions.

Discussion

Activation, inactivation, decay kinetics and pharmacology of the current elicited at low threshold and also encountered on the soma (Bossu et al. 1989), are similar to those of the $I_{Ca,T}$ identified in PC recorded in cerebellar slices (Crepel and Penit-Soria 1986). The corresponding channel has a conductance of 10 pS (see also Bossu, Fagni and Feltz 1989), as the T channel described on the soma of numerous neuronal preparations (Tsien et al. 1988), and also evidenced on the soma of cultured PC obtained from rat fetuses (Hirano and Hagiwara 1989). For the current elicited at a high threshold the situation is more complex. It is clear that such a current is generated mainly from the dendrite as proposed by Llinas and Sugimori (1980). Indeed it could not be characterized properly when recorded from somatic locations (Bossu et al. 1989). One question to be resolved is which type of Ca channel fits the properties of this Ca current. The 24 pS conducting channel activated with a high threshold in presence of BAY K 8644 could be considered as a L type Ca channel (Bossu, Fagni and Feltz 1989). Without BAY K 8644 in the pipette solution, the activity of this L channel is almost absent and then could not account for the high threshold current recorded from the dendrite. This could mean that this channel is localized to hot spots, as it happens on the soma (Hirano and Hagiwara 1989 but Llano et al. 1990). A second possibility is that this L channel normally quiescent is experimentally activated by dilution of a component in WCR condition. A third possibility is that an another distinct type of channel be involved, such as the 10-15 pS conducting channel opened over a large domain of membrane potentials. It was present in about 50% of the patches and displayed some voltage sensitivity. A channel with similar characteristics (10 pS in isotonic $BaCl_2$, open time sensitivity to membrane potential) has been isolated from cerebellar extracts (Llinas et al. 1989). Finally we cannot exclude that a N type Ca channel (such as the one observed on the somatic membrane of neurons (Tsien et al. 1988) is also present. The latter may be localized to the finest branches of the dendrite not accessible for patch-clamp studies. At the present time, one is missing a proper pharmacology of the Ca channels to favor one or the other of these possibilities.

References

Bossu JL, Dupont JL, Feltz A (1989) Calcium currents in rat cerebellar Purkinje cells maintained in culture. Neuroscience 30: 605-617
Bossu JL, Fagni L, Feltz A (1989) Voltage-activated calcium channels in rat Purkinje cells maintained in culture. Pflügers Archiv 414: 92-94
Crepel F, Debono M., Flores R. (1987) α-Adrenergic inhibition of rat cerebellar Purkinje cells in-vitro: a voltage-clamp study. J Physiol 383: 487-498
Crepel F, Penit-Soria J (1986) Inward rectification and low threshold calcium conductance in rat cerebellar Purkinje cells: an in-vitro study. J Physiol 372: 1-23
Hirano T, Hagiwara S (1989) Kinetics and distribution of voltage-gated Ca, Na and K channels on the somata of rat cerebellar Purkinje cells. Pflügers Arch 413: 463-469
Kapoor R, Jaeger CB, Llinas R (1988) Electrophysiology of the mammalian cerebellar cortex in organ culture. Neuroscience 26: 493-507
Llano I, Gähwiler BH, Marty A (1990) In: Neurobiology of the cerebellar system. Llinas r, Sotelo C (eds). University Press, Oxford
Llinas R, Sugimori M (1980) Electrophysiological properties of in vitro Purkinje cell dendrites in mammalian cerebellar slices. J Physiol 305: 197-213
Llinas R, Sugimori M, Lin JW, Cherksey B (1989) Blocking and isolation of a calcium channel from neurons in mammals and cephalopods utilizing a toxin fraction (FTX) from tunnel-web spider poison. Proc Natl Acad Sci USA 86: 1689-1693
Ross WN, Werman R (1987) Mapping calcium transients in the dendrites of Purkinje cells from the Guinea-pig cerebellum in-vitro. J Physiol 389: 319-336
Tank DW, Sugimori M, Connor J, Llinas R (1988) Spatially resolved calcium dynamics of mammalian Purkinje cells in cerebellar slice. Science 242: 773-777
Tsien RW, Lipscombe D, Madison DV, Bley KR, Fox Ap (1988) Multiple types of neuronal calcium channels and their selective modulation. Trends Neurosci 11: 431-437

Characterization of Chloride Channels in Pituitary Intermediate Lobe Cells: Sensitivity to Internal Calcium

O. Taleb, J. Trouslard, M. Hamann, P. Vanderheyden and P. Feltz

Laboratoire de Physiologie Générale (URA 309 CNRS), Université Louis Pasteur 21, rue Descartes, 67084 Strasbourg, France

Intermediate lobe (IL) cells display a pattern of oscillatory firing functionally related to the secretion of pro-opiomelanocortin-derived peptides (Taraskevich and Douglas 1982; Trouslard et al. 1989). These excitable cells are under multiple hypothalamic controls, i.e., a tonic inhibition mediated by dopamine and γ-aminobutyric acid (GABA) and positive control exerted by corticotropin releasing factor and thyrotropin releasing hormone (TRH).

Using the patch clamp technique, we characterized spontaneous and $GABA_A$-activated chloride currents and their regulation by intracellular calcium.

In the absence of any ligand, two different spontaneously active chloride channels were recorded. The first type was similar to $GABA_A$-activated channels in conductance levels (11, 23, and 31 pS) and halide permeability sequence ($I^- > Br^- > Cl^-$) (Taleb et al. 1988); however, a dominant short event of approximately 1 ms duration was typical for this spontaneous channel, whereas $GABA_A$-activated channel openings consisted of 4 ms event and a bursting activity of 19 ms (Taleb et al. 1987). Moreover, the convulsant, noncompetitive $GABA_A$ antagonist ter-butylbicyclophosphorothionate also inhibited the spontaneous chloride channel activity (Hamann et al. 1990), whereas it was not affected by the competitive agonists SR 42641 and bicuculline at doses at which they block $GABA_A$ receptors (Taleb et al. 1987). Interestingly, the maximal activity of GABA-activated channels as well as these spontaneous chloride channels appeared in the same concentration range of extracellular calcium, i.e., between 10^{-8} and 10^{-7} M, and was inhibited by 10^{-6} M (Fig. 1A) (Taleb et al. 1987). At such high concentrations of intracellular calcium (i.e., ≥ 0.75 μM), a second spontaneous chloride channel activity with a small conductance (2 - 3 pS) was observed (Taleb et al. 1988) (Fig. 1B). Analysis of the time constants, evaluated by measuring the relaxation consecutive to an instantaneous voltage jump on the whole population of channels present on the patch, indicated that the fast constant (10 ms) was voltage insensitive, whereas the slow one (75 ms) increased twofold over 120 mV (Taleb et al. 1988). In addition, the ratio of the slow to fast relaxations was 1 : 5 at -60 mV, compared to 1 : 1 at +80 mV.

What is the physiological role of chloride currents? Obviously, they are involved in modulation of the integrated electrical activity of IL cells. Taraskevitch and Douglas (1982) showed that activation of the $GABA_A$ receptor induced, on rat IL cells, brief bursts of action potentials followed by a long lasting inhibition. An inhibitory effect by calcium on GABA responses has been reported on sensory neurones (Behrends et al. 1988; Inoue et al. 1986). Similarly, our study, performed on neuroendocrine cells, shows that both GABA-evoked and the first type of spontaneous chloride activity can be inhibited at micromolar concentrations of internal calcium. Under these conditions, another spontaneous chloride channel activity (2 - 3 pS), similar to the one in exocrine glands and *Xenopus* oocytes, was recorded. High internal calcium is attained during rhythmic firing or in response to neurotransmitters such as TRH. In this context, IL cell possess all three types of voltage-dependent calcium channels (T, N, and L) (Taleb et al. 1986) that might be implicated in a complex mode of firing, i.e., sodium-dependent action potentials followed by short trains of calcium spikes (Fig. 2A) (Trouslard et al. 1989). Figure 2B illustrates a stimulatory effect of TRH on the firing frequency by mobilization of intracellular calcium through the phosphatidylinositol pathway (Trouslard et al. 1989). When high concentrations of calcium are attained, activation of

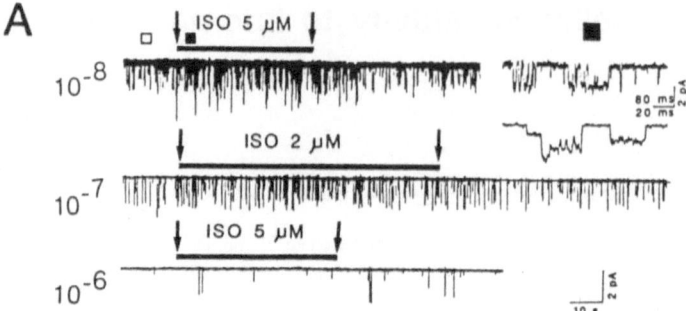

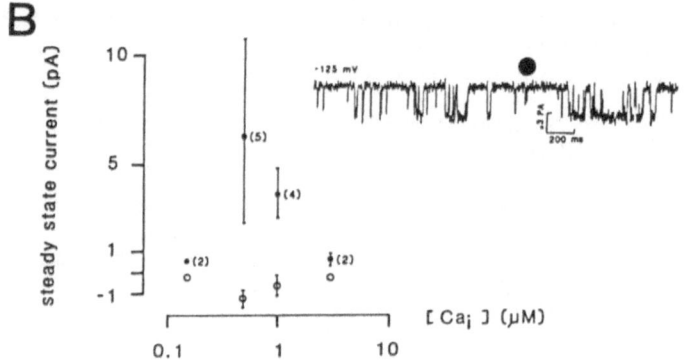

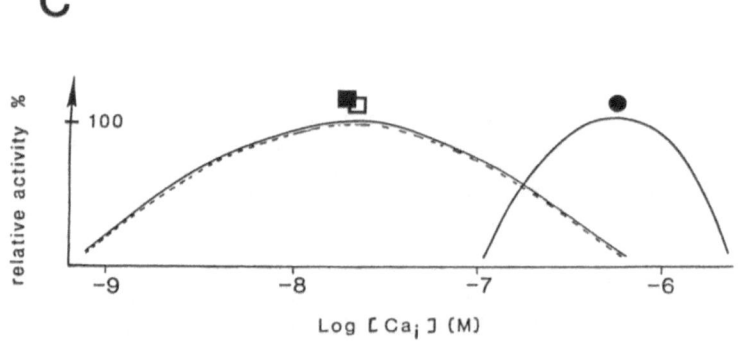

Fig. 1A-C. Effect of internal calcium on spontaneous (□), GABA-activated (■), and small conductance chloride channels (●). A Three successive outside/out patches were performed on the same cell using a different calcium concentration in each pipette (E_{Cl} = 0 mV, HP = -80 mV). The recording traces consist of spontaneous (□) and GABA$_A$-activated chloride channels evoked by isoguvacine, ISO (■). The two *upper right traces* show ISO-induced chloride channel bursting activity at larger time scales. B Relaxation of Cl⁻ currents on outside/out patches as function of internal calcium at -60 mV (○) and +80 mV (●). *Points* correspond to average values ± S.D., and the number of experiments is given in *brackets*. With increasing internal calcium from 0.1 μM to 3 μM, the plot of steady state values displayed a bell-shaped curve with a maximum at 0.5 μM. At the *right* a typical trace of this small conductance channel is shown (E_{Cl} = 2.5 mV, HP = -125 mV, [Ca] = 1 μM). C Schematic curves giving the different calcium dependences of the first spontaneous (□) and GABA-activated channels (■) as compared to small conductance chloride channels (●)

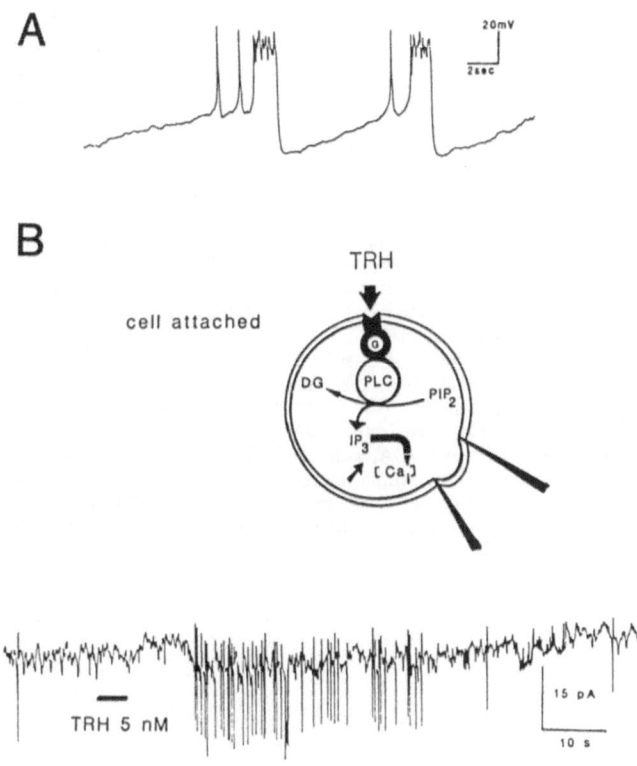

Fig. 2. A Whole cell recording of sodium-dependent action potentials (sensitive to tetrodotoxin, followed by short trains of calcium spikes. The extracellular medium was close to physiological conditions and intracellular calcium was buffered at 10^{-8} M (RP = -50 mV). **B** Effect of TRH on spiking frequency. The cell-attached configuration was chosen to avoid alteration of intracellular second messengers. TRH triggered action potentials under conditions at which spontaneous action potentials were abolished (pipette potential = -60 mV). The *upper panel* shows the mechanism of TRH action by stimulation of phospholipase C (*PLC*) inducing and increased breakdown of polyphosphoinositide (*PIP$_2$*) into diacylglycerol (*DG*) and inositol-3-phosphate (*IP$_3$*) and, ultimately, the release of intracellular calcium from the endoplasmic reticulum (Trouslard et al. 1989)

this small conductance chloride channel may contribute to maintaining the oscillatory pattern of firing. This results in a repolarization until the chloride gradient is reduced.

Chloride and calcium currents may exist in a state of dynamic equilibrium, each conductance regulating the other and contributing to the complex regulation of electrical activity of IL cells. One question that remains to be addressed is whether spontaneous chloride currents can substitute for GABAergic modulation of oscillatory firing.

Acknowledgements. This work was supported by INSERM (CRE 87-6011), DRET (89-34036), and by a FRM fellowship to P.V. (CNRS URA 309).

References

Behrends JC, Maruyama T, Tokutomi N, Akaike N (1988) Ca^{2+} mediated suppression of the GABA-response through modulation of chloride channels gating in frog sensory neurones. Neurosci Lett 86: 311-316

Hamann M, Desarmenien M, Vanderheyden P, Piguet P, Feltz P (1990) Electrophysiological study of TBPS-induced block of spontaneous chloride channels. Mol Pharmacol (in press)

Inoue M, Oomura Y, Yakushi T, Akaike N (1986) Intracellular calcium ions decrease the affinity of the GABA receptor. Nature 324: 156-158

Taleb O, Trouslard J, Demeneix BA, Feltz P (1986) Characterization of calcium and sodium currents in porcine pars intermedia cells. Neurosci Lett 66: 55-60

Taleb O, Trouslard J, Demeneix BA, Feltz P, Dupont J-L, Feltz A (1987) Spontaneous and GABA-evoked chloride channels on pituitary intermediate lobe cells and their internal Ca requirements. Pflügers Arch 409: 620-631

Taleb O, Feltz P, Bossu J-L, Feltz A (1988) Small-conductance chloride channels activated by calcium on cultured endocrine cells from mammalian pars intermedia. Pflügers Arch 412: 641-646

Taraskevich PS, Douglas WW (1982) GABA directly affects electrophysiological properties of pituitary pars intermedia cells. Nature 299: 733-734

Trouslard J, Demeneix BA, Feltz P (1989) Spontaneous spiking activity of porcine pars intermedia cells: effects of thyrotropin-releasing hormone. Neuroendocrinology 50: 33-43

Trouslard J, Loeffler JP, Demeneix BA, Feltz P (1989) Thyrotropin-releasing hormone stimulated porcine melanotrope cells in primary culture. Neurosci Lett 98: 234-239

Sprouting and Degeneration of Dendrites of the Identified Neuron B3 in the Buccal Ganglia of Helix Pomatia after Epileptic Activity

U. Altrup[1], A. Lücke[2], A. Lehmenkühler[2] and E.-J. Speckmann[1,2]

[1] Institut für Experimentelle Epilepsieforschung, Hüfferstr. 68, D-4400 Münster, FRG
[2] Institut für Physiologie, Robert-Koch-Str. 27a, D-4400 Münster, FRG

Introduction

Neuropathological studies have shown that epilepsies are often associated with a loss of neurons in parts of the nervous system (Pfeiffer 1963). In experimental animal studies it was demonstrated that prior to cell death there are changes in the shape of neurons with a decreased number of dendritic spines and multiple dendritic bulgings (see Ribak 1986). The aim of this investigation was to obtain more information on the processes which link epileptic activity with morphological changes in neurons. For this purpose, neuronal fibers of *Helix pomatia* (neuron B3, buccal ganglia) were studied under control and epileptic conditions. It will be shown that the sequence of morphological changes associated with epileptic activity consists of dendritic sprouting followed by degeneration and cell death.

Materials and Methods

The buccal ganglia of *Helix pomatia* (Altrup 1987; Schulze et al. 1975) including the left and right B3 neurons were separated from the animal and mounted in an experimental chamber. The chamber was continuously perfused with a bath solution containing: 130 mM NaCl, 4.5 mM KCl, 9 mM CaCl$_2$ and 5 mM Tris-Cl (pH: 7.4; 20 °C). Epileptic activity was induced by adding pentylenetetrazol (PTZ, 40 mM; see Lücke et al. 1989) to the control bath solution. The experimental protocol was as follows: (1) The dye cobalt-lysine was pressure injected into the soma of neuron B3 within ca. 30 to 60 min into recording of its membrane potential. (2) In the test experiments PTZ was applied after either 40 min (n=29) or 10 h (n=5). (3) Ganglia were washed with bath solution for 60 min. (4) The samples were fixed for histological analysis (see Altrup and Peters 1982).

Results and Discussion

Under control conditions (without adding of PTZ), the dendrites of neuron B3 were distributed typically in the neuropile of the ganglia when compared with those of neurons B1 and B2 (Altrup 1987) and B4 (Peters and Altrup 1984). Dendrites ran in several bundles into the medial lobe of the ipsilateral ganglion giving the impression of a crab's claw (Fig. 1A). Normally, they do not enter the bucco-buccal commissure and the layer of somata. Single dendrites were branched to a small extent only; their diameters decreased slightly with increasing distance to the soma and they terminated in a "growth cone" displaying some filopodia (Fig. 2A). In general, the shape and intraganglionic distribution of dendrites of neuron B3 proved sufficiently constant to be used as a control.

After epileptic activity of 40 min, sprouting and degeneration were found (Figs. 1B, 2B and C). Sprouting was apparent: (i) by the increased number of dendrites (Fig. 1B); (ii) by the appearance of dendrites in regions normally without fibers of neuron B3, as in the commissure and in the layer of neuronal somata, and; (iii) by the increased number of filopodia (Fig. 2B). Degeneration of

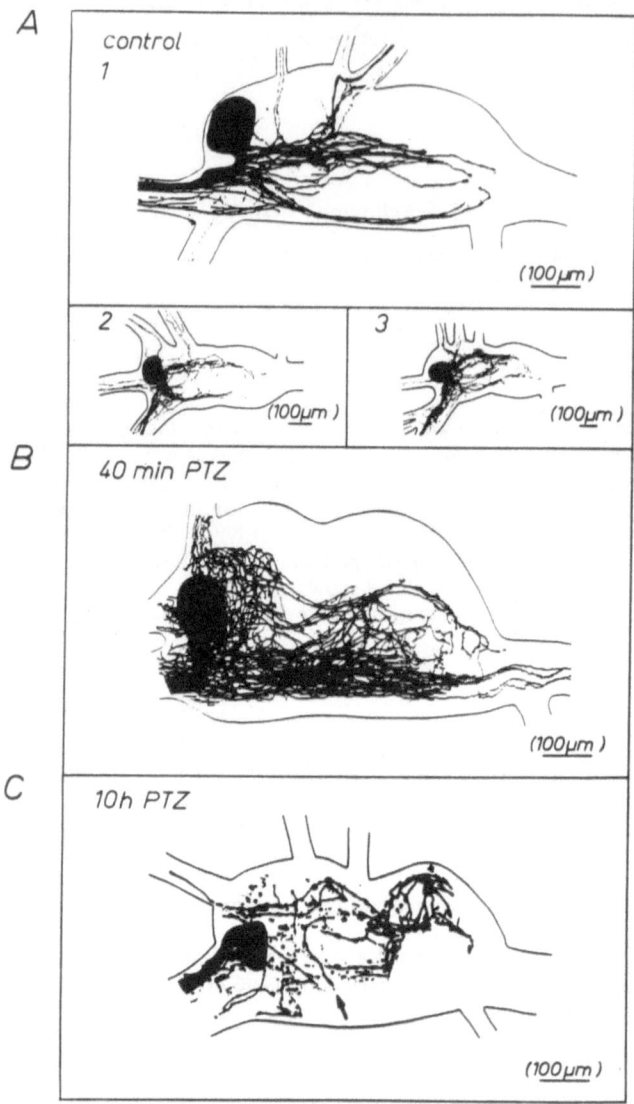

Fig. 1A-C. Pattern of dendrites of neuron B3 in the buccal ganglion of *Helix pomatia* in control experiments (A) and after adding of pentylenetetrazol (*PTZ*; **B,C**). Cobalt lysine staining; whole mount preparations; camera lucida drawings; calibrations not corrected for shrinkage. **A** Typical examples of neuron B3 under control conditions (1 to 3). **B** Neuron B3 after epileptic activity of 40 min induced by PTZ. **C** Neuron B3 after epileptic activity of 10 h induced by PTZ. The arrow marks an isolated dendrite situated above the layer of neuronal somata

dendrites was characterized by separation of whole dendrites from the rest of the cell (Fig. 2C) and in multiple isolations especially in the terminal regions of dendrites. Obviously changed dendrites could be situated beside ones of normal appearance in the same preparation. The described changes in the shape of dendrites were found in more than 50% of test experiments; however, comparable dendrites were also found in less than 20% of control preparations.

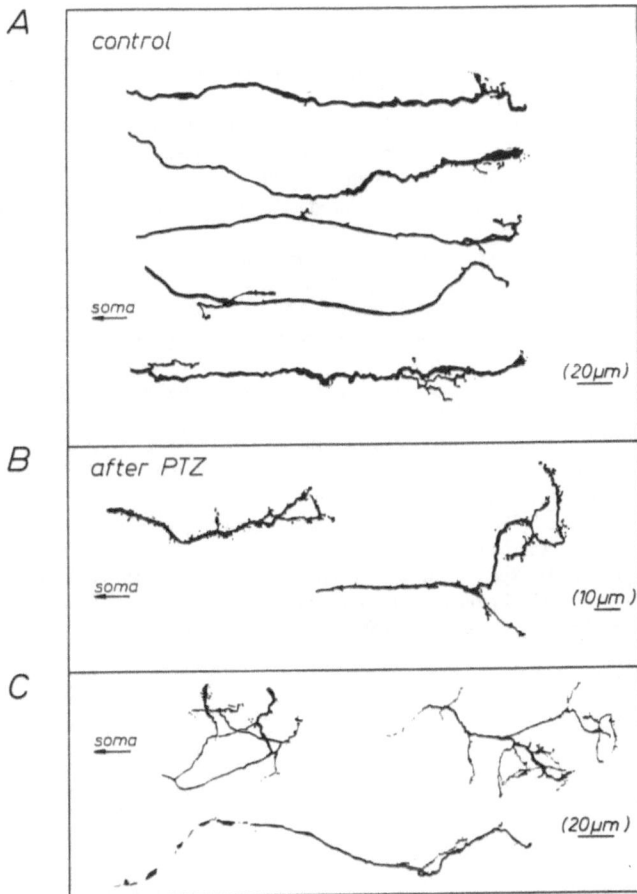

Fig. 2A-C. Shape of single dendrites of neuron B3 in the buccal ganglia of *Helix pomatia* in control experiments (A) and after adding of pentylenetetrazol (*PTZ*; B,C). Cobalt lysine staining; whole mount preparations; camera lucida drawings; calibrations not corrected for shrinkage

After 10 h of epileptic activity degenerations prevailed. They consisted in the above described separations and multiple isolations (Figs. 1C and 2C). As described for the effects of 40 min of epileptic activity, some dendrites were found in regions normally without dendrites of neuron B3. For example, the isolated dendrite marked by the arrow in Fig. 1C is situated between the layer of somata and the perineurium.

As to the mechanisms underlying these changes in neuronal shape, it is unknown whether they are induced during the period of epileptic activity or during the washing of the preparation after epileptic activity. The changes might, however, be linked to calcium inward currents, which have been reported to occur during epileptic activity (see Speckmann et al. 1986). To study whether or not calcium enters neuron B3, the free calcium concentration was measured on the outer surface of its soma during epileptic activity and was found to decrease extracellularly during each paroxysmal depolarization shift (see Lücke et al. 1989). Since sprouting processes (Cooper and Schliwa 1985; Patel et al. 1985) as well as degenerations (Meldrum 1986) have been linked to intracellular calcium, the morphological changes following epileptic activity could be mediated by the calcium inflow.

64

References

Altrup U (1987) Inputs and outputs of giant neurons B1 and B2 in the buccal ganglia of *Helix pomatia*. Brain Res 414: 271-284

Altrup U (1987) Die Dendriten der identifizierten B3-Zelle in den Buccalganglien der Weinbergschnecke: Experimentell ausgelöste Formänderungen. In: Elsner N, Creutzfeld O (eds) New frontiers in brain research. Thieme, Stuttgart, p 80

Altrup U, Peters, M (1982) Procedure of intracellular staining in the snail *Helix pomatia*. J Neurosci Methods 5: 161-165

Cooper MS, Schliwa M (1985) Electrical and ionic controls of tissue cell locomotion in DC electric fields. J Neurosci Res 13: 223-244

Lücke A, Speckmann E-J, Lehmenkühler A, Altrup U (1989) Changes in free calcium concentration at the surface of identified snail neurons during pentylenetetrazol-induced paroxysmal depolarization shifts. Pflugers Arch 413: R4

Meldrum BS (1986) Cell damage in epilepsy and the role of calcium in cytotoxicity. Adv Neurol 44: 849-855

Patel NB, Xie Z-P, Young SH, Poo M-M (1985) Response of nerve growth cone to focal electric currents. J Neurosci Res 13: 245-256

Peters M, Altrup U (1984) Motor organization in pharynx of *Helix pomatia*. J Neurophysiol 52 389-409

Pfeiffer J (1963) Morphologische Aspekte der Epilepsien. Springer, Berlin Göttingen Heidelberg

Ribak CE (1986) Contemporary methods in neurocytology and their application to the study of epilepsy. Adv Neurol 44: 739-764

Schulze H, Speckmann E-J, Kuhlmann D, Caspers H (1975) Topography and bioelectrical properties of identifiable neurons in the buccal ganglion of *Helix pomatia*. Neurosci Lett 1: 277-281

Speckmann E-J, Schulze H, Walden J (eds) (1986) Epilepsy and calcium. Urban and Schwarzenberg, Munich

Temporal Coupling of Epileptic Discharges in the Buccal Ganglia of Helix Pomatia

M. Madeja[1], U. Altrup[2], A. Lehmenkühler[1] and E. -J. Speckmann[1,2]

[1] Institut für Physiologie, Robert-Koch-Strasse 27a, 4400 Münster, FRG
[2] Institut für Experimentelle Epilepsieforschung, Hüfferstrasse 68, 4400 Münster, FRG

Introduction

Mechanisms underlying temporal coupling of epileptic discharges were studied in an invertebrate nervous system. Such nervous systems offer the opportunity to investigate giant neurons which represent individual cells with constant functions and interconnections (see Altrup 1987; Cottrell and Macon 1974; Peters and Altrup 1984; Schulze et al. 1975). Thus, temporal coupling of neuronal discharges during epileptic activities can be studied in a neuronal network, the wiring of which is principally known.

Material and Methods

The experiments were performed on identified and unidentified neurons in the buccal ganglia of *Helix pomatia* (Peters and Altrup 1984; Schulze et al. 1975). The ganglia were separated from the animal and kept in an experimental chamber containing a conventional bath solution (Altrup U. 1987). The membrane potentials of several neurons were recorded simultaneously. Epileptic activity was induced by adding pentylenetetrazol (40 mM) or etomidate (0.5 mM) to the bath solution (Altrup et al. 1989; Speckmann and Caspers 1973). Extracellular K^+ concentration was measured by K^+-selective microelectrodes based on the potassium ionophore I-cocktail B (Fluka).

Results and Discussion

With application of pentylenetetrazol or etomidate paroxysmal depolarization shifts (PDS) could be elicited in several neurons (Altrup et al. 1989; Speckmann and Caspers 1973). Simultaneous recordings revealed that PDS were coupled in time within defined neuronal networks, whereas PDS were not coupled in neurons belonging to different networks. For example, PDS in neurons B3 belonging to one of these networks are shown in Fig. 1. The temporal coupling of PDS developed within 30 min after administration of the drug (Fig. 1A). With higher time frame resolution (Fig. 1B), a latency of up to 1 s became apparent and enabled distinction of a "leader" and a "follower neuron". The more depolarized neuron was generally the leader. When the membrane potential of one neuron B3 shifted to more negative values spontaneously or by current injection (Fig. 1C), a rectangular depolarization (amplitude of ca. 5 mV) was unmasked (Fig. 1C 2). The depolarization was coupled in time to the PDS in the contralateral neuron B3 and was labeled as a "coupling depolarization".

In order to analyze whether or not the coupling depolarization is mediated by chemical transmission, a bath solution containing high magnesium and low calcium (Mg^{2+}: 9 mM; Ca^{2+}: 2 mM) was applied. Coupling depolarizations and coupling of PDS (Fig. 2A) were (Jefferys and Haas 1982) unaffected by this. Furthermore, in neuronal networks in which PDS were found to be synchronized, the neurons were electrically coupled with a coefficient between 0.1 and 0.01. The lower values appeared in cases in which there was no direct coupling, e.g., in neurons B3 (Fig.

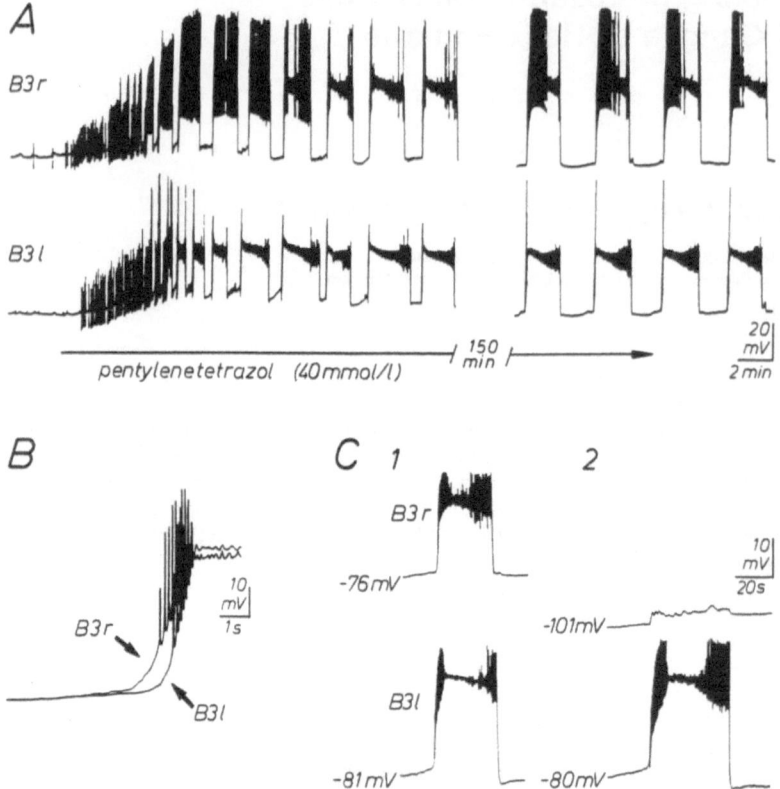

Fig. 1A-C. Paroxysmal depolarization shifts (*PDS*) simultaneously recorded in neurons B3 of the left (*l*) and right (*r*) buccal ganglion of *Helix pomatia* (pentylenetetrazol, 40 m*M*). **A** Development of coupling of PDS in time during administration of pentylenetetrazol (horizontal line). **B** Commencement of PDS in B3r and B3l (superimposed recordings). **C** Epileptic activities in neurons B3 before (*1*) and during hyperpolarization of neuron B3r (*2*) induced by intrasomatical current injection via a second microelectrode

2B). Neurons which were not coupled electrically (see Fig. 2B, neuron B2 vs B3) did not normally show synchronized PDS. Thus, coupling depolarization can be explained as an electrically transmitted PDS in defined networks.

The role of extracellular potassium ion concentration ($[K^+]_o$) in neuronal synchronization (Heinemann et al. 1977; Lux 1974) was estimated by measuring $[K^+]_o$ during epileptic activities using ion-selective microelectrodes (Fig. 2C). PDS in a motoneuronal network (one motoneuron is displayed in Fig. 2C 1) were accompanied by increases in $[K^+]_o$ which, in turn, did not synchronize PDS in neuron B3 (Fig. 2C 1) or in other neuronal networks, e.g., the network of neurons B2. Finally, the contribution of one neuron to $[K^+]_o$ was found to be very low (Fig. 2C 2). In general, $[K^+]_o$ did not synchronize PDS of different networks and it did not appear that K^+ release during the PDS of one neuron synchronized neighbouring cells.

In summary, epileptic discharges in the buccal ganglia of the snail *Helix pomatia* are coupled in time in different neuronal networks. As elementary mechanisms underlying this coupling, electrical contacts have to be taken into account, whereas changes in $[K^+]_o$ and chemical synaptic transmission appear to play minor roles.

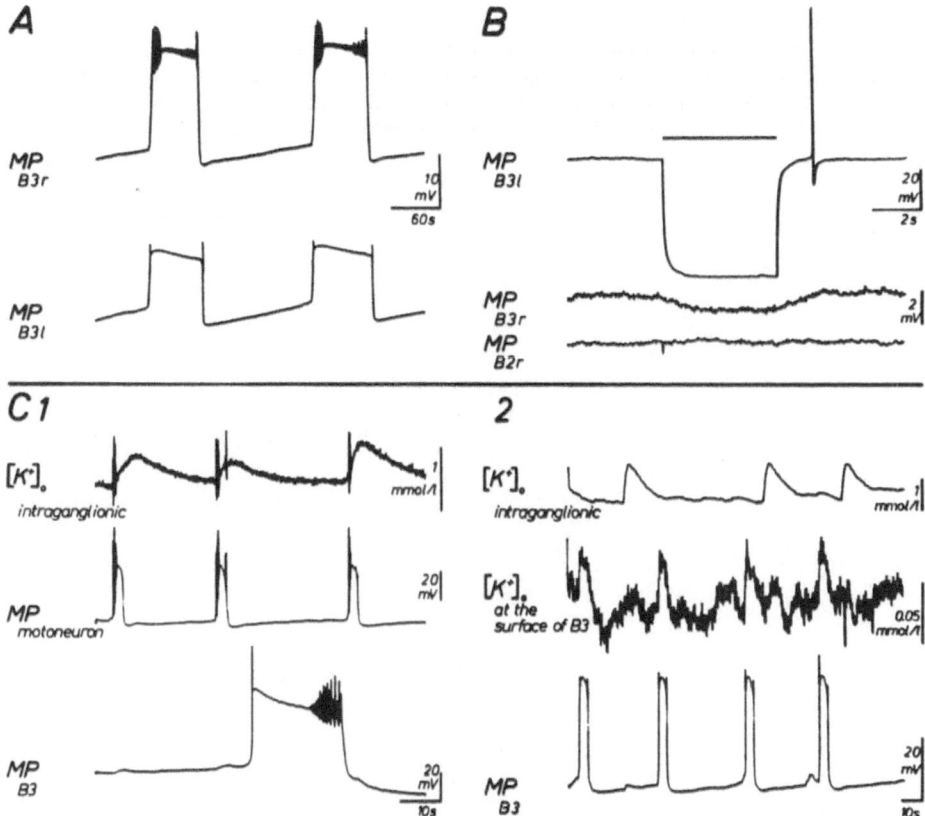

Fig. 2A-C. Membrane potential (*MP*) and extracellular K^+-concentration ($[K^+]_o$) in the buccal ganglia of *Helix pomatia* with respect to temporal coupling of paroxysmal depolarization shifts (*PDS*). *l*, left; *r*, right. **A** PDS in simultaneous recordings of neurons B3 in high Mg^{2+}, low Ca^{2+} bath solutions (9 m*M* $MgCl_2$, 2 m*M* $CaCl_2$; 40 m*M* pentylenetetrazol). **B** Membrane potentials of neurons B2 and B3 during injection of a hyperpolarizing current (horizontal bar; 10 nA) into neuron B3l. **C** $[K^+]_o$ recorded in the neuropile of the ganglion (*1,2* top) and on the surface of neuron B3 (*2* middle) simultaneously with the MP of neuron B3 (*1,2* bottom) and with a motoneuron (*1* middle); etomidate 0.5 m*M*

References

Altrup U (1987) Inputs and outputs of giant neurons B1 and B2 in the buccal ganglia of *Helix pomatia*: an electrophysiological and morphological study. Brain Res 414: 271-284

Altrup U, Lehmenkühler A, Speckmann E-J (1989) Paroxysmal depolarizations induced by etomidate in identified neurons (*Helix pomatia*). Pflügers Arch 413: R23

Cottrell GA, Macon JB (1974) Synaptic connexions of two symmetrically placed giant serotonin-containing neurones. J Physiol 236: 435-464

Heinemann U, Lux HD, Gutnick MJ (1977) Extracellular free calcium and potassium during paroxysmal activity in the cerebral cortex of the cat. Exp Brain Res 27: 237-243

Jefferys JGR, Haas HL (1982) Synchronized bursting of CA1 hippocampal pyramidal cells in the absence of synaptic transmission. Nature 300: 448-450

Lux HD (1974) The kinetics of extracellular potassium: Relation to epileptogenesis. Epilepsia 15: 375-393

Peters M, Altrup U (1984) Motor organization in pharynx of *Helix pomatia*. J Neurophysiol 52: 389-409

Schulze H, Speckmann E-J, Kuhlmann D, Caspers H (1975) Topography and bioelectrical properties of identifiable neurons in the buccal ganglion of *Helix pomatia*. Neurosci Lett 1: 277-281

Speckmann E-J, Caspers H (1973) Paroxysmal depolarization and changes in action potentials induced by pentylenetetrazol in isolated neurons of *Helix pomatia*. Epilepsia 14: 397-408

II BIOCHEMICAL MECHANISMS OF EPILEPTOGENIC ACTIVITY

Mechanisms Underlying Enhanced Synaptic Activation of NMDA Receptors During Hippocampal Seizures

Y. Yaari[1], M. S. Jensen[2] and A. Konnerth[3]

[1] Department of Physiology, Hebrew University School of Medicine, Jerusalem, Israel
[2] PharmaBiotec, Institute of Physiology, Aarhus University, Aarhus, Denmark
[3] Laboratory of Cellular Neurophysiology, Max-Planck-Institute for Biophysical Chemistry, Göttingen, FRG

Introduction

Fast synaptic excitation in the hippocampus is mediated by glutamate or analogue excitatory amino acid (EAA) transmitters acting at kainate/quisqualate postsynaptic receptors (Mayer and Westbrook 1987). Recent evidence indicates that hippocampal glutamatergic synapses also contain N-methyl-D-aspartate (NMDA) receptors (Monaghan and Cotman 1985). These receptors mediate only a small component of glutamatergic excitatory postsynaptic potentials (EPSPs) recorded under normal conditions (Andersean et al. 1988). Presumably most NMDA receptor channels are blocked by extracellular Mg^{2+} (Mg^{2+}_o) at normal resting membrane potentials and contribute only little to the excitatory postsynaptic currents (EPSCs) underlying the EPSPs. Indeed, when Mg^{2+}_o is washed out, the NMDA receptor mediated component of glutamatergic EPSPs is much increased (Coan and Collingridge 1985).

The Mg^{2+}_o block of NMDA receptor channels is strongly voltage dependent and is relieved by depolarization of the neuron (Mayer et al. 1984; Nowak et al. 1984). Since neurons in the discharge zone of an epileptic seizure depolarize profoundly for many seconds (Ajmone Marsan 1961), it could be expected that synaptic activation of NMDA receptors would be enhanced during seizure episodes. We tested this notion using an in vitro model of focal hippocampal epilepsy.

NMDA EPSPs Are Enhanced During Seizures

Exposing standard rat hippocampal slices to high K^+ solution induces an epileptic focus of ictal paroxysms in the CA1 field of approximately 40% of the slices (Jensen and Yaari 1988; Jensen and Yaari 1990; Traynelis and Dingledine 1988; Yaari and Jensen 1988). Each ictal episode characteristically consists of a tonic phase of sustained neuronal depolarization accompanied by repetitive, highly synchronized discharge. This is followed by a variable clonic phase of intermittent giant bursts (Fig. 1). CA1 pyramidal cells depolarize approximately 10 mV during the tonic phase of seizure (the actual transmembrane depolarization is about 15 mV due to concurrent extracellular negativity of approximately 5 mV).

Glutamatergic EPSPs in CA1 pyramidal cells acquire a slow component when evoked during the tonic phase of seizure (Fig. 1). This slow component was reversibly blocked by the NMDA receptor antagonist DL-5-aminophosphonovaleric acid (APV; 50 µM), indicating that it is mediated by synaptic activation of NMDA receptors (Fig. 1B,b). By contrast, EPSPs evoked between seizures did not contain a prominent APV-sensitive component. Thus synaptic activation of NMDA receptors is strongly facilitated during the tonic phase of seizure.

72

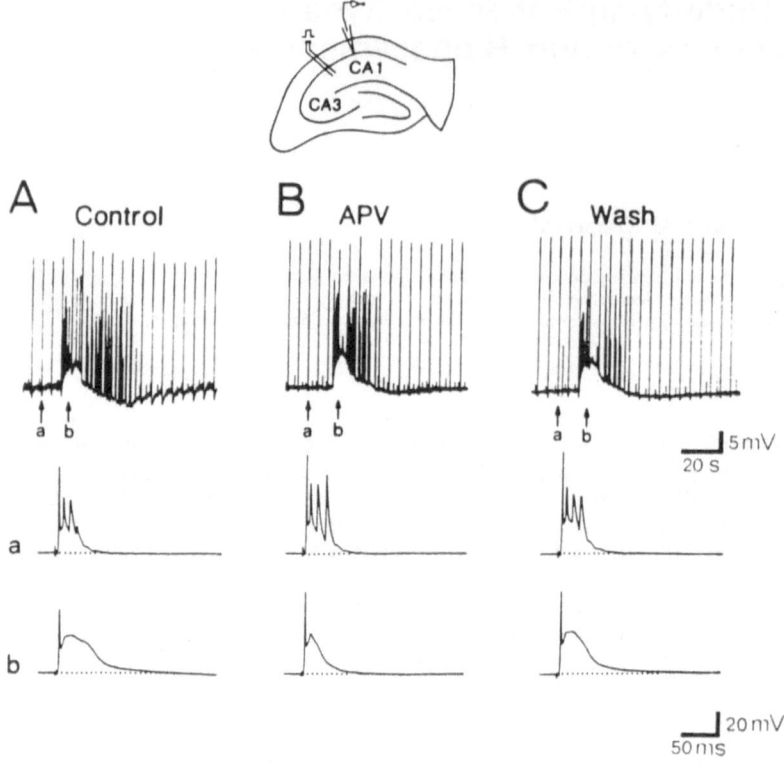

Fig. 1A-C. Appearance of a slow, APV-sensitive EPSP component during sustained paroxysms induced by K⁺ in standard rat hippocampal slices.

Above, intracellular recordings from a CA1 pyramidal cell. A constant stimulus (3 V) was applied to the stratum radiatum every 5 s. The preparation was perfused with saline containing 7.5 mM K⁺. The neuron resting potential in these conditions was -60 mV. Points *a* and *b* are selected EPSPs occurring, respectively, about 10 s before and during a tonic seizure. A Control. The EPSP evoked during the tonic seizure (*b*) contains a slow component not seen in the preictal EPSP (*a*). B Activity 30 min after adding 50 μM APV. The slow EPSP component is suppressed (*b*). C Activity 30 min after APV washout. The slow EPSP component reappears (*b*)

Membrane Voltage Modulates the Size and Kinetics of NMDA EPSCs

The appearance of an NMDA-mediated EPSP component (NMDA EPSC) during tonic seizures is likely to be a consequence of postsynaptic depolarization. We tested this possibility by determining the relationship between membrane voltage and NMDA EPSCs in hippocampal neurons (dentate granule cells and CA1 pyramidal cells) in thin hippocampal slices (Edwards et al. 1989). NMDA EPSCs (sensitive to block by APV) were isolated by blocking the non-NMDA (i.e. quisqualate and kainate) component of glutamatergic EPSCs with 5 μM 6-cyano-7-nitroquinoxaline-2,3-dione (CNQX) and recorded using the whole-cell configuration of the patch-clamp technique (Fig. 2).

Slow NMDA EPSCs, albeit small, were seen at normal resting potentials (approximately ca. -70 mV) and in 1 mM Mg²⁺ₒ. Depolarization had a dual effect on these currents. First, at potentials between -70 and -30 mV the size of NMDA EPSCs increased with depolarization, conferring a region of negative slope conductance to the I-V relationship. Second, at all potentials examined (from -90 to +50 mV), depolarization impeded the decay of the NMDA EPSCs. Washing the slices with nominally Mg²⁺ₒ-free saline markedly enhanced the amplitude of NMDA EPSCs at negative membrane voltage and shifted the region of negative slope conductance to more negative potentials. This treatment also slowed the decay of NMDA EPSCs but did not abolish its voltage sensitivity (Fig. 2).

In contrast to NMDA EPSCs, non-NMDA-mediated EPSCs (isolated by blocking the NMDA-mediated component with 50 μM APV) did not display a conspicuous voltage sensitivity.

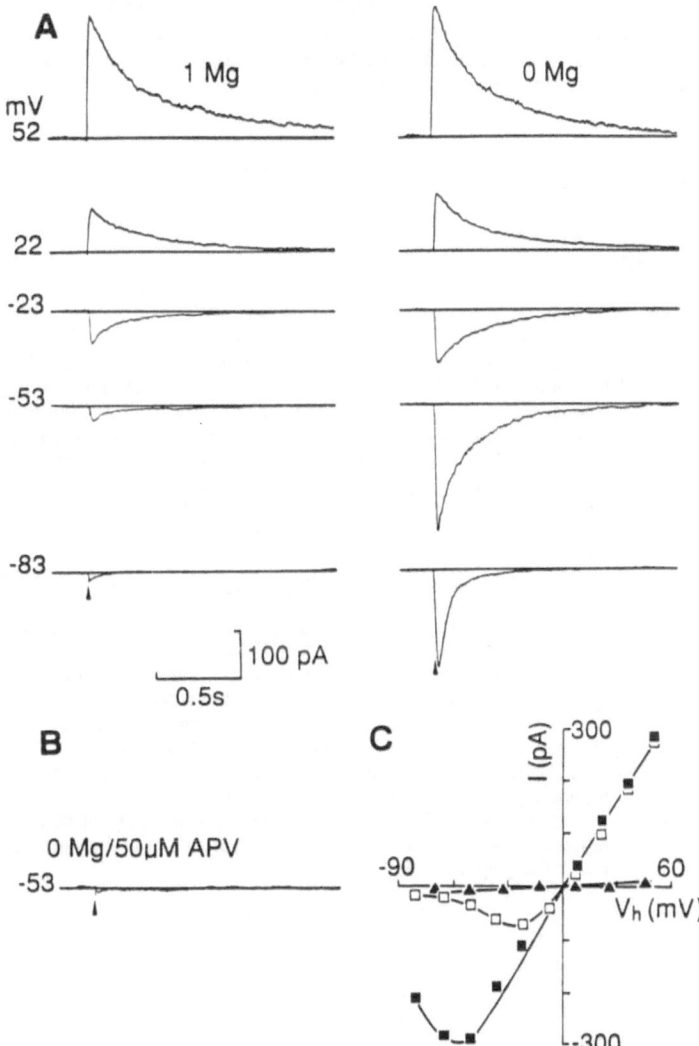

Fig. 2A-C. Effects of membrane voltage on NMDA EPSCs in dentate granule cells in thin rat hippocampal slices. NMDA EPSCs were pharmacologically isolated by blocking non-NMDA EPSCs with 5 μM CNQX. Inhibitory currents were blocked with 10 μM bicuculline methiodide. The NMDA EPSCs were evoked by stimulating afferent fibres near the apical outer border of the granule cell layer and recorded using the tight-seal whole-cell configuration of the patch-clamp technique. A NMDA EPSCs in 1 mM and 0 Mg^{2+}o evoked at different holding membrane potentials. Time of stimulation is indicated by the *black triangle* on the lower traces. Each displayed EPSC represents an average of five responses. **B** Effect of APV. An exemplary synaptic response evoked in 0 Mg^{2+}o after exposing the preparation to 50 μM APV. **C** I-V relation of NMDA EPSCs under different experimental conditions: in 1 mM Mg^{2+}o (*open squares*), in 0 Mg^{2+}o (*closed squares*) and in APV (*triangles*)

Conclusions

The data indicate that the NMDA-mediated EPSC component of glutamatergic EPSCs is enhanced in size and duration upon depolarization. The primary underlying mechanism most likely is removal of the Mg^{2+}_{o} block of NMDA receptor channels by depolarization. However, a Mg^{2+}_{o}-independent voltage sensitivity may contribute to the slowing of NMDA EPSC decay with depolarization. These mechanisms would be complementary in producing a slow, NMDA receptor-mediated EPSP component in glutamatergic synapses activated during sustained postsynaptic depolarization, as occurs in CA1 hippocampal pyramidal cells during the tonic phase of seizure. Whether additional mechanisms (e.g., facilitated glutamate release) contribute to the seizure enhanced activation of NMDA receptors has yet to be determined.

Acknowledgements. Supported by Israeli National Academy of Sciences, Danish MRC, ETP, GIF and Deutsche Forschungsgemeinschaft.

References

Ajmone Marsan C (1961) Electrographic aspects of "epileptic" neuronal aggregates. Epilepsia 2: 22-38

Andersean M, Lambert JDC, Jensen MS (1988) Effects of new non-NMDA antagonists on synaptic transmission in the in vitro rat hippocampus. J Physiol 414: 317-336

Coan EJ, Collingridge GL (1985) Magnesium ions block an N-methyl-D-aspartate receptor mediated component of synaptic transmission in rat hippocampus. Neurosci Lett 53: 21-26

Edwards FA, Konnerth A, Sakmann B, Takahashi T (1989) A thin slice preparation for patch clamp recordings from neurones of the mammalian central nervous system. Pfluegers Arch 414: 600-612

Jensen MS, Yaari Y (1988) The relationship between interictal and ictal paroxysms in an in vitro model of focal hippocampal epilepsy. Ann Neurol 24: 591-598

Jensen MS, Yaari Y. Role of glutamate receptors in potassium-induced paroxysms in rat hippocampal slices. This book

Mayer ML, Westbrook GL (1987) The physiology of excitatory amino acids in the vertebrate central nervous system. Prog Neurobiol 28: 197-276

Mayer ML, Westbrook GL, Guthrie PB (1984) Voltage dependent block by Mg^{2+} of NMDA responses in spinal cord neurones. Nature 309: 261-263

Monaghan DT, Cotman CW (1985) Distribution of N-methyl-D-aspartate receptors sensitive L-[3H]-glutamate binding sites in rat brain. J Neurosci 5: 2909-2919

Nowak L, Bregestovski P, Ascher P, Herbert A, Prochiantz A (1984) Magnesium gates glutamate-activated channels in mouse central neurones. Nature 307: 462-465

Traynelis SF, Dingledine RJ (1988) Potassium induced spontaneous electrographic seizures in the rat hippocampal slice. J Neurophysiol 59: 259-276

Yaari Y, Jensen MS (1988) Nonsynaptic mechanisms and interical-ictal transitions in the mammalian hippocampus. In: Dichter M A (ed) Mechanisms of epileptogenesis: from membrane to man. Plenum, New York, pp 183-198

Release of Intracellularly Stored Ca^{2+}
in Hippocampal Neurons by NMDA Receptor Activation

I. Mody[1], K. G. Baimbridge[2], J. A. Shacklock[2] and J. F. MacDonald[3]

[1] Department of Neurology and Neurological Sciences, Stanford University Medical Center, Stanford, CA 94305, USA
[2] Department of Physiology, University of British Columbia, Vancouver, B.C., V6T 1W5, Canada
[3] Department of Physiology and The Playfair Neuroscience Unit, Toronto Western Hospital, 399 Bathurst Street, Toronto, Ontario, M5T 2S8, Canada

Introduction

The excitatory cell-to-cell communication in the mammalian central nervous system (CNS) is mediated mainly by the two dicarboxylic amino acids, glutamate and aspartate (Watkins and Evans 1981; Mayer and Westbrook 1987b). Specific receptors for the excitatory amino acids (EAAs), in particular the N-methyl-d-aspartate (NMDA) receptor subtype, are critical in normal and pathological neuronal function (Watkins and Evans 1981; Mayer and Westbrook 1987b; Choi 1988). It is the increase in $[Ca^{2+}]_i$ following activation of NMDA receptors, that appears to trigger or mediate the altered neuronal excitability during long-term potentiation, epileptiform activity, spreading depression or neurotoxicity (Mayer and Westbrook 1987b; Choi 1988). There is considerable evidence that the extracellular origin of this $[Ca^{2+}]_i$ rise is extracellular; the combined influx of extracellular Ca^{2+} through NMDA- (MacDermott et al. 1986; Mayer and Westbrook 1987a; Ascher and Nowak 1988) and voltage-gated channels increases the intracellular concentration of the cation. This mechanism may not be the sole means of elevating $[Ca^{2+}]_i$. Neurons possess intracellular Ca^{2+} storage compartments of significant capacity which can be released by neurotransmitters (Smith et al. 1983; Neering and McBurney 1984; Alkon and Rasmussen 1988; Lipscombe et al. 1988). The quisqualate receptor of the EAA receptor family, has already been linked to the release of Ca^{2+} from intraneuronal organelles through an inositol 1,4,5,-trisphosphate (IP3)-dependent mechanism (Murphy and Miller 1988). Furthermore, caffeine has been shown to release Ca^{2+} from a distinct neuronal Ca^{2+} pool (Thayer and Miller 1988; Palade et al. 1989), an effect blocked by dantrolene Na (Thayer and Miller 1988).

An increase in free intracellular calcium concentration ($[Ca^{2+}]_i$) links excitatory amino acid (EAA) receptor activation to long-term neuronal excitability changes and neurotoxicity (Mayer and Westbrook 1987b; Choi 1988). The increase occurs when EAAs depolarize neurons to permit Ca^{2+} influx through voltage-dependent channels, or when N-methyl-d-aspartate (NMDA) channels open to allow Ca^{2+} entry (MacDermott et al. 1986; Mayer and Westbrook 1987a; Ascher and Nowak 1988). Using fura-2 microspectrofluorimetry in cultured rat hippocampal neurons, we have found that a third mechanism contributes to the rises in $[Ca^{2+}]_i$ induced by EAAs. Our study reveals that a significant portion of the Ca^{2+} responsible for the increase in $[Ca^{2+}]_i$ produced by activation of NMDA receptors is derived from an intraneuronal storage site. This storage site can be depleted in the absence of extracellular Ca^{2+} and cannot be restored by a rapid and transient reestablishment of the extra/intracellular Ca^{2+} gradient. The NMDA-sensitive intraneuronal Ca^{2+} pool is also susceptible to block by dantrolene-Na, an effective antagonist of sarcoplasmic Ca^{2+} release (Van Winkle 1976; Desmedt and Hainaut 1977). Thus, the source of Ca^{2+} contributing to the NMDA-induced $[Ca^{2+}]_i$ elevation is most likely the endoplasmic reticulum, which is known to function as a Ca^{2+} reservoir in nerve cells (Smith et al. 1983; Neering and McBurney 1984; Alkon and Rasmussen 1988).

Methods

Hippocampal neurons dissociated from embryonic rats (E18) were grown in cell culture (Banker and Cowan 1977) for 6-12 days on poly-d-lysine and laminin coated 18-mm diameter glass coverslips. Pluronic F-127 (BASF Wyandotte Corp. or Molecular Probes) and bovine serum albumin were used to facilitate loading (Poenie et al. 1986) of the neurons (for 1 hr at 37°C) with fura-2-AM (Molecular Probes). The cultures were washed for 30 min in control medium containing: 144 mM Na isethionate or NaCl, 5 mM KCl, 5.6 mM d-glucose 10 mM HEPES 1.8 mM CaCl2, 1-2 μM glycine, and 0.5-2 μM tetrodotoxin. EAA receptor agonists were added (25-100 μl; 300-500 μM) directly into a the laminar flow of medium (4 ml/min) entering a small volume (< 400 μl) flow-through chamber resulting in an estimated two to three fold dilution at their final site of action. Dantrolene Na (Norwich-Eaton), butanedione-monoxime (BDM; Sigma) and d-2-amino-5-phosphonovaleric acid (AP5; CRB) were dissolved in extracellular medium and administered by perfusion. Whenever perfusates were changed, the agonists were dissolved in the respective solutions. Resting or stimulated intraneuronal Ca^{2+} concentration ($[Ca^{2+}]_i$) was determined by measuring the ratio of fura-2 fluorescence at 510 nm following excitation at 350 and 380 nm (Grynkiewicz et al. 1985) every 1.8-5 s using a Jenalumar microscope equipped for epifluorescence. Individual cell bodies were isolated with an adjustable circular diaphragm and viewed through an oil immersion lens (*100). The fluorescent light was diverted either to the eyepieces or a photomultiplier tube.

After A/D conversion, the signal was stored on a computer for later calculation of $[Ca^{2+}]_i$ using the formula: $[Ca^{2+}] = K_D(F_0/Fs)(R-R_{min})/(R_{max}-R)$. Calibration was done in neurons exposed to the calcium ionophore Br-A23187 in the absence of external Ca^{2+} (with 10 mM EGTA) or in the presence of 1.8 mM Ca^{2+} (Grynkiewicz et al. 1985).

Results

We have investigated the possibility that the large NMDA-receptor mediated increases in $[Ca^{2+}]_i$ may involve Ca^{2+} release from intracellular storage sites. In the absence of extracellular Mg^{2+}, specific agonists and antagonists of the various EAA receptor/channels were used to activate only the NMDA receptor mediated component of the $[Ca^{2+}]_i$ increase. Additions of 1.8 mM Mg^{2+} to the perfusate or to the amino acid solutions effectively blocked the Ca^{2+} responses produced by glutamate (3.25-25 μM), aspartate (25-100 μM) and NMDA (100-200 μM; e.g., Fig. 1.). Because

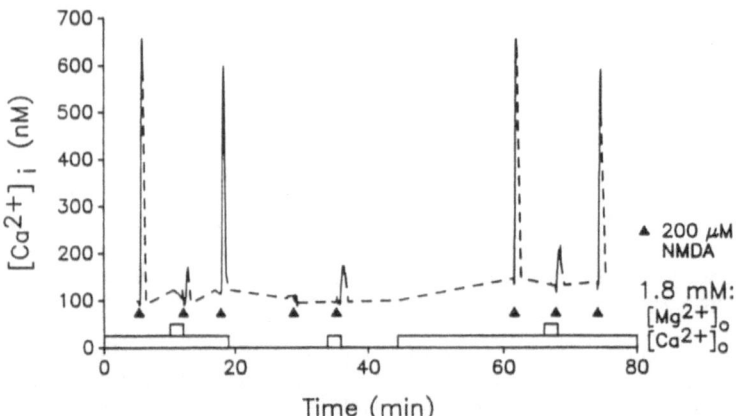

Fig. 1. The rises in $[Ca^{2+}]_i$ produced by NMDA receptor activation cannot be evoked in the absence of $[Ca^{2+}]_o$. Under such experimental conditions, Ca^{2+} responses are absent even when normal $[Ca^{2+}]_o$ is transiently restored. The Ca^{2+} responses induced by 50 μl 200 μM NMDA in the absence of $[Mg^{2+}]_o$ are significantly reduced by prior addition of 1 ml 1.8 mM Mg^{2+} to the flow-through chamber. However, in the absence of $[Ca^{2+}]_o$, a similar volume and concentration of Ca^{2+} fails to restore the NMDA-induced rises in $[Ca^{2+}]_i$ abolished by omission of extracellular Ca^{2+}. The NMDA-induced Ca^{2+} responses return to control level only following continuous perfusion of Ca^{2+} (1.8 mM) containing solutions, presumably indicating the necessity for the reloading of an intracellular Ca^{2+} pool. Horizontal bars (abscissa) indicate the presence (or absence) of 1.8 mM Ca^{2+} or Mg^{2+} in the extracellular solution. Note, however, that omission of Ca^{2+} from the solution with no added EGTA does not ensure a Ca^{2+}-free solution. It is estimated that a 0 mM Ca^{2+} solution contains 10-50 μM Ca^{2+} (Yaari et al. 1983)

Mg^{2+} is a specific voltage-dependent blocker of NMDA channels (Nowak et al. 1984; Mayer et al. 1984), it appears that these EAAs raise $[Ca^{2+}]_i$ through activation of NMDA-type receptors. This finding was further substantiated by the blocking effect of the specific competitive NMDA receptor antagonist AP5 (not shown).

Increases in $[Ca^{2+}]_i$ cannot be elicited by NMDA in the absence of extracellular Ca^{2+} ($[Ca^{2+}]_o$; Murphy et al. 1987). This finding has been attributed to the lack of available Ca^{2+} to enter through open NMDA- or voltage-gated channels. We tested this hypothesis in experiments where Ca^{2+} responses to NMDA or glutamate were abolished by omission of $[Ca^{2+}]_o$. We transiently (< 1 min) restored the Ca^{2+} gradient prior to and during agonist applications (Fig. 1.). This experimental manipulation, although temporarily reestablishing the extra/intracellular Ca^{2+} gradient is most likely insufficient to replenish the already depleted intraneuronal Ca^{2+} storage compartments. If extracellular Ca^{2+} entry were the sole source for the EAA-induced increases in $[Ca^{2+}]_i$, a full restoration of the NMDA-induced Ca^{2+} responses would be expected during the temporarily restored normal Ca^{2+} gradient. As shown in Fig. 1., this was clearly not the case. The NMDA receptor-mediated Ca^{2+} responses regained their full amplitudes only after exposing the neurons to normal (1.8 mM) $[Ca^{2+}]_o$ for 7-10 min. These findings suggest that the increases in $[Ca^{2+}]_i$ produced by NMDA receptor activation in control medium are only partially due to Ca^{2+} entry from the extracellular compartment and are supplemented by another component. Such a component may result from release of Ca^{2+} from storage sites within the neurons (Smith et al. 1983; Neering and McBurney 1984; Alkon and Rasmussen 1988; Murphy and Miller 1988). Most relevant to the present findings is the observation that such intracellular storage sites can be depleted when small cells are exposed to extracellular solutions nominally devoid of Ca^{2+} (Lipscombe et al. 1988; Murphy and Miller 1988). Moreover, these exchangeable intraneuronal Ca^{2+} stores recover relatively slowly when the normal Ca^{2+} gradient is reestablished (Murphy and Miller 1988).

We examined further the possibility that, in response to NMDA receptor activation, Ca^{2+} influx is complemented by the release of Ca^{2+} from an intraneuronal compartment. Analogous to the sarcoplasmic reticulum of muscle (Martonosi 1984), the endoplasmic reticulum of neurons may serve as an important intracellular Ca^{2+} reservoir and as such may store and release significant amounts of Ca^{2+} (Smith et al. 1983; Neering and McBurney 1984; Alkon and Rasmussen 1988; Lipscombe et al. 1988). There are no compounds known specifically to interact with the neuronal endoplasmic reticulum, but it has been established that Ca^{2+} release from the sarcoplasmic reticulum is blocked by dantrolene Na or BDM (Van Winkle 1976; Desmedt and Hainaut 1977; Fryer et al. 1988). We used both drugs in an attempt to block the intraneuronal release of Ca^{2+}. Dantrolene (10-20 μM) strongly antagonized the increases in $[Ca^{2+}]_i$ induced by glutamate or NMDA (Fig. 2). Similarly, the Ca^{2+} responses were significantly reduced by 10-20 mM BDM in a reversible and dose-dependent manner (not shown). The ranges of drug concentrations used in our experiments are known effectively to inhibit the release of sarcoplasmic Ca^{2+} (Van Winkle 1976; Desmedt and Hainaut 1977; Fryer et al. 1988). Dantrolene Na or BDM inhibited the NMDA-induced increases in $[Ca^{2+}]_i$, without a significant effect on resting $[Ca^{2+}]_i$. Figure 2B shows the results of such an experiment where NMDA-evoked increases in $[Ca^{2+}]_i$ were averaged in 14 neurons bathed in control medium and in another 11 neurons of the same culture while being exposed to 20 μM dantrolene.

Discussion

The possible effects of drugs known to inhibit the release of Ca^{2+} from the sarcoplasmic reticulum have not yet been examined in detail on neurons of the mammalian CNS. Although dantrolene prevents the caffeine-induced release of intracellularly stored Ca^{2+} in neurons (Kuba 1980; Thayer and Miller 1988), its effect on neuronal EAA responses is not known. In our present experiments, dantrolene could have inhibited the NMDA-induced increases in $[Ca^{2+}]_i$ simply by antagonism at the NMDA receptor/channel site, or alternatively by blocking voltage-dependent Ca^{2+} conductances. We have considered and examined both of these possibilities by using whole-cell voltage-clamp recordings in cultured hippocampal neurons (MacDonald et al. 1989). Dantrolene had no

78

A.

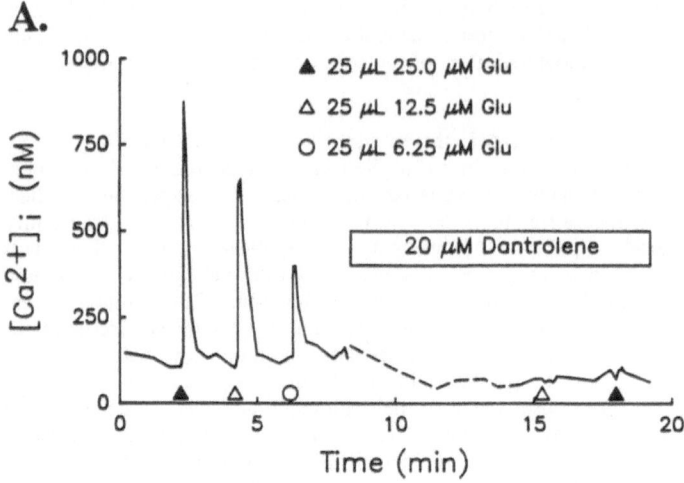

▲ 25 μL 25.0 μM Glu
△ 25 μL 12.5 μM Glu
○ 25 μL 6.25 μM Glu

20 μM Dantrolene

B.

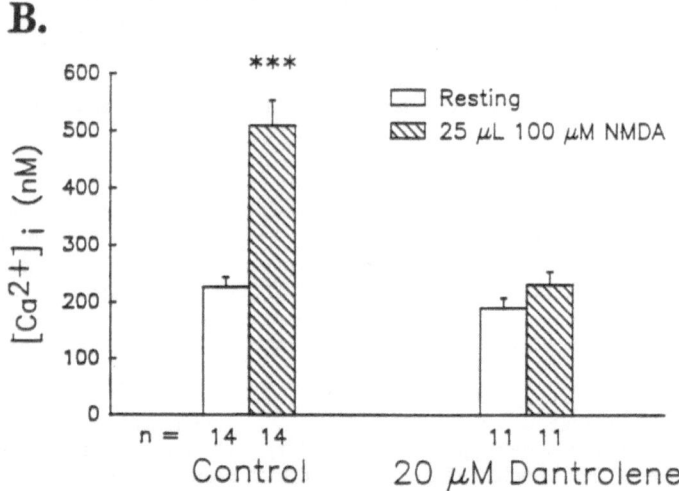

***

☐ Resting
▨ 25 μL 100 μM NMDA

n = 14 14   11 11
Control   20 μM Dantrolene

Fig. 2A,B. Inhibition of NMDA receptor mediated elevations in $[Ca^{2+}]_i$ by 20 μM dantrolene. **A** The increases in $[Ca^{2+}]_i$ produced by three different concentrations of glutamate (Glu) are effectively blocked by dantrolene. In the presence of dantrolene, a rapid Ca^{2+} entry through NMDA channels cannot be excluded, but such response may have been beyond the resolution of our sampling. **B** In a different set of experiments, NMDA was used as a specific agonist for the NMDA type of EAA receptors. Open bars, resting $[Ca^{2+}]_i$; shaded bars, intracellular Ca^{2+} responses evoked by applications of 25 μl of 100 μM NMDA in 25 different cells of the same culture in the absence (control; n=14) or in the presence of dantrolene (n=11). The average $[Ca^{2+}]_i$ increase produced by NMDA in the control perfusate was significantly different from all other groups (*** p<0.001; post-ANOVA Tukey test; F=24.7; error bars, SE)

significant effects on the amplitude or reversal potential of ionic currents produced by NMDA receptor activation (Mody and MacDonald, unpublished). Based on these findings, the possibility of a receptor/channel block mediating the effect of dantrolene on NMDA-induced increases in $[Ca^{2+}]_i$ can be excluded. Dantrolene also failed to reduce the voltage-dependent Ca^{2+} currents recorded in hippocampal neurons (Mody, Salter and MacDonald, unpublished). Analogous to its site of action in the muscle, the site of dantrolene block of Ca^{2+} responses to glutamate/NMDA is likely to be the neuronal endoplasmic reticulum. Taken together, our findings reveal the release of intracellularly stored Ca^{2+} by activation of NMDA receptors. It is important to note that our present experiments and previous studies (MacDermott et al. 1986; Murphy et al. 1987) have shown the blocking effect of NMDA receptor antagonists or channel blockers on increases in $[Ca^{2+}]_i$ produced by glutamate or NMDA. Thus, the opening of the NMDA channel could be required for the activation of the Ca^{2+} release mechanism. Possibly resembling the Ca^{2+}-dependent Ca^{2+} release process in muscle (Martonosi 1984), a limited influx of Ca^{2+} through the NMDA channel may trigger the intracellular Ca^{2+} release. Clearly, the possibility of such a Ca^{2+}- or voltage-dependent Ca^{2+} release from the neuronal endoplasmic reticulum needs to be investigated further. Whatever the intracellular source of the NMDA-released Ca^{2+} may be, it would appear to be rapidly depleted in the absence of $[Ca^{2+}]_o$ and, based on its sensitivity to dantrolene, is unlikely to be identical with the IP3-dependent intraneuronal Ca^{2+} pool (Thayer and Miller 1988; Palade et al. 1989).

In summary, our findings indicate that the bulk of the rise in $[Ca^{2+}]_i$ produced by activation of NMDA receptors involves the release of intraneuronal Ca^{2+} from a dantrolene-sensitive pool. This mechanism could be of considerable importance in long-term changes in neuronal excitability or neurotoxicity. Release of caged intracellular Ca^{2+} is sufficient to produce lasting potentiation of the synaptic responses in the hippocampus (Malenka et al. 1988), and dantrolene Na, presumably by blocking NMDA-induced release of intracellularly stored Ca^{2+}, effectively blocks long-term potentiation (Obenaus et al. 1989). Considering the role of intraneuronal Ca^{2+} elevation in NMDA receptor-mediated neurotoxicity (Choi 1988), our findings could lead to new strategies designed to prevent excessive cellular damage following ischemia/hypoxia or neuronal injury. In conclusion, three distinct but related mechanisms contribute to the increase in cytosolic Ca^{2+} by glutamate: i) entry of Ca^{2+} through the NMDA channel itself, which may in turn trigger, ii) the release of intraneuronally stored Ca^{2+} and finally, iii) influx of Ca^{2+} through the voltage-dependent Ca^{2+} channels opened by membrane depolarization.

Acknowledgements

Supported partly by NINCDS Grant NS12151 and the Hume Faculty Scholarship to I. Mody, and grants from the Canadian MRC to K.G. Baimbridge and J.F. MacDonald. The technical assistance of S. Atmaja is greatly appreciated. We would like to thank Norwich-Eaton for the generous gift of dantrolene Na.

References

Alkon DL, and Rasmussen H (1988) A spatial-temporal model of cell activation. Science 239: 998-1005.
Ascher P, and Nowak L (1988) The role of divalent cations in the N-methyl-d-aspartate responses of mouse central neurones in culture. J Physiol (Lond) 399: 247-266.
Banker GA, and Cowan WM (1977) Rat hippocampal neurons in dispersed cell culture. Brain Res 126: 397-425.
Choi DW (1988) Calcium-mediated neurotoxicity: relationship to specific channel types and role in ischemic damage. Trends in Neurosci 11: 465-469.
Desmedt JE, and Hainaut K (1977) Inhibition of the intracellular release of calcium by dantrolene in barnacle giant muscle fibres. J Physiol (Lond) 265: 565-585.
Fryer MW, Gage PW, Neering IR, Dulhunty AF, and Lamb GD (1988) Paralysis of skeletal muscle by butanedione monoxime, a chemical phosphatase. Pflügers Arch 411: 76-79.
Grynkiewicz G, Poenie M, and Tsien RY (1985) A new generation of Ca^{2+} indicators with greatly improved fluorescence properties. J Biol Chem 260: 3440-3450.
Kuba K (1980) Release of calcium ions linked to the activation of potassium conductance in a caffeine-treated sympathetic neurone. J Physiol (Lond) 298: 251-269.

Lipscombe D, Madison DV, Poenie M, Reuter H, Tsien RW, and Tsien RY (1988) Changes in calcium distribution in sympathetic neurons resulting from activity of calcium stores and calcium channels. Neuron 1: 355-365.

MacDermott AB, Mayer ML, Westbrook GL, Smith SJ, and Barker JL (1986) NMDA-receptor activation increases cytoplasmic calcium concentration in cultured spinal cord neurones. Nature 321: 519-522.

MacDonald JF, Mody I, and Salter MW (1989) Regulation of N-methyl-d-aspartate receptors revealed by intracellular dialysis of murine neurones in culture. J Physiol (Lond) 414: 17-34.

Malenka RC, Kauer JA, Zucker RS, and Nicoll RA (1988) Postsynaptic calcium is sufficient for potentiation of hippocampal synaptic transmission. Science 242: 81-84.

Martonosi AN (1984) Mechanism of Ca^{2+} release from sarcoplasmic reticulum of skeletal muscle. Physiol Rev 64: 1240-1320.

Mayer ML, and Westbrook GL (1987a) Permeation and block of N-methyl-d-aspartatic acid receptor channels by divalent cations in mouse cultured central neurones. J Physiol (Lond) 394: 501-527.

Mayer ML, and Westbrook GL (1987b) The physiology of excitatory amino acids in the vertebrate nervous system. Progr Neurobiol 28: 197-276.

Mayer ML, Westbrook GL, and Guthrie PB (1984) Voltagedependent block by Mg^{2+} of NMDA responses in spinal cord neurones. Nature 309: 261-263.

Murphy SN and Miller RJ (1988) A glutamate receptor regulates Ca^{2+} mobilization in hippocampal neurons. Proc Natl Acad Sci USA 85: 8737-8741.

Murphy SN, Thayer SA, and Miller RJ (1987) The effects of excitatory amino acids on intracellular calcium in single mouse striatal neurons in vitro. J Neurosci 7: 4145-4158.

Neering IR, and McBurney RN (1984) Role for microsomal calcium storage in mammalian neurones? Nature 309: 158-160.

Nowak L, Bregestowski P, Ascher P, Herbet A, and Prochiantz A (1984) Magnesium gates glutamate-activated channels in mouse central neurones. Nature 307: 462-465.

Obenaus A, Mody I, and Baimbridge KG (1989) Dantrolene-Na (Dantrium) blocks induction of long-term potentiation in hippocampal slices. Neurosci Lett 98: 172-178.

Palade P, Dettbarn C and Volpe P (1989) Inhibitors of inositol 1,4,5-trisphosphate-induced Ca^{2+} release from brain microsomes. Biophys J 55: 216a.

Poenie M, Alderton J, Steinhardt R, and Tsien RY (1986) Calcium rises abruptly and briefly throughout the cell at the onset of anaphase. Science 233: 886-889.

Smith SJ, MacDermott AB, and Weight FF (1983) Detection of intracellular Ca^{2+} transients in sympathetic neurones using arsenazo III. Nature 304: 350-352.

Thayer SA, and Miller RJ (1988) Mobilization, influx and buffering of Ca^{2+} in neurons. Biophys J 53: 430a.

Van Winkle WB (1976) Calcium release from skeletal muscle sarcoplasmic reticulum: site of action of dantrolene sodium? Science 193: 1130-1131.

Watkins JC, and Evans RH (1981) Excitatory amino acid transmitters. Annu Rev Pharmacol Toxicol 21: 165-204.

Yaari Y, Konnerth A, and Heinemann U (1983) Spontaneous epileptiform activity of CA1 hippocampal neurons in low extracellular calcium solutions. Exp Brain Res 51: 153-156.

NMDA Receptor-Mediated EPSP in Developing Rat Neocortex

N. Kato[*], M. S. Braun, A. Artola and W. Singer

Max-Planck-Institute for Brain Research, D-6000 Frankfurt-M. 71, FRG
[*] Present address: Institute for Brain Research, Kyoto University, 606 Kyoto, Japan

Introduction

NMDA receptors are known to play a crucial role in neural plasticity and cell death. Both phenomena occur more readily in developing animals than in adults, suggesting a predominance of NMDA receptor-mediated mechanisms in the young animal's brain. Such predominance is supported by several lines of evidence. NMDA receptors are denser in the kitten's visual cortex than in the adult's (Bode-Greuel and Singer 1989). Similarly, glutamate binding sites are denser in the visual cortex of 2-week-old rats than in that of 2-month-old rats (Schliebs et al. 1986). In the hippocampus of young rats the number of NMDA binding sites transiently increases at 1-2 weeks of age without change in affinity (Tremblay et al. 1988). NMDA receptor-mediated response contribute more to light-evoked activation of neurones in the visual cortex of young kittens than in that of adult cats (Tsumoto et al. 1987; Fox et al. 1989).

To our knowledge there are no electrophysiological studies on the development of NMDA receptor-mediated EPSPs (NMDA EPSPs) in the brain of young animals. Therefore we examined the postnatal change of NMDA receptor-mediated EPSPs in slices of rat visual cortex.

Methods

Slices were obtained from albino rats aged 12 days to 4 weeks. A submerged-type chamber was used for recording. A set of stimulating electrodes consisting of two parallel tungsten wires was inserted into the white matter underneath the recording site. Intracellular recordings were obtained from cells in layer II-IV with a glass micropipette filled with 3 M potassium acetate. Postsynaptic potentials (PSP) were evoked by subthreshold (70% - 80% of intensity required for spike generation) stimulation of the white matter at low frequency (0.06 - 0.07 Hz).

Results

Nearly all cells (98.0%) showed response properties characteristic of regular-spiking cells (McCormick et al. 1985). Only stable recordings from cells with no spontaneous firing, resting membrane potential (V_{m}) negative to -55 mV, and overshooting spikes were included in this analysis. NMDA receptor-mediated EPSPs were isolated by subtracting from the composite PSP the PSP component recorded after bath application of 25 nM APV (Fig. 1). The peak amplitude, onset latency, and time-to-peak of the NMDA receptor-mediated EPSPs were determined.

The peak amplitude of the NMDA receptor-mediated EPSPs was significantly larger ($p < 0.01$) in 2- and 3-week-old rats than in adults (Fig. 1), but there was no longer a significant difference between 4-week-old rats and adults. The time-to-peak was significantly longer ($p < 0.05$) than in adults at 2 weeks but not at 3 weeks of age. Other response parameters showed no significant developmental changes (Table 1).

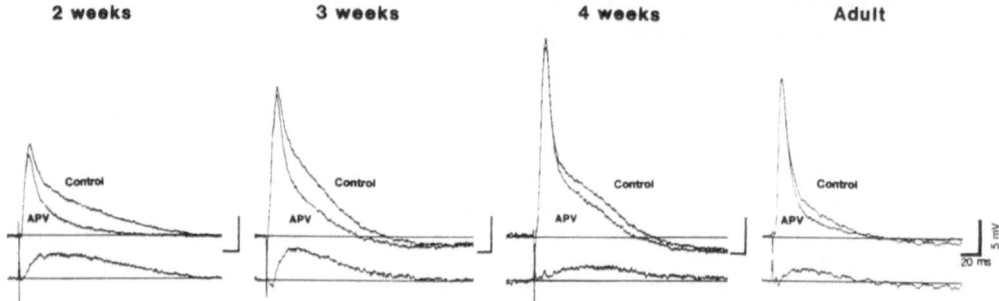

2 weeks   **3 weeks**   **4 weeks**   **Adult**

Fig. 1. NMDA receptor-mediated EPSPs recorded in slices from rats at different ages. *Upper traces,* evoked EPSPs before *(control)* and after APV application *(APV)*. All the traces are the average of five recordings

Table 1. Development of several parameters of the NMDA receptor-mediated EPSP

	Aged 2 weeks (n = 9)	Aged 3 weeks (n = 6)	Adults (n = 8)
Peak amplitude (mV)	3.2 ± 0.69[a]	3.4 ± 1.2[a]	2.0 ± 0.7
Onset latency (ms)	2.5 ± 3.6	2.9 ± 3.0	9.0 ± 6.8
Time-to-peak (ms)	38.3 ± 12.3[b]	35.9 ± 13.6	26.4 ± 9.4

[a] $p < 0.01$
[b] $p < 0.05$

In adult rat, reduction of the $GABA_A$ receptor-mediated initial IPSP (iIPSP) leads to a drastic enhancement of NMDA receptor-mediated EPSPs both in the visual cortex (Artola and Singer (1989) and in the hippocampus (Herron et al. 1985). Thus, one reason for the larger NMDA receptor-mediated EPSPs in slices from young rats could be reduced inhibition. Hence we tried to obtain an estimate of the strength of $GABA_A$ receptor-mediated iIPSPs in young rats.

We elicited PSPs at different levels of membrane potential (V_m) and observed in all cells examined a negative deflection of the PSP at 20 - 22 ms poststimulus at V_m more positive than -50 to -70 mV, indicating the presence of an iIPSP. The difference (V_r) between the maximum of this deflection and the V_m before stimulation (Fig. 2d) was measured. These values were plotted against the V_m for each cell and separately for the three age groups (Fig. 2a, b, c). The slope of this function is an indirect measure of the membrane current at the peak of the $GABA_A$ receptor-mediated initial IPSP. The average slope was -0.71 ± 0.13 (n = 5) at 2 weeks, -0.50 ± 0.07 (n = 4) at 3 weeks of age, and -0.50 ± 0.09 (n = 18) in adults. Since these slopes depend critically on the membrane resistance (R_m), we calculated the response conductance (Connors et al. 1988) in order to obtain a more direct measure for the iIPSP independent to R_m. The response conductance at about 20 ms poststimulus was 66.0 ± 27.0 (n = 5; nS) at 2 weeks of age and 63.9 ± 28.0 (n = 8; nS) in adults. These results indicate that the iIPSP is roughly as pronounced in young rats as it is in adults.

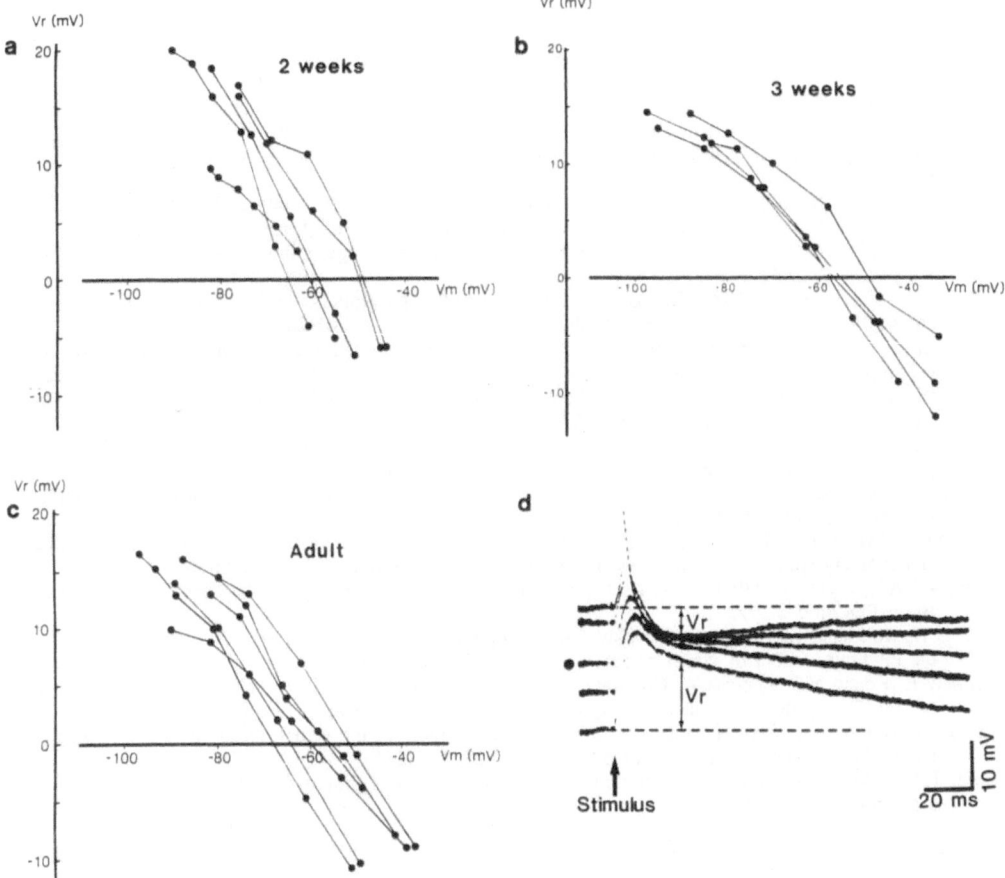

Fig. 2A-C. Graphs showing iIPSP amplitudes at different levels of membrane potential (V_m). **D** Evoked responses recorded from a cell at different levels of V_m. *Lines, arrows* show how the measurement was carried out. Potential deflections (V_r) were measured from the V_m to the bottom of the deflection at about 20 - 22 ms poststimulus at which point the deflection was largest. *Filled circle,* resting V_m

Discussion

Our results showed that (1) the size of NMDA receptor-mediated EPSPs is larger in the visual cortex of 2- and 3-week old rats than in that of adults, and that (2) the iIPSP is already pronounced on 2-week-old rats as in adults. This suggests that in the cortex of young rats NMDA receptor-dependent mechanisms per se are more prominent and/or NMDA receptor-mediated EPSPs are less sensitive to the suppressive effect of the iIPSP than in adults.

NMDA receptors are known to play a key role in neural plasticity as well as in learning and memory. In the visual cortex two forms of neural plasticity, ocular dominance plasticity and long-term potentiation, have been reported to depend on NMDA receptors (Artola and Singer, 1987, 1990; Kleinschmidt et al. 1987; Gu et al. 1989). Both phenomena are known to be age dependent (Wiesel and Hubel 1963; Hubel and Wiesel 1970; Kato et al. 1988; Perkins and Teyler 1988). The present experiments suggest that this age dependence may be attributed to the developmental decline in the NMDA receptor-mediated EPSP.

References

Artola A, Singer W (1987) Long-term potentiation and NMDA receptors in rat visual cortex. Nature 33: 649-652

Artola A, Singer W (1990) The involvement of N-methyl-D-aspartate receptors in induction and maintenance of long-term potentiation in rat visual cortex. Eur J Neurosci 2: 254-269

Bode-Greuel KM, Singer W (1989) The development of NMDA-receptors in cat visual cortex. Dev Brain Res 46: 197-204

Connors BW, Malenka RC, Silva LR (1988) Two inhibitory postsynaptic potentials, and GABA$_A$ and GABA$_B$ receptor-mediated responses in neocortex of rat and cat. J Physiol (Lond) 406: 443-468

Fox K, Sato H, Daw N (1989) The location and function of NMDA receptors in cat and kitten visual cortex. J Neurosci 9: 2443-2454

Gu Q, Bear MF, Singer W (1989) Blockade of NMDA-receptor prevents ocularity changes in kitten visual cortex after reversed monocular derivation. Dev Brain Res 47: 281-288

Herron CE, Williamson R, Collingridge GL (1985) A selective N-methyl-D-aspartate antagonist depress epileptiform activity in rat hippocampal slices. Neurosci Lett 61: 255-260

Hubel DH, Wiesel TN (1970) The period of susceptibility to the physiological effects of unilateral eye closure. J Physiol (Lond) 206: 419-436

Kato N, Artola A, Singer W (1988) Susceptibility of visual cortical neurones to undergo long-term potentiation decreases with age. Eur J Neurosci [Suppl]: 31

Kleinschmidt A, Bear MF, Singer W (1987) Blockade of "NMDA" receptors disrupts experience=dependent plasticity of kitten striate cortex. Science 238: 355-358

McCormick DA, Connors BW, Lighthall JW, Prince DA (1985) Comparative electrophysiology of pyramidal and sparsely spiny stellate neurons of the neocortex. J Neurophysiol 54: 782-806

Perkins IV AT, Teyler TJ (1988) A critical period for long-term potentiation in the developing rat cortex. Brain Res 439: 222-229

Schliebs R, Kullmann E, Bigl V (1986) Development of glutamate binding sites in the visual structures of the rat brain. Effect of visual pattern deprivation. Biomed Biochim Acta 45: 495-506

Trembley E, Roisin MP, Represa A, Charriaut-Marlangue C, Ben-Ari Y (1988) Transient increased density of NMDA binding sites in the developing rat hippocampus. Brain Res 461: 393-396

Tsumoto T, Hagihara K, Sato H, Hata Y (1987) NMDA receptors in the visual cortex of young kittens are more effective than those of adult cats. Nature 327: 513-514

Wiesel TN, Hubel DH (1963) Single-cell responses in striate cortex of kittens deprived of vision in one eye. J Neurophysiol 26: 1003-1017

Role of Glutamate Receptors in Potassium-Induced Paroxysms in Rat Hippocampal Slices

M. S. Jensen[1] and Y. Yaari[2]

[1] PharmaBiotec Research Center, Institute of Physiology, University of Aarhus, Denmark
[2] Department of Physiology, Hebrew University School of Medicine, Jerusalem, Israel.

The epileptic cortex generates two patterns of abnormal neuronal discharges (Ajmone-Marsan 1961): (a) Brief synchronous bursts which are responsible for the interictal spikes in the EEG. These paroxysms originate at a defined focus and project therefrom to other brain regions only to a limited extent.(b) Sustained paroxysms (ictal episodes) which tend to spread and recruit large brain areas into epileptic seizures.

Epileptiform activity consisting of both sustained (ictal) and intervening brief (interictal) paroxysms can be induced in rat hippocampal slices simply by moderately elevating the potassium concentration ($[K^+]_0$) in the perfusing saline (Yaari and Jensen 1987; Yaari et al. in this volume). The underlying mechanisms responsible for the two types of paroxysms in this model are not been resolved, but appear to be different in nature (Jensen and Yaari 1988). Thus, we have investigated the role of glutamatergic synapses in generating paroxysmal discharges in an in vitro model of hippocampal epilepsy.

Potassium-Induced Paroxysms in Hippocampal Slices

Perfusing hippocampal slices with high-K^+ (7 - 8.5 mM) saline induced interictal bursts in CA3 and CA1 fields, recurring regularly every 1 - 2 sec. Cutting the pathways connecting the two hippocampal fields prevented interictal bursting in CA1, but not in CA3. Thus, interictal bursts arise in a CA3 "spiking" focus and project therefrom to the synaptically connected CA1 (Jensen and Yaari 1988).

In approximately 40% of the slices, interictal "spiking" in CA1 was disrupted by sustained (20 - 120 sec) paroxysms recurring every 1 - 3 min. These ictal episodes typically consisted of a tonic-clonic sequence (Fig. 1A), termed in this way because their appearance resembles electrical activity recorded in vivo during tonic-clonic motor seizures. During the tonic phase, CA1 hippocampal pyramidal cells (HPCs) depolarized and fired repetitively and in synchrony. Consequently, $[K^+]_0$ in the pyramidal layer of CA1 rose to about 11 mM (Yaari and Jensen 1987), causing the extracellular field to shift negatively. The tonic phase was followed by a clonic phase, consisting of a long series of intermittent giant bursts triggered by the "spiking" focus in CA3 (Fig. 1B).

Ictal paroxysms developed also in surgically isolated (and interictally silent) CA1 fields. These paroxysms consisted of a tonic phase only (Fig. 2Aa); however, stimulating afferent fibres in the dendritic layer (stratum radiatum) at 0.5 or 1 Hz to mimic interictal input from CA3 restored the clonic phase (Fig. 2Ab) (Jensen and Yaari 1988). Interestingly, clonic bursts could similarly be produced by stimulating the axons of CA1 HPCs in the alveus (Fig. 2Ac).

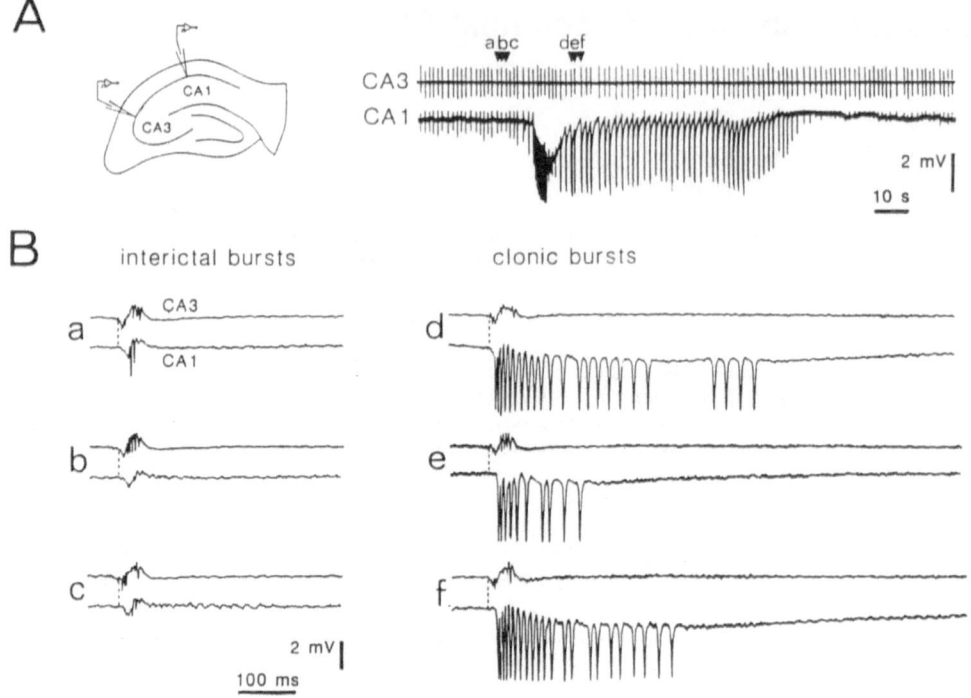

Fig. 1A,B. Epileptiform activity induced by high-K⁺ saline in hippocampal slice. **A** *Left*, experimental setup. Extracellular microelectrodes (1M NaCl; 15 MΩ) were positioned in CA3 and CA1 pyramidal layer of an intact slice. *Right*, interictal bursts and an ictal episode induced in the slice after increasing [K⁺] in the saline from 3.5 to 7 mM. **B** Excerpts from A before (*a,b,c*) and after (*d,e,f*) the sustained tonic paroxysmal discharge in CA1 are shown in B on an expanded time scale.
Methods: Hippocampal slices (500 μM) from 9 to 13 week old rats were kept at 35°C and perfused with oxygenated (95% O_2 - 5% CO_2) saline (in mM): NaCl, 124; KCl, 3.5 - 8.5; NaH_2PO_4, 1.25; $MgSO_4$, 1.2; $CaCl_2$, 1.2; $NaHCO_3$, 26 and D-glucose, 10. All drugs used were added directly the perfusing saline. Conventional techniques for stimulation and recording were used.

Role of Glutamatergic Synaptic Transmission

Glutamate appears to be the major excitatory transmitter in the hippocampus. Glutamate activates both NMDA (N-methyl-D-aspartate) and non-NMDA (i.e. kainate and quisqualate) receptors in hippocampal neurons. Specific non-NMDA receptor antagonists, such as 6-cyano-7-nitroquinoxaline-2,3-dione (CNQX; 15 μM) (Honore et al. 1988), block fast synaptic excitation of CA1 and CA3 HPCs (Andreasen et al. 1988). Although NMDA receptors are abundant in area CA1 (Monaghan et al. 1983), they normally contribute only sparsely to formation of excitatory postsynaptic potentials (EPSPs) in CA1 HPCs.

The specific NMDA receptor antagonist 2-amino-5-phosphonovaleric acid (APV; 50 μM) did not suppress interictal "spiking" in CA3, and in CA1 was unable to block the spontaneous interictal bursts. In contrast the non-NMDA receptor antagonist CNQX (15 μM) promptly abolished their generation in the CA3 subfield, and consequently also impeded interictal activity in CA1. Interictal discharges in area CA1 evoked by afferent fiber stimulation were blocked by CNQX, but not by APV.

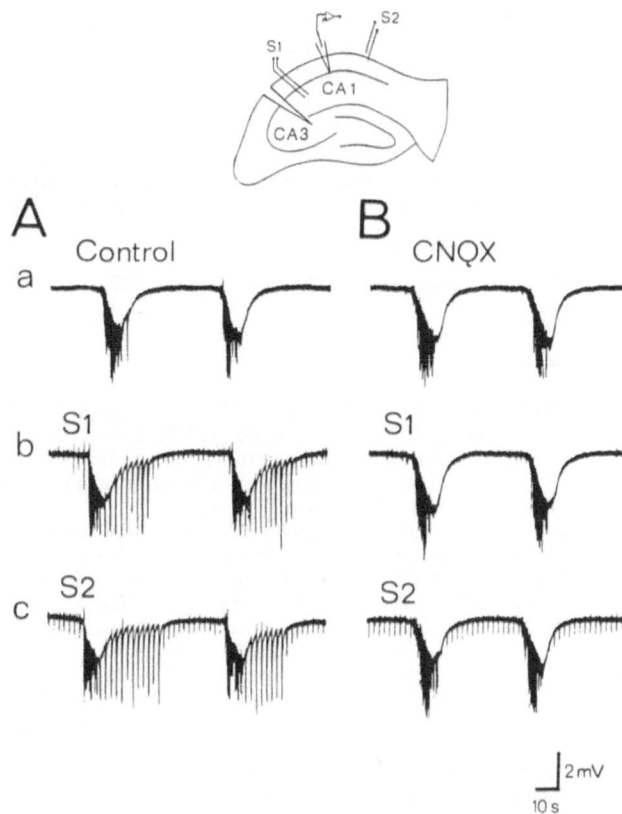

Fig. 2A,B. Brief bursts in CA1 are blocked by CNQX (15 μM). *Top* experimental setup. CA1 was isolated from CA3. Stimulating electrodes were positioned in stratum radiatum (CA2) and alvius fibers (CA1) for ortho- and antidromic stimulation of CA1 HPCs. Extracellular electrode (1M NaCl; 10 MΩ) was positioned 200 μM deep, in stratum pyramidale CA1. A Tonic paroxysms induced by elevating [K⁺]ₒ to 7.5 mM in the isolated CA1 field. Orthodromic (3 V; 0.5 Hz) and antidromic (2.5 V; 0.5 Hz) evoked brief paroxysms. B CNQX does not affect the generation of tonic paroxysms but blocks both ortho- (*b*) and antidromic (*c*) evoked interictal and clonic bursts.

Furthermore, APV had no apparent effects on the frequency or intensity of the ictal episodes (Yaari et al. in this volume). By contrast, CNQX (15 μM) promptly abolished interictal bursts in CA3 and in CA1, as well as the clonic phase of ictal episodes. Likewise, CNQX blocked clonic paroxysms evoked by ortho- or antidromic stimulation (Fig. 2bc). However, CNQX, alone or in combination with APV, did not interfere with the generation of tonic paroxysms (Fig. 2a).

Conclusions

These data suggest that K⁺-induced brief and sustained paroxysms are fundamentally different. The spontaneous interictal bursts generated in CA3 are blocked by CNQX and evolve therefore through local synaptic excitation mainly mediated by non-NMDA receptors.

Although brief paroxysms do not arise spontaneously in CA1 when [K⁺]₀ is elevated, they can be evoked very easily in this field by ortho- or antidromic volleys. The sensitivity of these paroxysms

88

to CNQX indicates that recurrent synaptic excitation, mediated by non-NMDA receptors, also underlies the genesis of brief paroxysms in CA1. Accordingly, the appearance of giant clonic paroxysms suggests that recurrent synaptic excitation is particularly enhanced during ictal episodes.

Excitatory neuronal interactions underlying the sustained paroxysms in CA1 do not involve activation of NMDA or non-NMDA receptors. They are generated most probably by nonsynaptic mechanisms (e.g., $[K^+]_0$ accumulation and ephaptic coupling) (Jensen and Yaari 1988; Yaari and Jensen 1987) which are more powerful in CA1 than in CA3.

Acknowledgement. Supported by The Danish MRC, ETP and the Israeli National Academy of Science.

References

Ajmone-Marsan C (1961) Electrographic aspects of "epileptic" neuronal aggregates Epilepsia 2: 22-38
Andreasen M, Lambert JDC, Jensen MS (1988) Effects of new non-NMDA antagonists on synaptic transmission in the in vitro rat hippocampus. J Physiol 414: 317-336
Honore T, Davies SN, Drejer J, et al. (1988) Quinoxalinediones: potent competitive non-NMDA glutamate receptor antagonists. Science 241: 701-703
Jensen MS, Yaari Y (1988) The relationship between interictal and ictal paroxysms in an in vitro model of focal hippocampal epilepsy. Ann Neurol 24: 591-598
Monaghan DT, Holets VR, Toy DW, Cotman CW (1983) Anatomical distributions of four pharmacologically distinct ^{3}H-L-glutamate binding sites. Nature 306: 176-179
Traynelis SF, Dingledine RJ (1988) Potassium induced spontaneous electrographic seizures in the rat hippocampal slice. Neurophysiol 59: 259-276
Yaari Y, Jensen MS (1987) Soc Neurosci Abstr 13: 1155
Yaari Y, Jensen MS (1987) Nonsynaptic mechanisms and interictal-ictal transitions in the mammalian hippocampus. In: Dichter MA (ed) Mechanisms of epileptogenesis: From membrane to man. Plenum Press, New York, pp.183-198
Yaari Y, Jensen MS, Konnerth A, Mechanisms underlying enhanced synaptic activation of NMDA Receptors during hippocampal seizures. in this volume

Excitatory Amino Acids-Induced Synaptic Plasticity in Hippocampus: Relation to the Age of Animals

O. A. Krishtal[1], A. V. Petrov[1], S. V. Smirnov[1], Ye. Sh. Yanovsky[1] and M. C. Nowycky[2]

[1] A. A. Bogomoletz Physiology Institute, Kiev, USSR
[2] Department of Anatomy, Medical College of Pennsylvania, Philadelphia, PA 19129, USA

Introduction

The excitatory synaptic input to pyramidal CA1 neurons from the Schaffer collateral-commissural pathway is operated by glutaminergic mechanisms, which normally respond only to non-N-methyl-D-aspartate (non-NMDA) blockers during low frequency stimulation (Collingridge et al. 1983). Intensive stimulation of this pathway leads to a transient activation of NMDA receptors and accompanying elevation of intracellular Ca^{2+} concentrations, which, in turn, is thought to be a necessary step in the formation of long-term potentiation (LTP), a stimulus-induced form of synaptic plasticity (Bliss and Lomo 1973; Collingridge and Bliss 1987). Recent reports indicate that under certain conditions the involvement of NMDA receptors during low-frequency stimulation may become persistent. *In vivo* this has been observed in experiments on kindled animals (Mody and Heinemann 1987). *In vitro* a similar observation was made in hippocampal slices continuously perfused with EAA (Krishtal et al. 1988). Some aspects of such EAA-induced synaptic plasticity are investigated in this study.

Material and Methods

Slice Preparation

Experiments were performed on hippocampal slices of rats (3 - 4 week, 8 - 12 week, and 16 - 20 week old animals). After isolation, the hippocampus was cut in slices 200 - 500 μm thick in the presence od ice-cold preincubation solution. During preincubation and recording, the slices were kept in oxygenated (95% O_2, 5% CO_2) solution which was composed of (in mM): NaCl - 124, KCl - 5, $MgSO_4$ - 2, NaH_2PO_4 - 1.24, $NaHCO_3$ - 26, $CaCl_2$ - 2.5, glucose - 20. The pH was buffered at the level of 7.4. Experiments were carried out at room temperature and at 32°C. The slices were incubated for 2 h before recordings.

Perfusion and Recording

Slices were placed on platinized metal net in a 0.7 ml chamber. Each drug or component was added to 30 ml of standard solution, which was recycled through the chamber at a rate of 5 ml/min during the period indicated. For electrical stimulation (0.5 - 1 Hz, 20 V, 0.2 ms) monopolar wire nichrom electrode (100 μm thick) was positioned onto the surface of the slice. Pyramidal neurons from the CA1 region were activated by stimulation the Schaffer collateral-commissural pathway. For field potential recording 50 μm thick nichrom wire was used. The noninsulated tip of wire was placed onto the surface of a slice in the CA1 cell body region. For intracellular recording glass pipettes filled with 3 M buffered (pH 7.5) potassium acetate were used. Resistance of microelectrodes was 40 - 80 MΩ. A ring with metal platinized net was used for slice fixation.

Experimental Brain Research Series 20
© Springer-Verlag Berlin · Heidelberg 1991

Results

Figure 1 illustrates the phenomenon in slices obtained from 3 - 4 weeks old rats and demonstrated its extremely long-lasting nature. As has been previously reported (Bernstein and Fisher 1985; Collingridge et al. 1983; Fagni et al. 1983; Fagni et al. 1983; Krishtal et al. 1988), bath application of Glu or Asp greatly suppresses Schaffer collateral-commissural evoked PSPs, without affecting other inputs (Bernstein and Fisher 1985). If perfusion of EAA is maintained, PSP amplitudes are largely or completely restored (Krishtal et al. 1988). These recovered responses display altered pharmacological properties. If the slice is perfused with Glu, the recovered PSPs become sensitive to NMDA antagonists. On washout of Glu, synaptic transmission is again insensitive to NMDA antagonists as in the initial state. However, there is an indication of some long-lasting changes, since during subsequent Glu application PSPs are only partially and briefly suppressed. Similar behavior was obtained with Asp perfusion, but the pharmacology of the PSPs present under Asp remains unclear, since the responses are insensitive to both NMDA antagonists and γ-D-glutamylglycine a nonspecific antagonist (Krishtal et al. 1988). Enhanced uptake, which probably participates in synaptic plasticity, may be responsible for the lack of total inhibition with later repeated application of Glu and Asp, but cannot explain the change in receptor pharmacology.

Neurons with stable (25 min) resting potentials (more then 65 mV) were selected for studying the action of Glu on the resting potential and EPSP values. All neurons under investigation produced persistent EPSPs and action potentials as response to Schaffer collateral-commissural pathway stimulation. Recordings generally lasted for 1.5 - 2 h.

Glu produced hyperpolarization of neuronal membrane which could be observed within the whole period of Glu application and could not be removed by Glu washout. The hyperpolarization value was 8 - 11 mV in different neurons. Short-term neuronal membrane depolarization (Bernstein and Fisher 1985) was observed only at the initial period of Glu application (Fig. 2). As concerns intracellularly recorded EPSP, Glu induces its 10 - 20 min decrease followed by a gradual recovery to the value of 60 - 80% of initial amplitude (Fig. 2).

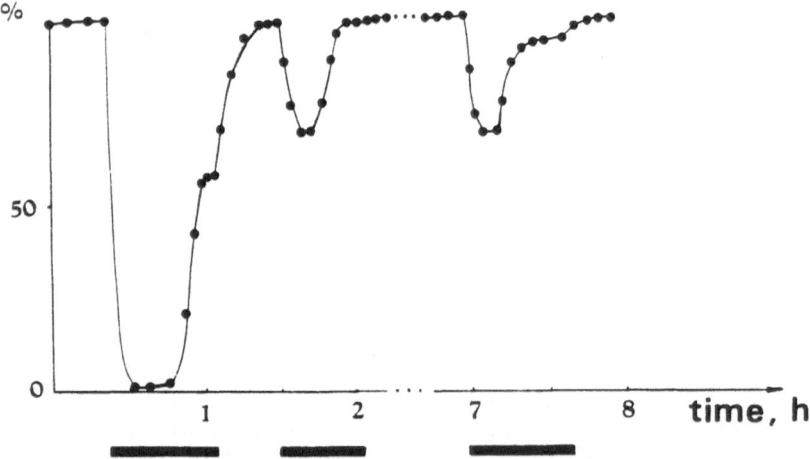

Fig. 1. Effects of excitatory amino acid perfusion on the hippocampal synaptic transmission. Excitatory postsynaptic field potentials (PSPs) were recorded extracellularly from CA1 cell body region following stimulation of the Schaffer collateral-commissural pathway. Data are plotted as percent of control peak amplitude. Bath application of Glu (0.5 mM, *filled bar*) produces an essentially complete inhibition of synaptic transmission within 5 min, which is followed by a slower recovery of PSP amplitudes to a plateau level. On removal of Glu, PSP amplitudes return to their original levels. The existence of a persistent, modified state of synaptic transmission is indicated by subsequent perfusion of Glu, which produces only a transient, partial inhibition to the plateau level. These changes last for many hours (*third filled bar*)

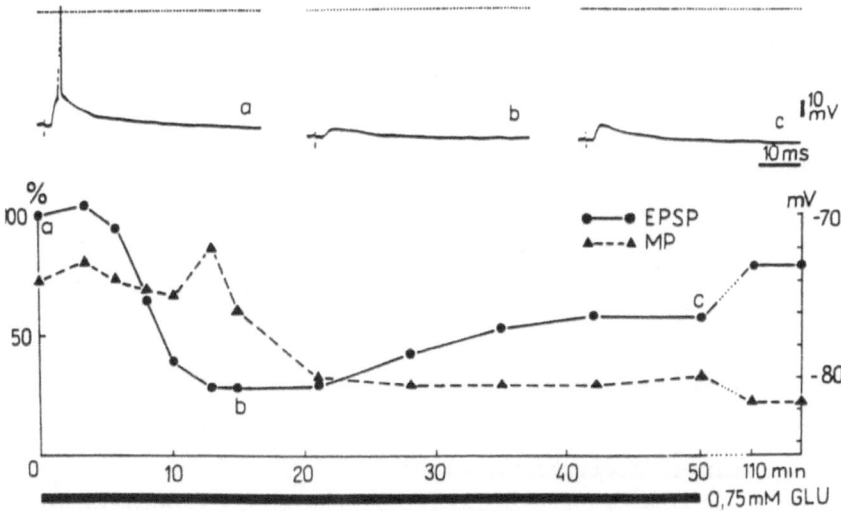

Fig. 2. Time course of intracellularly recorded EPSP (●) and membrane potential (▲). Application of 0.75 mM Glu is shown by the *filled bar*. Hyperpolarization is irreversible for at least 2 h. *Inset, a, b,* and *c* show intracellularly recorded EPSPs at times indicated in the plot below

Field potential in slices obtained from older animals (8 - 12 weeks old) do not produce biphasic effects: no recovery of synaptic transmission could be obtained in the case of EAA application on such slices. The recovery ability strongly correlates with the sensitivity to muscarinic receptor agonist Cch. With the increase in the animals' age, the sensitivity to Cch falls and ability of the stimulus-evoked PSPs to recover also falls. Such results suggest that phosphoinositide (PI) metabolism, activated by muscarinic agonists, may play a role in described phenomena.

Recently it was reported that synaptic transmission from mossy fibers to CA3 neurons in the hippocampus is largely inhibited by ω-CgTX (Kamiya et al. 1988), a specific and irreversible blocker of neuronal N- and L-type Ca^{2+} channels (McCleskey et al. 1987). We have confirmed this finding for CA1 neurons excited by stimulation of the Schaffer collateral-commissural pathway and explored whether the toxin is effective in the other postulated "states" of synaptic transmission (Krishtal et al. 1988). As in the study of Kamiya et al. (1988), in the initial 'non-NMDA' state (Krishtal et al. 1988), ω-CgTX blocked PSPs to about 20 - 30% of control levels within 30 min of perfusion. This block was largely, though not completely, irreversible. Figure 3 demonstrates an experiment in which the slice first was subjected to a prolonged bath application of Glu. As expected, PSPs were initially strongly inhibited and, after a lag period, reappeared. In this transformed state, synaptic transmission loses its sensitivity to ω-CgTX (Fig. 3). Prolonged application of ω-CgTX at concentrations that were maximally effective in control conditions (see also Kamiya et al. 1988, Fig. 2) decreases PSPs by a maximum of 20%, with no block in some experiments ($n = 5$). The similar insensitivity to ω-CgTX was obtained with PSPs reappearing after Asp perfusion ($n = 2$). After removal of Glu, sensitivity to ω-CgTX was restored, but the time course of toxin block (shown in Fig. 3) was considerably slower than the sum of the times needed to remove Glu and to produce ω-CgTX block in the initial state. This argues against simple competition between the EAA and ω-CgTX for a binding site. It appears that during EAA perfusion ω-CgTX cannot bind to Ca^{2+} channels: if the toxin was added to the slice in the presence of Glu and then both were washed out, PSPs were not blocked ($n = 2$, data not shown).

When applied in the initial state, ω-CgTX also irreversibly blocked the ability of the synaptic mechanism to undergo subsequent transition to the transformed state; we made 4 attempts using Glu and 2 with Asp after washing out ω-CgTX. This is not due solely to an inhibition of calcium

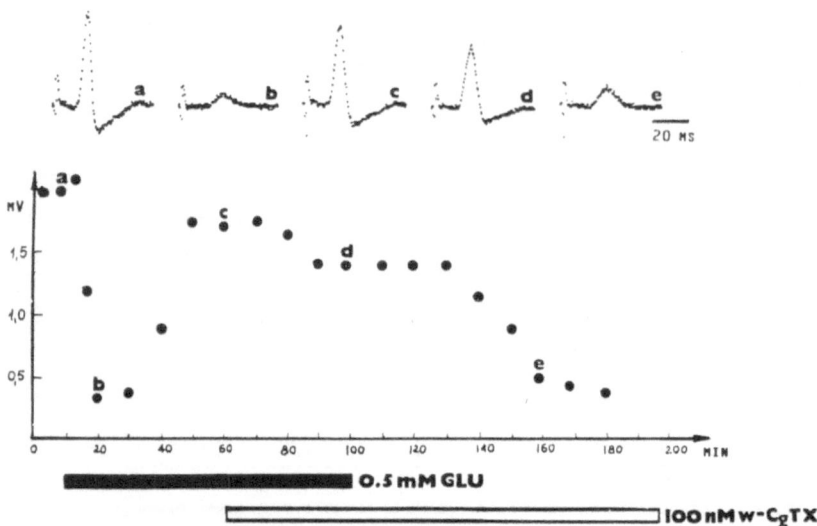

Fig. 3. Hippocampal synaptic transmission loses sensitivity to ω-CgTX during excitatory amino acid perfusion. ω-CgTX does not block synaptic transmission during the EAA induced transformed state. *Inset, a - e* show representative PSPs at times indicated in the plot. PSPs which recovered during continuous Glu perfusion were not blocked by 40 min of concurrent administration of ω-CgTX. Following washout of Glu (which takes place for 5 - 10 min), ω-CgTX blocked synaptic transmission, but only after a delay of approximately 40 min. This block, as in the initial state, was irreversible (data not shown)

entry following ω-CgTX block to Ca^{2+} channels. Complete suppression of PSPs and Ca^{2+} entry by 10 mM Mg^{2+} had no effect on the transforming process. If Glu was perfused concurrently with Mg^{2+}, the slice was shown to be in the NMDA-sensitive state (Krishtal et al. 1988) as soon as the divalent cation was washed out ($n = 2$).

We tested another modulator of synaptic transmission which undergoes changes in potency depending on the biochemical status of the preparation. Adenosine is usually a powerful, reversible inhibitor of excitatory synaptic transmission in the hippocampus (Dolphin and Ascher 1983; Dunwiddie and Hoffer 1980; Schubert and Mitzdorf 1979) as well as at many other synapses. Following pre-exposure of a hippocampal slice to the muscarinic agonist, Cch, adenosine becomes completely ineffective in blocking PSPs (Worley et al. 1987). A similar phenomenon was observed in our experimental conditions. In the initial state, adenosine (20 μM) fully blocked PSPs in a completely reversible manner ($n = 15$, not shown). If the slice was exposed to EAAs, adenosine had no effect on the recovered PSPs obtained under Glu or Asp ($n = 9$ and 7, respectively). After EAA washout, sensitivity to adenosine (as in the case of ω-CgTX) recovers.

Recently, Worley et al. (1988) proposed that the sensitivity of synaptic transmission to adenosine is regulated by muscarinic agonists secondary to stimulation of PI metabolism. EAA also increase PI turnover (Sladeczek et al. 1988). To determine if this second messenger pathway is involved in the EAA-modified state, we tested several compounds which interact with the PI system. So far, we have no evidence that PI turnover plays some role in producing the transitions between states. Li^+, which inhibits PI turnover, was ineffective at concentrations similar to those used by Worley et al. (1987), (0.5 mM) prior and during Glu application ($n = 3$). Staurosporine (6 nM), a potent protein kinase C (PKC) inhibitor, was ineffective as to the occurrence or kinetics of transition ($n = 2$). Ca^{2+} release from internal stores by caffeine application (10 mM) was also ineffective ($n = 3$). So, the role of muscarinic agonist-stimulated PI turnover remains unclear and the mechanism or pathways by which the transitions between states occur await further work.

Discussion

We can only speculate on the ω-CgTX resistance of synaptic transmission in the EAA-transformed state. Of the three types of calcium channels (high-threshold N and L types and low-threshold T type that had been described in neurons (Nowycky et al. 1985)) only the first two could be blocked by ω-CgTX in a specific and irreversible manner. Existing evidence suggests that N-type channels are the most important entry pathway for calcium during synaptic transmission (Hirning et al. 1988; Kamiya et al. 1988). Thus, one possible explanation is that the ω-CgTX insensitive, T-type channels underlie synaptic transmission in the transformed by EAA but not initial state. Unfortunately, no truly specific blockers of T-type channels are available at present.

A more likely explanation is that in the transformed state, ω-CgTX cannot bind to the calcium channels responsible for transmission, perhaps because of some conformational change in the channels. High-threshold calcium channels undergo conformational changes as a consequence of ion permeation (Pietrobon et al. 1985), membrane potential (Brown et al. 1986), and G-protein binding (Scott and Dolphin 1987). Recently, it was reported that the toxin blocks neuronal calcium channels in their calcium passing state, but not if they become permeable to sodium in the absence of calcium (Carbone and Lux 1988). In the EAA-transformed state, a conformational change in the channels could be induced by altered levels of a second messenger system, resulting, for example, in different degrees of phosphorylation or G-protein binding. The toxin's irreversible block of transitions between states argues for this hypothesis, since once the channels are bound they may be immobilized from further conformational changes. On the other hand, exposure to the toxin in the EAA-transformed state does not inhibit synaptic transmission after Glu washout, suggesting that the toxin does not bind to channels in this state.

Parallels between the *in vitro* experiments described here and *in vivo* work can be drawn qualitatively, although not quantitatively. While the concentration of EAA was 0.5 mM in the perfusing solution, the actual concentration within the slice is estimated to be much lower than in the bath because of bulk uptake (Gartwaithe 1985). LTP studies have shown substantial liberation of Glu and Asp from slices even if only selective pathways are stimulated (Bliss et al. 1986). Thus, conditions of intense stimulation of EAA pathways may lead to large increases of Glu and Asp concentrations, locally and transiently. These high concentration transients may induce the transitions described here much more quickly than the Glu and Asp application in slice experiments.

An *in vivo* correlate to these findings is the observation of acquired NMDA sensitivity of low-frequency synaptic transmission following kindling (Mody and Heinemann 1987). NMDA-sensitive responses to single volleys also have been observed in slices undergoing epileptic activity (Slater et al. 1985) or in the presence of convulsant drugs (Dingledine et al. 1986; Herron et al. 1985). The significance of the different states may be that (1) excitatory synaptic transmission will be unimpaired even if neurons are repeatedly subjected to Glu and Asp, and (2) this transmission will exhibit long-lasting plastic changes. These changes are profound, involving not only changes in EAA receptor activity, but also changes in the properties of calcium channels which form part of the machinery for transmitter release.

Acknowledgements. This study was supported by the Ukrainian Academy of Sciences and by ICI.

References

Bernstein J, Fisher RS (1985) Excessive glutamate as an inhibitor of excitatory transmission in rat hippocampal slice. Neurosci Lett 61: 19-24
Bliss TVP, Douglas RM, Errington ML, Lynch MA (1986) Correlation between long-term potentiation and release of endogenous amino acids from dentate gyrus of anaesthetized rats. J Physiol (Lond) 377: 391-408

94

Bliss TVP, Lomo T (1973) Long-lasting potentiation of synaptic transmission in the dentate area of the anaesthetized rabbit following stimulation of the perforant path. J Physiol (Lond) 232: 331-356

Brown AM, Kunze DL, Yatani A (1986) Dual effects of dihydropyridines on whole cell and unitary calcium currents in single ventricular cells of guinea-pig. J Physiol (Lond) 379: 415-514

Carbone E, Lux HD (1988) Omega-conotoxin blockade distinguishes Ca from Na permeable states in neuronal calcium channels. Pflügers Arch 413: 14-22

Collingridge GL, Bliss TVP (1987) NMDA receptors - their role in long-term potentiation. Trends Neurosci 10: 288-293

Collingridge GL, Kehl SJ, McLennan HJ (1983) The antagonism of amino-acid induced excitations of rat hippocampal CA1 neurons in vivo. J Physiol (Lond) 334: 19-31

Collingridge GL, Kehl SJ, McLennan HJ (1983) Excitatory amino acids in synaptic transmission in the Schaffer collateral-commissural pathway of the rat hippocampus. J Physiol (Lond) 334: 33-46

Cotman CW, Iversen LL (1987) Excitatory amino acids in the brain - focus on NMDA receptors. Trends Neurosci 10: 263-265

Dingledine R, Hynes MA, King GL (1986) Involvement of N-methyl-D-aspartate receptors in epileptiform bursting in the rat hippocampal slice. J Physiol (Lond) 380: 175-189

Dolphin AC, Ascher ER (1983) An adenosine agonist inhibits and cyclic AMP analogue enhances the release of glutamate but not GABA from slices of rat dentate gyrus. Neurosci Lett 43: 49-54

Dunwiddie TV, Hoffer BJ (1980) Adenine nucleotides and synaptic transmission in the in vitro rat hippocampus. Br J Pharmacol 69: 59-68

Fagni L, Baudry M, Lynch G (1983) Desensitization to glutamate does not affect synaptic transmission in rat hippocampal slices. Brain Res 261: 167-171

Fagni L, Baudry M, Lynch G (1983) Classification and properties of acidic amino acid receptors in hippocampus. J Neurosci 8: 1538-1546

Fox AP, Nowycky MC, Tsien RW (1987) Kinetic and pharmacological properties distinguishing three types of calcium currents in chick sensory neurones. J Physiol (Lond) 394: 149-172

Gartwaithe J (1985) Cellular uptake disguises action of L-glutamate on N-methyl-D-aspartate receptors. Br J Pharmacol 85: 297-307

Herron CE, Williamson R, Collingridge GL (1985) A selective N-methyl-D-aspartate antagonist depresses epileptiform activity in rat hippocampal slices. Neurosci Lett 61: 255-260

Hirning LD, Fox AP, McCleskey EW, Olivera BM, Thayer SA, Miller RJ, Tsien RW (1988) Dominant role of N-type Ca^{2+} channels in evoked release of norepinephrine from sympathetic neurons. Science 239: 57-61

Kamiya H, Sawada S, Yamamoto C (1988) Synthetic ω-CgTX blocks synaptic transmission in the hippocampus in vitro. Neurosci Lett 91: 84-88

Krishtal OA, Smirnov SV, Osipchuk YuV (1988) Changes in the state of the excitatory synaptic system in the hippocampus on prolonged exposure to excitatory amino acids and antagonists. Neurosci Lett 85: 82-88

McCleskey EW, Fox AP, Feldman DH, Cruz LJ, Olivera BM, Tsien RW, Yoshikami D (1987) ω-Conotoxin: direct and persistent blockade of specific types of calcium channels in neurons but not muscle. Proc Natl Acad Sci USA 84: 4327-4331

Mody I, Heinemann U (1987) NMDA receptors of dentate gyrus granule cells participate in synaptic transmission following kindling. Nature (Lond) 326: 701-704

Nowycky MC, Fox AP, Tsien RW (1985) Three types of neuronal calcium channel with different calcium agonist sensitivity. Nature (Lond) 316: 440-443

Pietrobon D, Prod'hom B, Hess P (1985) Conformational changes associated with ion permeation in L-type calcium channels. Nature (Lond) 333: 373-376

Schubert P, Mitzdorf U (1979) Analysis and quantitative evaluation of the depressive effect of adenosine on evoked potentials in hippocampal slices. Brain Res 172: 186-190

Scott RH, Dolphin AC (1987) Activation of a G protein promotes agonist responses to calcium channel ligands. Nature (Lond) 330: 760-762

Sladeczek F, Recasens M, Bockaert J (1988) A new mechanism for glutamate receptor action: phosphoinositide hydrolysis. Trends Neurosci 11: 545-549

Slater NT, Stelzer A, Galvan M (1985) Kindling-like stimulus patterns induce epileptiform discharges in the guinea pig in vitro hippocampus. Neurosci Lett 6: 25-31

Worley PF, Baraban JM, McCarren M, Snyder SH, Alger BE (1987) Cholinergic phosphatidylinositol modulation of inhibitory. G protein-linked, neurotransmitter actions: electrophysiological studies in rat hippocampus. Proc Natl Acad Sci USA 8: 3467-3471

Worley PF, Heller WA, Snyder SH, Baraban JM (1988) Lithium blocks a phosphoinositide-mediated cholinergic response in hippocampal slices. Science 23: 1428-1429

Increase in Glutamate Responses
by GABA in Neocortical and Archicortical Structures

J. Walden[1], E. -J. Speckmann[1,2], D. Bingmann[3] and H. Straub[1]

[1] Institut für Physiologie, Universität Münster, D-4400 Münster, FRG
[2] Institut für Experimentelle Epilepsieforschung, Universität Münster, D-4400 Münster, FRG
[3] Institut für Physiologie, Universität Essen, D-4300 Essen, FRG

Introduction

The excitatory transmitter, glutamate, and the inhibitory transmitter, gamma-aminobutyric acid (GABA), have been implicated in the pathophysiology of epileptic activity (Nistico et al. 1986). Since glutamate may contribute to the expression and spread of epileptic discharges and GABA to their restriction, the aim of the present investigation was to test the influence of GABA on the action of the glutamate subreceptor agonist N-methyl-D-aspartate (NMDA). The experiments were performed in the motor cortex of the rat (in vivo studies), in neocortical organotypic tissue cultures from newborn rats, and in hippocampal slice preparations from guinea pigs (in vitro studies) using changes in field and membrane potentials as indicators of transmitter action.

Methods

In Vivo Studies.
The experiments were carried out in anesthetized (120 mg/kg phenobarbital or 1.5 g/kg urethane) and artificially ventilated rats. The motor cortex was exposed and covered by agar dissolved in artificial CSF or superfused with CSF. Field potentials (DC potentials) were measuted from different cortical laminae by means of micropipettes placed against a reference point on the frontal nasal bone (Walden et al. 1989).

In Vitro Studies.
The experiments were carried out in organotypic neocortical explants of newborn rats (Bingmann et al. 1988) and in CA3 neurons from the hippocampal slice (Bingmann and Speckmann 1986). Membrane potential recordings were made with single electrodes filled with 2 M potassium-methylsulfate.

Administration of Transmitters.
For application of transmitters a 3-barrelled micropipette was glued in parallel with the shank of the recording microelectrode. Transmitters were ejected by pressure or iontophoresis. For iontophoresis one barrel of the application pipette was filled with 2 M NaCl for current neutralization. The duration of drug administration ranged from 10 to 120 s in the in vivo preparations and 10 to 180 ms in the in vitro preparations. All drugs were aadministered at constant intervals (Hamon and Heinemann 1986).

Results and Discussion

In Vivo Studies.
NMDA induced negative and GABA positive cortical field potential changes (CFP; Fig. 1). The amplitudes of GABA elicited positive CFPs were comparatively small. When GABA was applied up to 300 s before the ejection of NMDA, the amplitude of the negative CFPs induced by the excitatory amino acid was increased (Fig. 1A; 26 out of 32 experiments). The same effect could

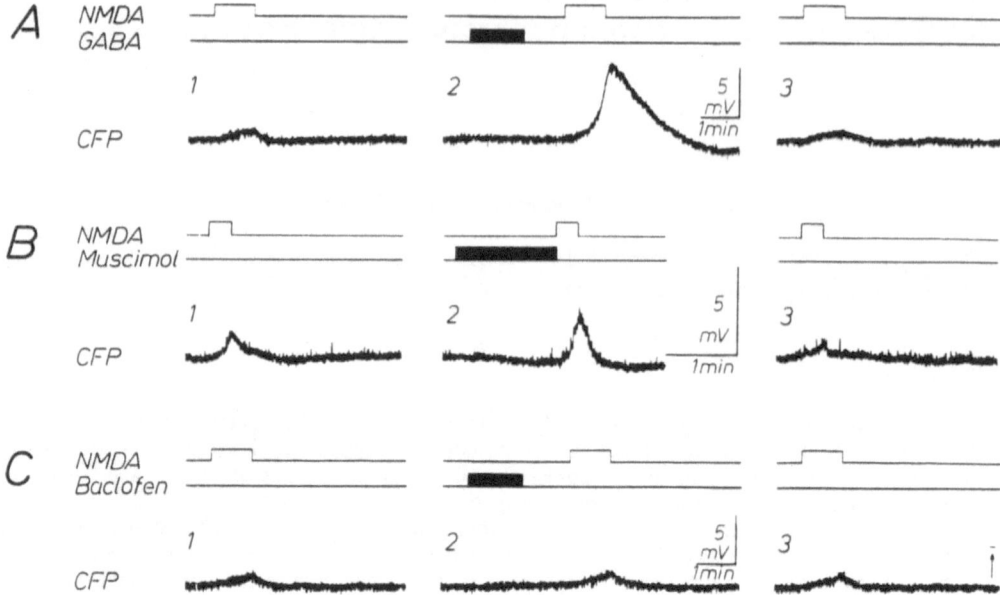

Fig. 1A-C. Augmentation of NMDA-induced CFPs by preapplication of GABA and muscimol but not by baclofen. Pressure application; concentrations of the solutions ejected: 10^{-4} M NMDA, $10^{-6}M$ GABA and muscimol, 10^{-5} M baclofen. A NMDA induced CFP 5 min before (*1*), 1 min after (*2*) and 8 min after (*3*) a GABA preapplication. B NMDA elicited CFP 5 min before (*1*), immediately after (*2*), and 7 min (*3*) after a muscimol preapplication. C NMDA evoked CFP 5 min before (*1*), ca. 1 min after (*2*), and 5 min after (*3*) a baclofen preapplication (From Walden et al. 1989).

also be observed, with minimal reactions in the field potential, after GABA ejection (Fig. 1A). The increase in excitatory amino acid induced CFP by GABA was more pronounced with NMDA (relative increase up to ca. 9 from its initial value) than with quisqualate (relative increase up to ca. 5 from its initial value) and glutamate (relative increase up to ca. 2 from its initial value). The augmentation was long-lasting and was found even up to 10 min after the end of the GABA application. When GABA and NMDA were applied simultaneously the NMDA response usually was abolished. Furthermore, under these conditions the CFP elicited by a subsequent NMDA application was augmented for about 10 min (Walden et al. 1989). Since GABA may act on different receptor sites, the $GABA_A$ receptor agonist, muscimol, and the $GABA_B$ receptor agonist, baclofen, were substituted for GABA and applied before NMDA. An augmentation of the NMDA response was found to occur after muscimol but not after baclofen (muscimol:four out of five experiments; baclofen: six out of six experiments). The relative increase in NMDA- induced CFPs by muscimol was up to ca. 6 from its initial value.

In Vitro Studies.
In both neocortical organotypic tissue culture (Fig. 2) and hippocampal slice preparations (CA3 neurons), local application of NMDA elicited depolarizations. GABA induced membrane potential changes which were pre- dominantly hyperpolarizing in nature (Bingmann et al. 1989; Müller and Misgeld 1989). Depolarizations induced by the excitatory amino acid were increased in amplitude up to three times the initial value when GABA was given before NMDA (Fig. 2; neocortical neurons: 8 out of 11; hippocampal neurons: 6 out of 7). The augmentation, which did not depend on the actual membrane potential, lasted up to 10 min. Simultaneous application of NMDA and GABA led to a decrease or abolition of NMDA reactions. After this depressed NMDA response there was, however, a marked increase in the NMDA-induced depolarization lasting up to 10 min

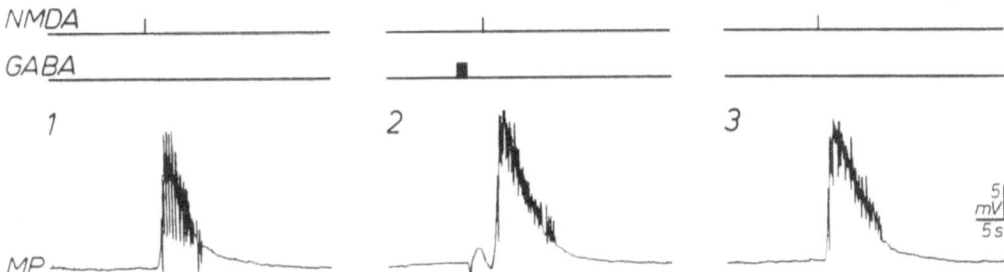

Fig. 2 Augmentation of NMDA-induced depolarization in neocortical neurons. Organotypic neocortical tissue cultures; newborn rat. Pressure application. Concentrations of the solutions ejected: 10^{-7} M NMDA, 10^{-5} M GABA. NMDA induced depolarizations 180 s before (*1*), 3 s after (*2*), and 120 s after (*3*) a GABA preapplication.

(see in vivo studies). Tetrodotoxin blocked action potentials, but but did not influence the GABA-induced augmentation effect. As to the elementary mechanisms underlying augmentation of the NMDA response by GABA, the following explantations may be considered: GABA might act by an allosteric modulation of the receptor for excitatory amino acids. Such a mechanism is postulated for glycine, which augments NMDA effects (Johnson and Ascher 1987). In this context, it has to be taken into account that responses to quisqualate were also enhanced by GABA, which has not been reported for glycine. Furthermore, glutamate and its analogues enhance hydrolysis of inositol phospholipids (Sugiyama et al. 1987). Since GABA has been described to upregulate the hydrolysis of phosphoinositides (Yu and Chuang 1988), the present data suggest that the augmentation of the NMDA response by GABA is based on a common intracellular pathway of GABA and glutamate action.

References

Bingmann D, Speckmann E-J (1986) Actions of pentylenetetrazol on CA3 neurons in hippocampal slices of guinea pigs. Exp Brain Res 64: 94-104

Bingmann D, Speckmann E-J, Baker R E, Ruijter J, de Jong B M (1988) Differential antiepileptic effects of the organic calcium antagonists verapamil and flunarizine in neurons of organotypic neocortical explants of newborn rats. Exp Brain Res 72: 439-442

Bingmann D, Baker R, Ruijter J M, Walden J, Speckmann E-J, Straub H (1989) Polyphasic membrane potential changes induced by GABA in neocortical and hippocampal neurons (in vitro). Eur J Neurosci [Suppl] 2: 110

Hamon B, Heinemann U (1986) Effects of GABA and bicuculline on N-methyl-D-aspartate and quisqualate induced reductions in extracellular free calcium in area CA1 of the hippocampal slice. Exp Brain Res 64: 27-36

Johnson J W, Ascher P (1987) Glycine potentiates the NMDA response in cultured mouse brain neurons. Nature 325: 529-531

Müller W, Misgeld U, Lux H-D (1989) Gamma-amino-butyric acid-induced ion movements in the guinea pig hippocampal slice. Brain Res 484: 84-191

Nistico G, Morselli P L, Lloyd K G, Fariello R G, Engel J (eds) (1986) Neurotransmitters, seizures and epilepsy III. Raven, New York

Sugiyama H, Ito I, Hirono C (1987) A new type of glutamate receptor linked to inositol phospholipid metabolism. Nature 325: 531-533

Walden J, Speckmann E-J, Bingmann D (1989) Augmentation of glutamate responses By GABA in the rat's motorcortex in vivo. Neurosci Lett 101: 209-213

Yu O, Chuang D-M (1988) GABA pretreatment enhances glutamate mediated phosphoinositide hydrolysis in neurons. Eur J Pharmacol 158: 179-180

The Action of Neuropeptides on Transmitter Release at the Neuromuscular Synapse

M. Abdul-Ghani, H. Meiri and R. Rahamimoff

Department of Physiology, Hebrew University, Hadassah Medical School, P.O. Box 1172, Jerusalem 91010 Israel

Introduction

Synaptic transmission is a plastic phenomenon which is regulated by intrinsic and extrinsic factors. Among the intrinsic factors, the frequency of activation is probably most relevant to the topic of this conference (for a review see Erulkar et al. 1982). The frequency of activation greatly changes the number of quanta of transmitter liberated (see Rahamimoff 1979; Nussinovitch and Rahamimoff 1988) and thus affects the probability of successful synaptic transmission. The frequency of activation also affects preferentially the amount of a peptide liberated. For example, stimulation of the splenic nerve releases both noradrenaline and neuropeptide Y at low frequencies of 2 Hz. Increasing the frequency of stimulation to 20 Hz produces a small increase in the amount of noradrenaline released, while it augments the release of neuropeptide Y more than sixfold (Lundberg and Hökfelt 1986). If other neuropeptide-releasing neurons behave in a similar way, it is expected that during high frequency epileptic activity there may be a substantial release of neuropeptides. It is thus interesting to examine the action of such neuropeptides on basic processes of synaptic transmission such as transmitter release.

The purpose of this article is therefore to summarize in brief the action of neuropeptides on quantal transmitter release at one of the simplest and best studied synapses: the neuromuscular junction (see Katz 1969).

The Action of Neuropeptides on Synaptic Transmission

Many neuropeptides affect the amplitude of the synaptic potential at the neuromuscular junction. Most of the peptides examined augment this amplitude (Fig. 1). The main advantage of the neuromuscular synapse is that one can easily perform a quantal analysis and determine whether the action of the peptide is presynaptic or postsynaptic in nature. Such an analysis was performed for a number of neuropeptides (see Table 1) and showed unequivocally that the augmenting action of ACTH and vasopressin is presynaptic in nature. Both augment the number of quanta liberated by the nerve impulse without having any significant postsynaptic effect.

All the peptides augmenting transmitter release that appear in Table 1 are similar in their site of action, the presynaptic nerve terminal, but differ substantially in the duration of their effect. Some neuropeptides, such as VIP, calcitonin and LH-RH induce a monotonous increase of release, which decays following removal of the peptides (Gold 1982; Akasu 1986; Abdul-Ghani et al. 1986b). ACTH, α-MSH and Arg-vasopressin exert a completely different effect. The enhancement of acetylcholine secretion induced by these peptides, lasts for a long time after their wash from the bathing medium (Johnston et al. 1983; Abdul Ghani et al. 1990).

Since many peptides produce a similar augmenting effect on transmitter release, the question arises whether this may be due to some unspecific effect. This does not seem to be the case, since small alterations in the Arg-vasopressin molecule cause a loss of the peptide action (Abdul-Ghani et al. 1990). Not all peptides augment quantal liberation of acetylcholine. Inhibition of release was

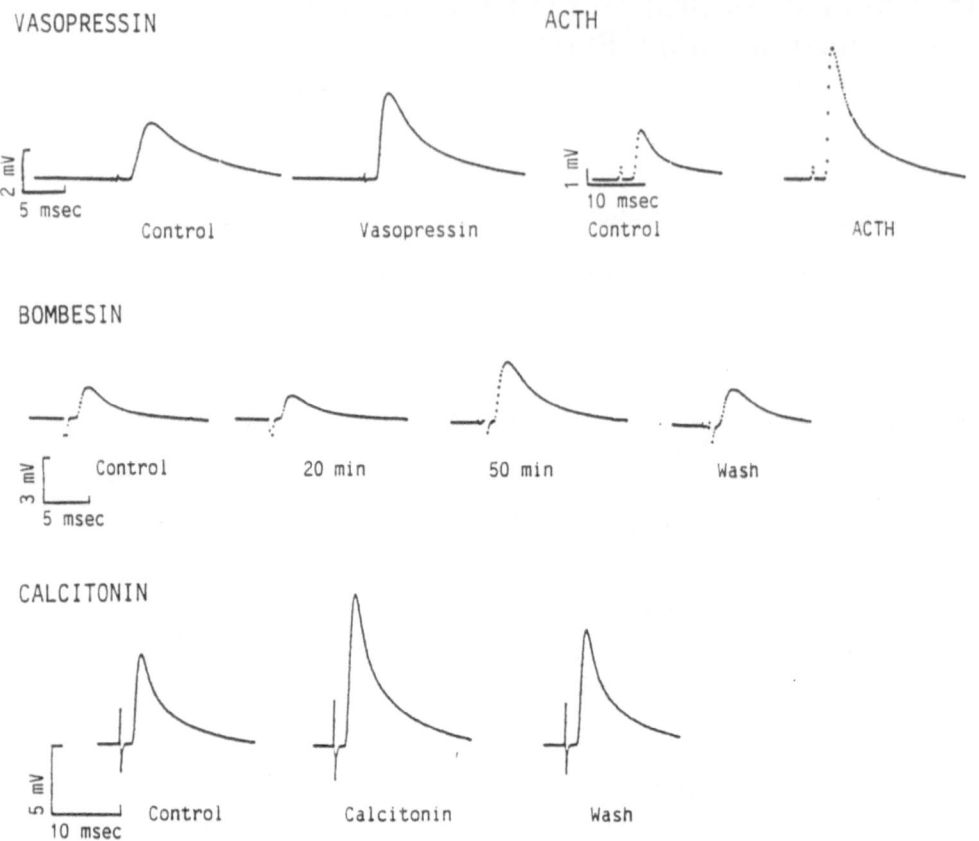

Fig. 1. The augmenting effect of neuropeptides on synaptic transmission at the neuromuscular junction. Intracellular recording at the frog neuromuscular junction. Every trace is the average of 200 responses (100 for ACTH). Peptide concentration 10 μM, 3.9 μM ACTH, 6.33 μM calcitonin, 1.33 μM bombesin

observed following exposure of the nerve terminal to Met-enkephalin, and the release returned to its original level after wash (Bixby and Spitzer 1983). Partially reversible inhibition of release was also induced by substance P when the nerve terminal was presented with a high concentration of this peptide; however, at lower concentrations the substance exhibited a more complex influence: an initial decrease followed by an augmentation (Steinacker 1977). A similar biphasic effect was also exerted by bombesin (Abdul-Ghani et al. 1986a).

Discussion and Conclusions

This article summarizes the effects of a number of neuropeptides on synaptic transmission at the neuromuscular junction. The striking effect is that all of these peptides act on the presynaptic nerve terminal, where they alter the number of quanta liberated. Most of the peptides examined do not have a substantial postsynaptic effect. Therefore if they have a modulatory role in synaptic transmission, it is mainly presynaptic in nature.

The concentrations of the peptides in most of the articles quoted here are high compared to their circulating levels. However, if they are released in the confined space of the synaptic gap, the local concentrations may be much higher than the circulating levels. While some peptides such as

Table 1. The effect of neuropeptides on transmitter release at the neuromuscular junction

Peptide	Existence in nerve terminal	Effect on transmitter release	Maximal effect	Reference
ACTH	Unknown	Enhancement	106%	Johnston et al. (1983)
Calcitonin	Unknown	Enhancement	130%	Abdul-Ghani et al. (1986)
VIP	Cholinergic	Enhancement	14%[a]	Gold (1982)
Enkephalin	Sympathetic	Reduction	40%	Bixby and Spitzer (1983)
Vasopressin	Sympathetic	Enhancement	199%	Abdul-Ghani et al. (1990)
Substance-P	CNS	Biphasic	80%/500%[b]	Steinacker (1977)
Bombesin	Unknown	Biphasic	58%/204%[b]	Abdul-Ghani et al. (1986)
LHRH	Sympathetic	Enhancement	34%[a]	Akasu (1986)
α-MSH	NMJ	Enhancement		Johnston et al. (1983); Oren et al. (1989)

[a] Mean effect.
[b] Numerator, decrease; denominator, increase.

enkephalin have an inhibiting effect on quantal liberation, most of them augment the number of quanta liberated by the nerve impulse. The actions of some peptides is of a short duration, such as those of calcitonin, whereas other peptides have long lasting effects which outlast the exposure by hours. Thus they may serve as a cellular switch of synaptic efficiency.

The cellular and molecular mechanism of the action of these peptides on the nerve terminals remains unknown. It is tempting to speculate that some of the effects are due to an alteration in the concentration of the main determinant of transmitter release - calcium ions (Augustine et al. 1988), such as alteration in the calcium channel activity. But one should of course remember that this need not be the case, since at least one peptide, ACTH, is known to be able to increase spontaneous transmitter release in the virtual absence of extracellular calcium (Johnston et al. 1983).

Whatever the cellular mechanism may turn out to be, it is quite likely that peptides may participate in the normal modulation of synaptic activity or in pathophysiological states such as small cell carcinoma of the lung (Sorenson et al. 1981) or epilepsy.

Acknowledgements. The unfailing secretarial help of Mrs. Marsha Rapp is greatly appreciated. The original work in this article was supported by B.S.F., M.D.A. and C.T.R. Equipment was purchased with the aid of Lower Saxony State Government and the Sir Zelman Cowen Interuniversity Fund.

102

References

Abdul-Ghani M, Meiri H, Rahamimoff R (1986a) Biphasic effect of bombesin on transmitter release from motor nerve terminals. Rev Clin Basic Pharm 6: 47s

Abdul-Ghani MA, Silberberg SD, Johnston MF, Kravitz EA, Meiri H, Rahamimoff R (1986b) The role of peptides released by small cell carcinoma of the lung on neuromuscular transmission. In: Rahamimoff R, Katz B (eds) Calcium, neuronal function and transmitter release Nijhoff, Boston, pp 593-599

Abdul-Ghani M, Meiri H, Rahamimoff R (1989) Vasopressin augments transmitter release from frog motor nerve terminals. Pflugers Arch 413 (suppl 1): R40

Abdul-Ghani MA, Meiri H, Rahamimoff R (1990) Vasopressin produces long lasting increase in transmitter release. Brain Res 515: 355-357

Akasu T (1986) Luteinizing hormone releasing hormone modulates the cholinergic transmission in frog neuromuscular junction. Jpn J Physiol 36: 25-42

Augustine GJ, Charlton MP, Smith SJ (1987) Calcium action in synaptic transmitter release. Ann Rev Neurosci 10: 633-693

Bixby JL, Spitzer NC (1983) Enkephalin reduces quantal content at the frog neuromuscular junction. Nature 301: 431-432

Erulkar SD, Meiri H, Rahamimoff R (1982) The ionic basis of high-frequency synaptic activation. In: Klee MR, Lux HD, Speckmann EJ (eds) Physiology and pharmacology of epileptogenic phenomena . Raven, New York, pp 197-205

Gold MR (1982) The effects of vasoactive intestinal peptide on neuromuscular transmission in the frog. J Physiol 327: 325-335

Johnston MF, Kravitz EA, Meiri H, Rahamimoff R (1983) Adrenocorticotropic hormone causes long-lasting potentiation of transmitter release from motor nerve terminals. Science 220: 1071-1072

Katz B (1969) The release of neural transmitter substances. Sherrington Lecture 10, Liverpool University Press

Lundberg JM, Hökfelt T (1986) Multiple co-existence of peptides and classical transmitters in peripheral autonomic and sensory neurons - Functional and pharmacological implications. Prog Brain Res 68: 241-262

Nussinovitch I, Rahamimoff R (1988) Ionic basis of tetanic and post-tetanic potentiation at a mammalian neuromuscular junction. J Physiol 396: 435-455

Oren N, Micevych P, Letinsky MS (1989) The presence of alpha-MSH-like immunoreactivity in the innervation of amphibian skeletal muscle. J Neurosci Res 23: 225-233

Rahamimoff R (1979) Principles of chemical signalling in the nervous system. In: JG Nicholls (ed) The role of intercellular signals: navigation, encounter, outcome. Dahlem Konferenz, Berlin, pp 15-40

Sorenson GD, Pettengill OS, Brinck-Johnson T, Cate CC, Mauter LH (1981) Hormone production by cultures of small cell carcinoma of the lung. Cancer 47: 1289-1296

Steinacker A (1977) Calcium dependent presynaptic action of substance P at the frog neuromuscular junction. Nature 267: 268-270

Cholecystokinin and Epileptogenesis in the Hippocampal CA3 Region

P. G. Aitken[1,2], D. B. Jaffe[1], and J. V. Nadler[2,3]

Departments of Cell Biology[1], Neurobiology[2], and Pharmacology[3] Duke University Medical Center, Durham, NC 27710 USA

Introduction

Cholecystokinin (CCK) is a neuroactive peptide that has been localized to excitatory afferent fibers in CA1 and the fascia dentata and to local circuit interneurons scattered throughout areas CA1 and CA3 (Fredens et al. 1984; Roberts et al. 1984) The CCK-containing interneurons appear to be a subset of the GABAergic inhibitory interneurons known as basket cells (Hendry and Jones 1985; Sloviter and Nilaver 1987). CCK-containing synaptic boutons make synaptic contact with both pyramidal and nonpyramidal neurons (Totterdell and Smith 1986; Harris et al. 1985). CCK is found in the mossy fibers of the guinea pig, but not of the rat (Gall 1984). The above findings suggest that CCK may be co-released with GABA and may influence pyramidal cell excitability directly, via direct synapses, and indirectly, via synapses on interneurons. Application of CCK to neurons generally causes excitation (Dodd and Kelly 1981; Bradwejn and de Montigny 1984; Jaffe et al. 1987). The present study further investigated electrophysiological effects of CCK and a CCK receptor antagonist, benzotript (BZT), in the hippocampal CA3 region.

Methods

Transverse slices 0.4 mm thick were cut from the middle one-third of hippocampi removed from 150-175 g Sprague-Dawley rats or 300-400 g Hartley guinea pigs and transferred immediately to an interface chamber at 35.5°C. They were superfused at 1.6 cc/min with artificial CSF (ACSF) of standard composition ($K^+ = 3.5mM$). Extracellular field responses were recorded in stratum pyramidale of CA3b in response to mossy fiber stimulation spanning the range from subthreshold to near maximal. Input/output (I/O) curves were generated by plotting the amplitude of the mossy fiber-evoked response vs stimulus intensity. The equally spaced stimulus values were converted to an ordinal scale in which the smallest stimulus used in each experiment was given a value of 1, the second smallest a value of 2, and so on.

Each experiment generated one I/O curve before drug application, one or more during drug application, and a final I/O curve after drug washout. The area under each I/O curve was calculated and a control value determined by taking the average area of the pre-drug and washout curves. The area under the experimental I/O curve was then expressed as a percent change from this control value. Spontaneous epileptiform activity, or bursts, were monitored by feeding the amplified signal into a chart recorder which reproduced each burst as a pen deflection.

Drugs were applied by dissolving them in the perfusion medium. CCK octapeptide (Peninsula Laboratories) was dissolved in saline, divided into aliquots, lyophilized, and stored at -70°C until needed. The concentrations of CCK and BZT were chosen to achieve maximum effects, based on published affinities of these compounds for CCK receptors. Sulphated CCK was used unless specified otherwise.

Results

The application of 100 nM CCK to either rat or guinea pig hippocampal slices evoked no spontaneous activity and had no effect on I/O curves. Note that in earlier experiments (Jaffe et al, 1987), identical techniques and CCK concentrations caused a significant excitatory I/O curve shift in the CA1 region of rat hippocampal slices.

Treatment with 200 μM BZT induced spontaneous bursting in both rat and guinea pig hippocampal slices. While quantitative comparisons were not performed, the BZT-induced burst discharges appeared like bursts induced by kainic acid or elevated potassium. Burst frequency was 0.53 ± 0.14 Hz in rat hippocampal slices and 0.55 ± 0.15 Hz in guinea pig hippocampal slices. The application of 50 μM BZT to only rat slices caused bursting at 0.32 ± 0.07 Hz.

Superfusion with 200 μM BZT had opposite effects on I/O curves in slices from rats and guinea pigs. In rats, BZT had a significant excitatory effect, causing an increase in the area under the I/O curve to 139% ± 13% of control ($F = 15.9$, df = 1,54, $p < 0.0002$). In guinea pigs, BZT had a significant inhibitory effect, causing a decrease in I/O curve area to 75.7 ± 8% of control ($F = 5.83$, df = 1.48, $p < 0.02$). In both cases, the effects of BZT reversed upon washout. Summary I/O curves are shown in Fig 1.

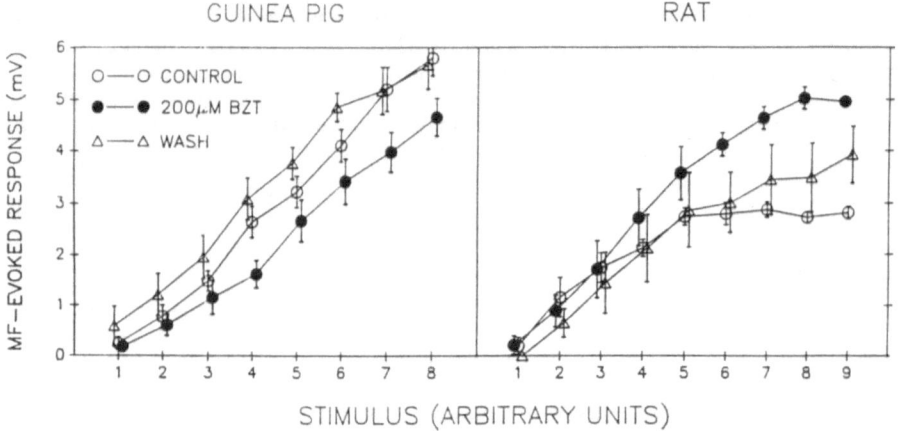

Fig. 1. Effects of benzotript on mossy fiber-CA3 input/output curves. BZT inhibits synaptic transmission in guinea pigs but facilitates it in rats. **1** = lowest stimulus intensity used in each experiment, **2** = next lowest, etc

In four experiments, rat hippocampal slices were pretreated with 200 nM sulphated CCK (CCK-S) or 200 nM unsulphated CCK (CCK-NS) for 30 min before being exposed to 50 μM BZT. Pretreatment with CCK-S (200 nM) either blocked the development of BZT-induced bursts, or reduced their frequency; CCK-NS had no effect. Burst frequencies are given in Table 1.

Discussion

That BZT, a CCK antagonist, causes spontaneous epileptiform activity implies that at least some of the CCK receptors in the CA3 region are normally occupied by endogenous CCK and that the tonic activation of these receptors inhibits bursting. Of course, these conclusions are valid only if BZT is acting solely as a specific antagonist of CCK receptors. That BZT is a CCK antagonist is

Table 1. The effects of CCK pretreatment on the latency and frequency of CA3 bursts evoked by BZT

Pretreatment	n	Burst latency	Final frequency Pre + BZT	BZT alone
200 nM CCK-S	5	38 ± 11*	0.08 ± 0.03	0.17 ± 0.03 **
200 nM CCK-NS	5	21 ± 3	0.24 ± 0.05	0.25 ± 0.06

* p < 0.01, Student's t test (compared to CCK-NS)
** p < 0.01, Student's t test (compared to pretreatment + BZT)

Slices were exposed to the pretreatment for 30 min, pretreatment + 50 µM BZT for 30 min, and 50 µM BZT alone for 30 min. Latency is time from initial BZT application to first burst. Frequencies were measured after 25 min in the indicated condition.

well established, but the possibility that it has other effects in the central nervous system remains open. The fact that CCK suppresses the actions of BZT is encouraging, but is not conclusive evidence of direct antagonism.

If most or all of the CCK receptors in the CA3 region are tonically occupied by endogenous CCK, it would explain our finding that exogenous CCK application has no effect on synaptic transmission. Yet it has been reported that iontophoretic CCK application causes potent activation of CA3 pyramidal cells *in situ* (Bradwejn and de Montigny 1984). This apparent discrepancy remains unresolved, but may be due to differences in the preparation and/or method of CCK delivery.

The species difference in the effects of BZT on mossy fiber synaptic transmission is difficult to explain. That BZT causes bursting in both species implies, as discussed above, that endogenous CCK is exerting a tonic inhibitory effect. The excitatory effect of BZT on synaptic transmission in rats is consistent with this scenario. In guinea pigs, however, the inhibitory effect of BZT on mossy fiber transmission would argue that endogenous CCK is tonically inhibiting bursting while also causing a facilitation of synaptic transmission. CCK is present in guinea pig, but not rat, mossy fibers (Gall 1984), suggesting that it may be a component of excitatory transmission between mossy fibers and CA3 pyramidal cells. BZT would then be expected to antagonize both excitatory transmission and tonic inhibition. The observed effects of BZT on I/O curves would be obtained if its antagonism of excitatory transmission "outweighs" its antagonism of tonic inhibition. Rat-guinea pig differences in CCK receptor distribution (Hill et al. 1987; Mantyh and Mantyh 1985) may also play a role; these reports do not, however, provide the spatial resolution required to detect differences within hippocampal subfields.

Acknowledgements. We thank George Somjen for helpful discussion and a critical reading of the manuscript. This work was supported by NIH grant NS 17771.

References

Bradwejn J, de Montigny C, (1984) Benzodiazepines antagonize cholecystokinin-induced activation of hippocampal neurones. Nature 312:363-364
Dodd J, Kelly JS, (1981) The actions of cholecystokinin and related peptides on pyramidal neurons of the mammalian hippocampus. Brain Res 205:337-350
Fredens K, Stengaard-Pedersen K, Larsson L-I, (1984) Localization of enkephalin and cholecystokinin immunoreactivities in the perforant path terminal fields of the rat hippocampal formation. Brain Res 304:255-263

Gall C, (1984) The distribution of cholecystokinin-like immunoreactivity in the hippocampal formation of the guinea pig: localization in the mossy fibers. Brain Res 306:73-83

Harris KM, Marshall PE, Landis DMD, (1985) Ultrastructural study of cholecystokinin-immunoreactive cells and processes in area CA1 of the rat hippocampus. J Comp Neurol 233:147-158

Hendry SHC, Jones EG, (1985) Morphology of synapses formed by cholecystokinin-immunoreactive axon terminals in regio superior of rat hippocampus. Neuroscience 16:57-68

Hill DR, Shaw TM, Woodruff GN, (1987) Species differences in the localization of 'peripheral' type cholecystokinin receptors in rodent brain. Neurosci Lett 79:286-289

Jaffe DB, Aitken PG, Nadler JV, (1987) The effects of cholecystokinin and cholecystokinin antagonists on synaptic function in the CA1 region of the hippocampal slice. Brain Res 415:197-203

Mantyh CR, Mantyh PW, (1985) Differential localization of cholecystokinin-8 binding sites in the rat vs. the guinea pig brain. Eur J Pharmacol 113:137-139

Roberts GW, Woodhams PL, Polak JM, Crow TJ, (1984) Distribution of neuropeptides in the nervous system of the rat: The hippocampus. Neuroscience 11:35-77

Sloviter RS, Nilaver G, (1987) Immunocytochemical localization of GABA-, cholecystokinin-, vasoactive intestinal polypeptide-, and somatostatin-like immunoreactivity in the area dentata and hippocampus of the rat. J Comp Neurol 256:42-60

Totterdell S, Smith AD, (1986) Cholecystokinin-immunoreactive boutons in synaptic contact with hippocampal pyramidal neurons that project to the nucleus accumbens. Neuroscience 19:181-192

In Vivo and In Vitro Studies of the Epileptic Focus Induced by the Interruption of a Cortical GABA Infusion

C. Silva-Barrat, J. Champagnat, S. Brailowsky, C. Menini and R. Naquet

Laboratoire de Physiologie Nerveuse, CNRS, 91198 Gif-sur-Yvette Cedex, France

Localized chronic infusions of GABA, when administered into the motor cortex of both the naturally photosensitive epileptic baboon and the amygdala-kindled rat, have two main effects. (a) During chronic (7 days) infusion, powerful anticonvulsant effects are observed with the disappearance of both EEG and clinical epileptic manifestations in baboons, but only clinical manifestations in kindled rats. (b) The interruption of the GABA infusion gives rise, in the infused cortical area, to an epileptic focus which does not depend on the previous epileptic predisposition of the animals since it can also be obtained in the same cortical areas in nonepileptic baboons and nonepileptic rats (Brailowsky et al. 1987, 1988; Fukuda et al. 1987; Silva-Barrat et al. 1988). The goal of the present paper is to summarize *in vivo* and *in vitro* studies concerning normal nonepileptic rats presenting the focus after the interruption of GABA infusion.

GABA infusion was performed by using intracortical cannulae implanted at the somatomotor level and connected to GABA-filled osmotic minipumps providing delivery of GABA (1 *M*) at a very slow rate (1 μl/h) during 7 days. Within 10 - 60 min after discontinuation of the infusion, the epileptic focus appeared as localized cortical EEG paroxysmal discharges associated with myoclonic twitches of the hind limbs. The paroxysmal activity then became bilateral and diffused to anterior cortical areas (Fig. 1). About 4 h after the appearance of the first discharges, the amplitude of the paroxysmal discharges decreased progressively, whereas their frequency increased and reached its maximal value. In the final stages, which lasted 24 - 72 h, clinical manifestations were absent (Brailowsky et al. 1989). Focal epileptic activity can thus be considered as a local status epilepticus lasting for hours or days and resulting from a disturbance of neurotransmitter balance consecutive to a prolonged intracortical excess of exogenous GABA.

Focal epilepsy, observed after GABA infusion, resembled in many, but not all, respects the EEG and clinical characteristics of some partial human epilepsies, especially somamotor status epilepticus. This is characterized by the persistence of localized segmental myoclonus and is referred to as epilepsia partialis continua or Kozhevnikov's syndrome. Anticonvulsant drugs that enhance GABAergic transmission, such as benzodiazepines, valproate, and barbiturates, when administered systemically at high doses, blocked the myoclonic twitches of rats, but had no effect on EEG focal paroxysmal discharges (unpublished data), just as is observed in human intractable focal epilepsies. Reinstallement of GABA infusion into the epileptic site was, however, effective in blocking EEG focal paroxysmal discharges completely (Brailowsky et al. 1989), indicating a subsensitivity of GABA receptors and that only large doses of GABA are effective.

After GABA interruption, the epileptic focus was characterized by neuronal depopulation and gliosis due to the infusion cannula. Nevertheless, the short latency (10 - 60 min) of the focus appearance would argue against a critical role for this lesion. Moreover, animals perfused intracortically with saline or taurine (1 *M*), using the same method as for GABA, presented similar histological changes but with no associated signs of epilepsy upon cessation of infusion.

In order to analyze the hypothesis of a subsensitivity of GABA receptors that is due to prolonged infusion of exogenous GABA and which explains the appearance of the focus, *in vitro* intracellular recordings in neocortical slices obtained from rats presenting the focus were performed (Silva-

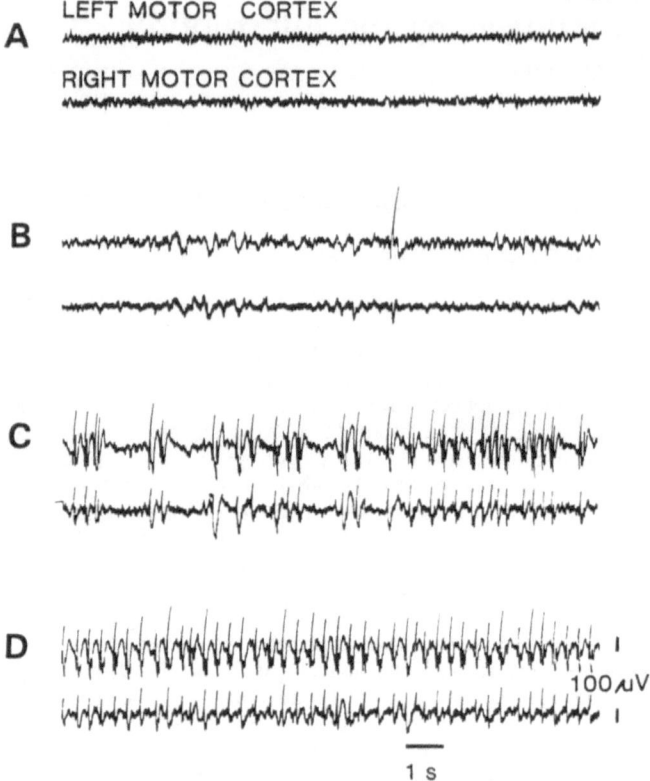

LEFT MOTOR CORTEX

A

RIGHT MOTOR CORTEX

B

C

D

100 μV

1 s

Fig. 1 A-D. EEG recordings in the motor cortex. **A** During GABA infusion, and then **B** 15 min, **C** 30 min, **D** 60 min after the interruption of GABA infusion. GABA infusion was performed in the left motor cortex where epileptic spikes appear and predominate

Barrat et al. 1989). Cells were recorded in the vicinity of the GABA infused site, i.e., near the lesion, and their activity was compared to that recorded at the same site in cortical slices obtained from saline-infused rats.

In slices obtained from epileptic rats, electrical stimulation of the white matter induced paroxysmal depolarization shifts (PDSs) in virtually all neurons in the vicinity of the lesion (A3 in Fig. 2). The application of GABA (1 - 10 μM) into the bath medium had no effect on these neurons, although it was effective in blocking action potentials in slices from saline-infused rats. In the slices obtained from epileptic rats, a population of neurons presented, in addition to synaptically-induced PDSs, voltage-dependent and cobalt-sensitive bursts of action potentials induced by depolarizing current injections (Fig. 2, A1, A2). These intrinsic bursting neurons represented 30% of the total recorded neurons but only 7% in slices from saline-infused rats. Intrinsic bursting neurons were unresponsive to high doses of GABA (100 μM).

The sensitivity of these neurons to isoguvacine, a specific GABA$_A$ agonist, was also tested (Silva-Barrat et al. 1989). Measurement of conductance changes (Fig. 2) confirmed the very low sensitivity of GABA receptors (ED$_{50}$ 100 times higher for intrinsic bursting cells than for cells presenting PDSs). This demonstrates a significant tolerance to GABA or to GABA$_A$ agonist of the entire population of neurons in slices from epileptic rats compared to slices from saline-infused rats. Moreover, tolerance was strikingly exaggerated in intrinsically bursting neurons compared to

109

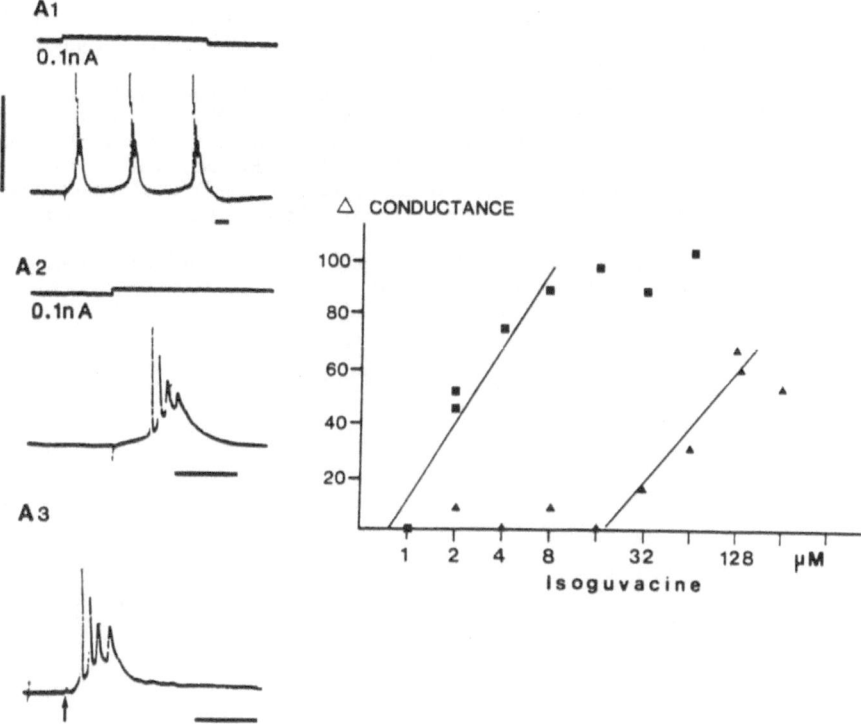

Fig. 2. Representative examples of an intrinsic bursting cell (*A1*). *A2*, Same discharge at a faster sweep speed. *A3*, PDS evoked in the same cell by white matter stimulation (*arrow*). Vm = 70 mV; calibration = 60 mV, 20 ms. *Right*, Dose-response curve for a cell presenting PDSs (*squares*) and for an intrinsic bursting cell (*triangles*) during isoguvacine application. *Ordinate*, % increase of conductance with respect to predrug values calculated by using current pulse injection. *Abscissa*, concentration (μM) of isoguvacine

neurons presenting PDSs. Some of mechanisms responsible for this tolerance seem, therefore, to be related to processes underlying generation of calcium-dependent intrinsic currents. It is known that the effectiveness of $GABA_A$ agonist is reduced by an increased intracellular calcium concentration or by changes in intracellular phosphorylation factors (Stelzer et al. 1988). Moreover, excessive neuronal uptake of exogenous GABA by presynaptic terminals would slow GABA synthesis. These factors, and a possible decrease in the number of active receptors, may account for the development of neuronal hyperexcitability and the epileptic focus.

This model of the epileptic focus may contribute to the understanding of basic mechanisms of GABA involvement in epilepsy and the action of GABA related drugs, such as barbiturates, benzodiazepines, and other anticonvulsants used for therapy of human epilepsies.

References

Brailowsky S, Kunimoto M, Menini C, Silva-Barrat C, Riche D, Naquet R (1988) The GABA-withdrawal syndrome: a new model of focal epileptogenesis. Brain Res 442: 175-179
Brailowsky S, Menini C, Silva-Barrat C, Naquet R (1987) Epileptogenic gamma-aminobutyric acid-withdrawal syndrome after chronic, intracortical infusion in baboons. Neurosci Lett 74: 75-80

Brailowsky S, Silva-Barrat C, Menini C, Riche D, Naquet R (1989) Effects of localized, chronic GABA infusions into different cortical areas of the photosensitive baboon, Papio papio. Electroencephalogr clin Neurophysiol 72: 147-156

Fukuda H, Brailowsky S, Menini C, Silva-Barrat C, Riche D, Naquet R (1987) Anticonvulsant effect of intracortical, chronic infusion of GABA in kindled rats. Exp Neurol 98: 120-129

Silva-Barrat C, Brailowsky S, Riche D, Menini C (1988) Anticonvulsant effects of localized chronic infusion of GABA in cortical and re·icular structures of baboons. Exp Neurol 101: 418-427

Silva-Barrat C, Champagnat J, Brailowsky S, Menini C, Naquet R (1989) Relationship between tolerance to $GABA_A$ agonist and bursting properties in neocortical neurons during GABA-withdrawal syndrome. Brain Res 498: 289-298

Stelzer A, Kay AR, Wong RKS (1988) $GABA_A$-receptor function in hippocampal cells is maintained by phosphorylation factors. Science 241: 339-341

Serotonergic Attenuation of Low Ca^{2+} Induced Bursting in the CA1 Region of the Rat Hippocampus

A. Obenaus[*], J. M. Klancnik and K. G. Baimbridge

Department of Physiology, University of British Columbia, Vancouver, British Columbia, Canada
[*] Present address: Mental Retardation Research Center, UCLA School of Medicine, 760 Westwood Plaza, Los Angeles, CA, 90024 USA

Introduction

Epileptiform activity may be induced in hippocampal slices by perfusion with medium containing low [Ca^{2+}] (Jefferys and Haas 1982; Taylor and Dudek 1982). This activity is characterized by a negative shift in the extracellular potential, often with superimposed population spikes. Such low Ca^{2+} induced bursting is initiated primarily in the CA1 pyramidal cell layer and is preceded by a maintained intracellular depolarization of CA1 hippocampal pyramidal cells (Haas and Jefferys 1984).

Several possible mechanisms may underlie the synchronization and spread of bursting activity, including loss of inhibitory chemical synaptic transmission, increased neuronal excitability due to increases in extracellular K$^+$ (Heinemann et al. 1986; Yaari et al. 1986), and decreases in the extracellular space (Lux et al. 1986). In addition, removal of extracellular Ca^{2+} may cause hyperexcitability by destabilization of the cell membrane surface (Frankenhaeuser and Hodgkin 1957) and alter membrane conductances which affect the ability of the cell to control intrinsic burst capabilities (Schwindt and Crill 1980; Storm 1987). This model of hyperexcitability occurs in the absence of synaptic transmission and is useful for testing the postsynaptic anticonvulsant properties of various compounds (Haas et al. 1984; Agopyan et al. 1985).

Previous investigations have suggested an inhibitory role for serotonin (5-hydroxytryptamine, 5-HT) in epileptogenesis (Waterhouse 1986). Haas et al. (1984) reported an inhibitory effect of 5-HT on the rate of low Ca^{2+} bursting in hippocampal slices. Recently, 5-HT receptors have been divided into a number of distinct subtypes (see Peroutka 1987). The hippocampus contains a very dense concentration of 5-HT$_{1A}$ receptors, which are localized on pyramidal cells (Pazos and Palacios 1985). In the present study we investigated the effects of 5-HT; the selective 5-HT$_{1A}$ receptor agonist, 8-hydroxy-2-(di-n-propylamino)tetralin (8-OH-DPAT); and the 5-HT$_2$ receptor agonist, 1-(2,5-dimethoxy-4-iodophenyl)-2-aminopropane (DOI) on low Ca^{2+} induced bursting.

Methods

Hippocampal slices (400 μm) were prepared by standard procedures. Recording electrodes were placed in the stratum pyramidale (SP) of the CA1 region and lowered to 75 - 150 μm below the surface of the slice. Normal (2 mM Ca^{2+}, 1.5 mM Mg^{2+}) artificial cerebrospinal fluid (ACSF) was then replaced with ACSF containing 0 mM Ca^{2+} and 1.7 mM Mg^{2+}. Perfusion with low Ca^{2+} ACSF abolished evoked population responses within 10 - 15 min and induced low Ca^{2+} bursts within 30 - 95 min.

After a 30 min baseline period of low Ca^{2+} induced bursting, 1, 5, 10, 20, or 50 μM 5-HT, 8-OH-DPAT or DOI was added to the low Ca^{2+} medium for a 10 min period (n = 6 at all doses). Each slice was exposed to only a single drug application, following which burst activity was recorded for 180 min. Control slices were treated similarly with drug-free low Ca^{2+} medium.

Experimental Brain Research Series 20
© Springer-Verlag Berlin · Heidelberg 1991

The data are expressed as a percentage of the initial baseline period of bursting (100% represents baseline), as the burst rate was highly variable from one slice to another (range 5 - 28 bursts/5 min interval). Statistical analysis included analysis of variance (ANOVA) and two-sample two-tailed t-tests.

Results

Data were collected from 96 hippocampal slices. Control hippocampal slices exhibited a slow initial increase in burst rate reaching 146% ± 22% of baseline after 60 min in low Ca^{2+} medium, followed by a decline to 72% ± 10% after 180 min (Figs. 1 and 2). Once initiated these bursts occurred at a frequency of 3.0 ± 2.3 per min (mean ± SD) and had an amplitude of 2 - 20 mV.

Perfusion of 5-HT was followed by a significant dose-related reduction in burst rate (ANOVA P < 0.02 - 0.005). Perfusion with 50 μM 5-HT produced the largest decrease in burst rate, 77% ± 13% at 60 min (P < 0.02) and 24% ± 2% at 180 min of baseline (P < 0.005) after drug perfusion.

Application of 8-OH-DPAT was followed by a significant decrease in burst rate at all doses used (ANOVA P < 0.0001) (Figs. 1 and 2). Perfusion of 1 μM 8-OH-DPAT was followed by a decrease in the burst rate to 81% ± 9% of baseline values at 60 min (P < 0.02) with a return to control values at 180 min (58% ± 10%). Although 50 μM 8-OH-DPAT produced the greatest decrease in burst rate after 60 min (48% ± 4% of baseline (P < 0.007), 10 μM 8-OH-DPAT produced the largest decrease after 180 min (19% ± 2%; P < 0.002; Fig. 2).

Application of the selective $5-HT_2$ agonist, DOI, resulted in a significant dose-related reduction in burst rate at 60 min (ANOVA P < 0.006); however, the long latency of this effect (30 - 60 min) suggest that the $5-HT_2$ receptor may not be involved.

No significant changes in burst amplitude were noted at any concentration of 5-HT, 8-OH-DPAT, or DOI.

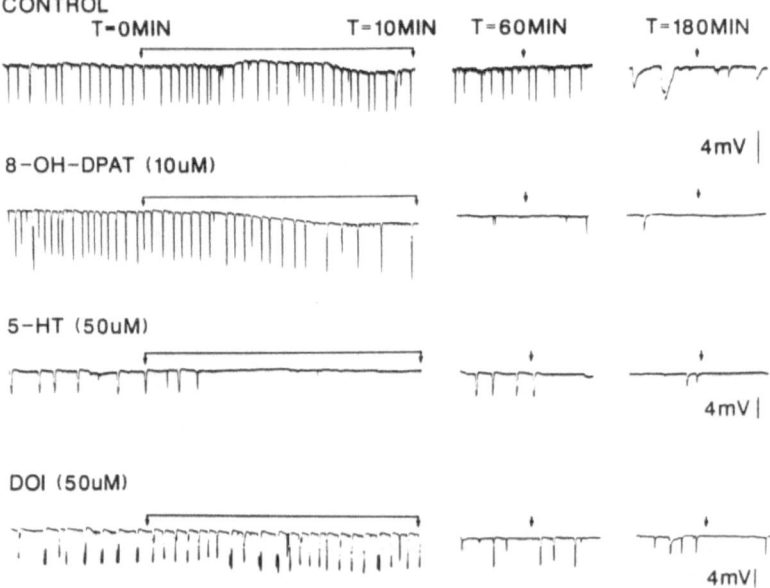

Fig. 1. Chart records of burst activity in the CA1 region after control, 8-OH-DPAT, 5-HT, and DOI drug perfusion

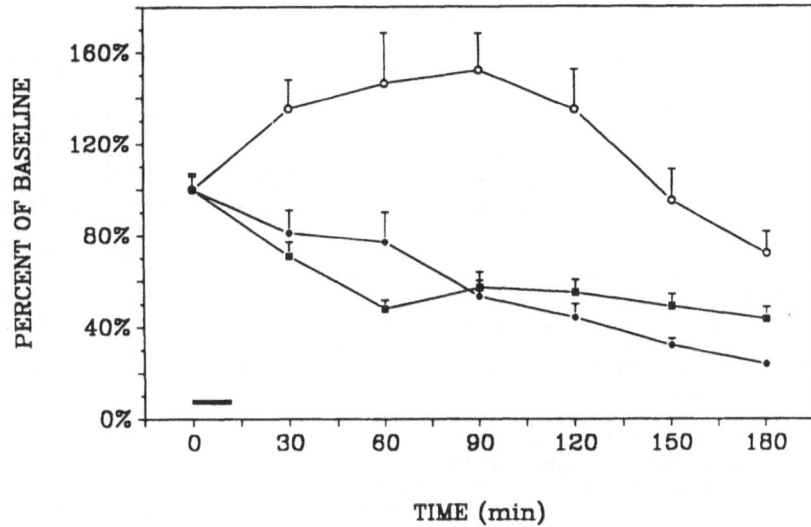

TIME (min)

Fig. 2. Application of 50 μM 8-OH-DPAT (■) and 5-HT (●) for 10 min (*solid horizontal bar*) resulted in a significant reduction in burst rate when compared to control (○) ACSF perfusion

Discussion

Both 5-HT and 8-OH-DPAT produced a dose-related decrease in burst rate, suggesting that this effect was mediated by activation of the 5-HT_{1A} receptor. Several underlying mechanisms are possible. Activation of the 5-HT_{1A} receptor by 5-HT and 8-OH-DPAT appears to hyperpolarize hippocampal pyramidal cells via an increased K^+ conductance (Andrade and Nicoll 1987), and such a mechanism may offset the depolarization observed in low Ca^{2+} conditions. Hippocampal 5-HT_{1A} receptors appear to mediate inhibition of adenylate cyclase (De Vivo and Maayani 1986) and decrease cAMP levels which could also act to reduce burst activity (Haas et al. 1984).

Serotonin and 8-OH-DPAT may also decrease bursting by reducing neuronal coupling in pyramidal cells, given that 5-HT application reduces coupling in *Helisoma* (Mercier and Kater 1986). This factor may be important since decreased $[Ca^{2+}]_i$ may facilitate neuronal coupling (Rao et al. 1987) and contribute in the synchronization of low Ca^{2+} bursts (Dudek et al. 1986).

The present results demonstrate that both 5-HT and 8-OH-DPAT can act to reduce low Ca^{2+} induced epileptogenesis and suggest that the 5-HT_{1A} receptor may mediate anticonvulsant effects in the CA1 region of the rat hippocampus. The long latency of the DOI effects indicates either non-specific effects or the possible involvement of second messenger systems.

References

Agopyan N, Avoli M, Rieb L, Tancredi V (1985) Depression of hippocampal low calcium field bursts by the antiepileptic drug valproic acid. Neurosci Lett 69: 57-62

Andrade R, Nicoll RA (1987) Pharmacologically distinct actions of serotonin on single pyramidal neurones of the rat hippocampus recorded *in vitro*. J Physiol (Lond) 394: 99-124

De Vivo M, Maayani S (1986) Characterization of the 5-hydroxytryptamine 1A receptor-mediated inhibition of Forskolin-stimulated adenylate cyclase activity in guinea pig and rat hippocampal membranes. J Pharmacol Exp Ther 238(1): 248-253

Dudek FE, Snow RW, Taylor CP (1986) Role of electrical interactions in synchronization of epileptiform bursts. In: Delgado-Escueta AV, Ward AA, Woodbury DM Jr, Porter RJ (eds) Basic mechanisms of the epilepsies. Molecular and cellular approaches. Raven, New York, pp 593-618

Frankenhaeuser B, Hodgkin AL (1957) The action of calcium on the electrical properties of squid axons. J Physiol (Lond) 137: 218-244

Haas HL, Jefferys JGR (1984) Low-calcium field burst discharges of CA1 pyramidal neurones in rat hippocampal slices. J Physiol (Lond) 354: 185-201

Haas HL, Jefferys JGR, Slater NT, Carpenter DO (1984) Modulation of low calcium induced field bursts in the hippocampus by monoamines and cholinomimetics. Pflügers Arch 400: 28-33

Heinemann U, Konnerth A, Pumain R, Wadman WJ (1986) Extracellular calcium and potassium concentration changes in chronic epileptic brain tissue. In: Delgado-Escueta AV, Ward AA, Woodbury DM Jr, Porter RJ (eds) Basic mechanisms of the epilepsies. Molecular and cellular approaches. Raven, New York, pp 641-661

Jefferys JGR, Haas HL (1982) Synchronized bursting of CA1 hippocampal pyramidal cells in the absence of synaptic transmission. Nature 300: 448-452

Lux HD, Heinemann U, Dietzel I (1986) Ionic changes and alterations in size of the extracellular space during epileptic activity. In: Delgado-Escueta AV, Ward AA, Woodbury DM Jr, Porter RJ (eds) Basic mechanisms of the epilepsies. Molecular and cellular approaches. Raven, New York, pp 619-639

Mercier AJ, Kater SB (1986) A role for serotonin in the regulation of electrotonic coupling in Helisoma. Soc Neurosci Abstr 12: 117

Pazos A, Palacios JM (1985) Quantitative autoradiographic mapping of serotonin receptors in the rat brain. I. Serotonin-1 receptors. Brain Res 346: 205-230

Peroutka SJ (1987) Serotonin receptors. In: Meltzer HY (ed) Psychopharmacology. The third generation of progress. Raven, New York, pp 303-311

Rao G, Barnes CA, McNaughton BL (1987) Occlusion of hippocampal electrical junctions by intracellular calcium injection. Brain Res 408: 267-270

Schwindt P, Crill W (1980) Role of a persistent inward current in motoneuron bursting during spinal seizures. J Neurophysiol 43: 1296-1318

Storm JF (1987) Action potential repolarization and a fast afterhyperpolarization in rat hippocampal pyramidal cells. J Physiol (Lond) 385: 733-759

Taylor CP, Dudek FE (1982) Synchronous neural afterdischarges in rat hippocampal slices without active chemical synapses. Science 218: 810-812

Waterhouse BD (1986) Electrophysiological assessment of monoamine synaptic function in neuronal circuits of seizure susceptible brains. Life Sci 39: 807-818

Yaari Y, Konnerth A, Heinemann U (1986) Nonsynaptic epileptogenesis in the mammalian hippocampus in vitro. II. Role of extracellular potassium. J Neurophysiol 56: 424-438

Excitability Changes Induced in Rat Neocortical Neurons by the Selective Blockade of a Low K_m, Ca^{2+}/Calmodulin - Independent cAMP - Phosphodiesterase

B. Sutor and G. ten Bruggencate

Institute of Physiology, University of Munich, Munich, FRG

Introduction

By analogy with the effects of neurotransmitters, the action of the intracellular second messenger cyclic adenosine-3',5'-monophosphate (cAMP) has to be terminated by an efficient and effectively regulated inactivation mechanism. Cyclic AMP is degraded by cAMP - phosphodiesterases (cAMP - PDE). Thus, the regulation of the activity of cAMP - PDE may be of crucial importance for normal neuronal excitability as well as for the development of abnormal neuronal behavior, including epilepsies. PDEs are a family of isozymes differing in their biochemical and pharmacological properties (Beavo 1988). The availability of selective inhibitors for a low K_m, Ca^{2+}/calmodulin-independent isozyme (e.g., denbufylline, see Nicholson et al. 1989) allows one to study the influence of this cAMP - PDE isozyme on the excitability of rat neocortical neurons *in vitro*.

Methods

Coronal slices (500 μm) were prepared from the frontal cortex of male Wistar rats (120 - 160 g). In the recording chamber, the slices were kept submerged in artificial CSF consisting of (in mM): 124 NaCl, 3 KCl, 1.25 NaH$_2$PO$_4$, 2.0.CaCl$_2$, 1.3 MgCl$_2$, 25 NaHCO$_3$, and 10 glucose (continuously gassed with 5% CO$_2$ in O$_2$; pH 7.4 at 32°C). Intracellular recordings were made from superficially located neurons (layer II and III) by means of glass microelectrodes filled with either 4 M potassium acetate (pH 7.2) or 3 M KCl. Postsynaptic potentials were evoked by electrical stimulation using a bipolar silver electrode positioned in cortical layer IV. Stimuli were applied at a frequency of 0.1 Hz. In the present study, the stimulus intensities (30 - 300 μA) are given as multiples of the threshold intensity (T) necessary to evoke an action potential. Denbufylline (DBF) was dissolved in dimethyl sulfoxide (DMSO) at concentrations of 10^{-2} M. From this stock solution, appropriate amounts were added to the bathing solution in order to obtain the desired drug concentration. The final solvent concentration never exceeded 0.1%. At this concentration, DMSO did not affect the neuronal electrophysiological properties.

Results

The results are based on intracellular recordings from 25 cortical neurons with a mean resting potential of -79.6 ± 3.2 mV (mean ± SD) and a mean input resistance of 26.5 ± 2.3 MΩ. When added to the bathing solution at concentrations between 10 and 100 nM, DBF increased the amplitude of excitatory postsynaptic potentials (EPSPs). Figure 1 depicts an example of DBF-induced changes of EPSPs. At low stimulus intensities (0.4 T, Fig. 1A), this neuron responded with an early EPSP (eEPSP, control trace). Following the application of DBF (100 nM, center trace), the amplitude of the eEPSP increased and a late EPSP (lEPSP) occurred. Upon an increase in stimulus strength to 0.5 T (Fig. 1B), an eEPSP followed by a small lEPSP was observed (see Sutor and Hablitz 1989). Again, DBF (100 nM) produced an enhancement in the amplitude of both EPSPs and, in addition, a decrease in the latency of the lEPSP (center trace). These DBF-induced

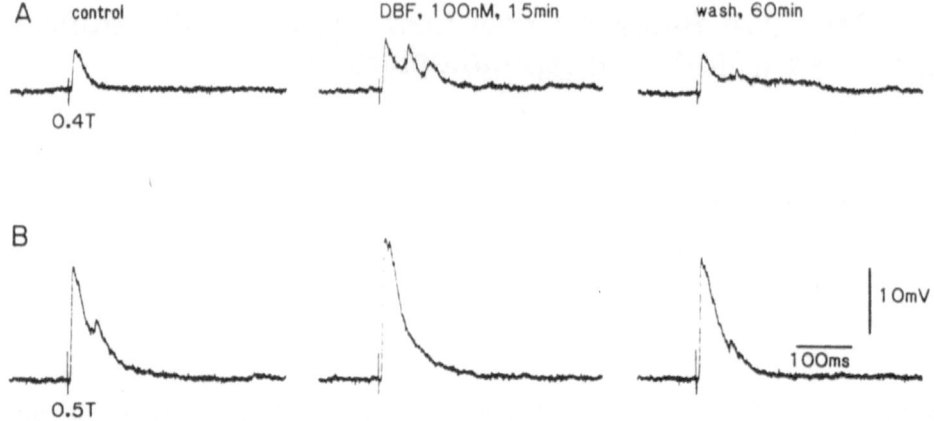

Fig. 1A,B. Actions of DBF (100 nM) on EPSPs in rat neocortical neurons. The resting potential was -85 mV. EPSPs were evoked by electrical stimulation of cortical layer IV using different stimulus strength (**A** and **B**). The time point of stimulation is indicated by the stimulus artifact. Stimulus intensities are given as multiples of 1 T

changes of EPSPs were reversible upon washout of the drug (Fig. 1A and B, right traces). At these low concentrations (10 - 100 nM) no significant DBF-induced effects on inhibitory postsynaptic potentials (IPSPs) were observed. However, at higher concentrations (1 - 10μM), DBF produced a marked prolongation of GABAergic, Cl⁻-dependent IPSPs (Fig. 2). At a stimulus strength of 0.5 T (Fig. 2A), an eEPSP was observed (control trace), the amplitude of which was enhanced following the application of DBF (10 μM). Stimulation with an intensity of 0.9 T (Fig. 2B) evoked a sequence of postsynaptic potentials consisting of an EPSP followed by a depolarizing, Cl⁻-dependent IPSP and a small hyperpolarizing, K⁺-dependent IPSP (control trace, see Howe et al. 1987). In the presence of DBF (10 μM), both the amplitude of the EPSP and the duration of the depolarizing IPSP increased (center trace; see also superimposed traces). At a stimulus intensity of 1 T (Fig. 2C), the DBF-induced effects were similar to those observed at 0.9 T. The action of DBF on the depolarizing IPSP was reversible upon washout of the drug.

DBF affected neither the membrane potential nor the input resistance (measured with hyperpolarizing current pulses). In the depolarizing direction, a slight enhancement of the steady state inward rectification (see Sutor and Zieglgänsberger 1987) was observed. The latter effect is probably responsible for the DBF-induced increase in direct excitability.

Discussion

These experiments demonstrate a concentration-dependent effects of the selective cAMP - PDE inhibitor, DBF, on synaptic potentials in rat neocortical neurons. At low concentrations, DBF facilitates excitatory synaptic transmission without affecting inhibitory transmission. At higher concentrations, DBF predominantly prolongs the duration of GABAergic, Cl⁻-dependent IPSPs. In this context, two other observations are of interest: (1) In cytochemical studies, we were able to show that the cAMP - PDE activity is predominantly located in the postsynaptic density of axospineous or axodendritic synapses of cortical neurons (see also Florendo et al. 1971) and (2) by measuring the effects of DBF on the total PDE activity in rat frontal cortex we found that at concentrations which effectively facilitate synaptic transmission (up to 10 μM), only a maximum of 10% of the total PDE activity was blocked. These results suggest that the low K_m, Ca^{2+}/calmodulin-independent cAMP - PDE might be of enormous importance for synaptic transmission in the neocortex. The ultrastructural localization of most of the PDE activity and the

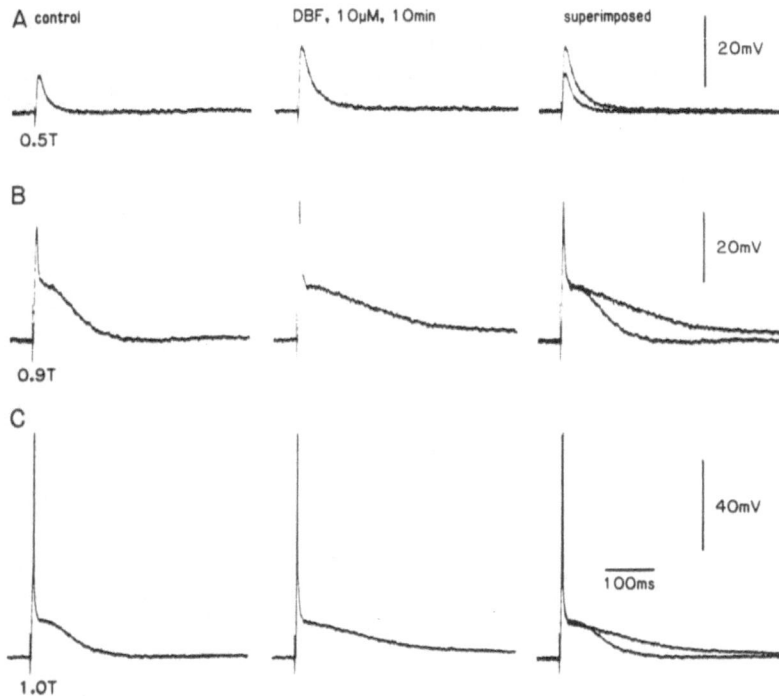

A control   DBF, 1 0µM, 1 0min   superimposed

20mV

0.5T

B   20mV

0.9T

C   40mV

100ms

1.0T

Fig. 2A-C. Actions of DBF (10 µM) on IPSPs in rat neocortical neurons. The resting potential was -81 mV. **A** At an intensity of 0.5 T, the stimulus produced an EPSP (control) the amplitude of which increased upon application of DBF. In this and the following two panels, the single recordings were superimposed and are shown on the right-hand side of the figure. **B** At a just subthreshold stimulus intensity (0.9 T), the stimulus evoked an EPSP/IPSP sequence. GABA-mediated, Cl⁻-dependent IPSPs display a depolarizing time course in rat neocortical neurons in vitro. **C** At 1.0 T, the stimulus induced a slowly decaying depolarizing IPSP following the synaptically evoked action potential. Note the enhancing effect of DBF on the IPSP duration

observation that the inhibition of only a small fraction of the total PDE activity leads to marked facilitation of synaptic transmission suggest that the low K_m cAMP - PDE is located at a "strategically" significant site and that this isozyme might be the determining factor in the regulation of intracellular cAMP concentration in the synaptic region.

Acknowledgements. This work was supported by the Deutsche Forschungsgemeinschaft (Su 104/1-1 and SFB 220, A9).

References

Beavo JA (1988) Multiple isozymes of cyclic nucleotide phosphodiesterase. Adv in Second Messenger and Phosphoprotein Res 22: 1-38

Florendo NT, Barrnett RJ, Greengard P (1971) Cyclic 3',5'-nucleotide phosphodiesterase: cytochemical localization in cerebral cortex. Science 173: 745-747

Howe JR, Sutor B, Zieglgänsberger W (1987) Baclofen reduces post-synaptic potentials of rat cortical neurones by an action other than its hyperpolarizing action. J Physiol 384: 539-569

Nicholson CD, Jackman SA, Wilke R (1989) The ability of denbufylline to inhibit cyclic nucleotide phosphodiesterase and its affinity for adenosine receptors and the adenosine re-uptake site. Br J Pharmacol 97: 889-897

118

Sutor B, Hablitz JJ (1989) EPSPs in rat neocortical neurons in vitro. I. Electrophysiological evidence for two distinct EPSPs. J Neurophysiol 61: 607-620
Sutor B, Zieglgänsberger W (1987) A low-voltage activated, transient calcium current is responsible for the time-dependent depolarizing inward rectification of rat neocortical neurons in vitro. Pflügers Arch 410: 102-111

Action of a Phorbol Ester
in the Presence of Two PKC Blockers

M. R. Klee[1] and J. Hoyer[2]

[1] Max-Planck-Institut für Hirnforschung, Abteilung Neurophysiologie, 6000 Frankfurt, FRG
[2] Institut für Neurophysiologie, Universität Wien, 1090 Vienna, Austria

Introduction

Agonist-induced phosphorylation is a process that underlies a variety of plasticity changes in neuronal functioning. Protein kinase C (PKC), present in *Aplysia* ganglia (DeRiemer et al. 1985), can be activated consistently by phorbol esters. It is possible to determine whether actions evoked by a phorbol ester are the consequence of activation of the kinase C or whether they express intrinsic activities of the phorbol ester itself by means of blockage of the kinases by using a specific blocker. This could also provide further confirmation that the phorbol ester activates kinase C, because the effects of phorbol esters on *Aplysia* bag cell and cat motoneurons have been shown to be identical to those of injection of PKC itself into the same neuron (DeRiemer et al. 1985; Zhang and Krnjevic 1987). We conducted a series of experiments with two well-known blockers of PKC: H-7 (isoquinolinylsulfonyl-methylpiperazine dihydrochloride; Hidaka et al. 1984; Hidaka and Hagiwara 1987) and staurosporine (Tamaoki et al. 1986). These two PKC blockers act on the catalytic domain of the PKC rather than on the regulatory domain (Huang 1989; Nakadake et al. 1988), where they would be interfering with the action of phorbol esters.

Experiments were carried out in neurons of the visceral and pleural ganglia of *Aplysia californica* using current and voltage clamp techniques. The phorbol ester (12,13-phorbol-diacetate, PDAc, Sigma) was diluted in water and added to artificial sea water (ASW) in concentrations of 0.2 - 22 μM; H-7 (Sigma) was diluted in water and added in concentrations of 5 - 50 μM while staurosporine (Boehringer) was diluted in dimethyl-sulfoxide and added to ASW to yield a final concentration of 1 μM.

Effects of PDAc

As reported elsewhere (Klee 1988; Klee and Hoyer 1989), the action of phorbol esters on neurons in the visceral and pleural ganglia can be described as follows. PDAc increases the spike amplitude (and hence the overshoot) as well as the slope of the rising phase of the action potential (AP). The membrane resting potential and resistance remain unchanged. In voltage clamp, the effect of PDAc is expressed as an increase in the sodium inward current; a reduction in the L-type calcium inward current is observed in most of the neurons. However, an increase in this current is observed in the lower voltage range (-30 to 0 mV) in bursting cells of the left upper quadrant (L-2 to L-6). The total outward current remains practically unchanged (or increases slightly) in contrast to the delayed, voltage-dependent potassium current observed in cobalt ASW which is slightly decreased. The transient potassium current, I_A, is reduced. These effects are almost completely irreversible.

Synaptic transmission and agonist-induced currents are also changed by PDAc. After approximately 10 - 15 min, PDAc induces an increase in synaptic noise. Excitatory postsynaptic potentials (EPSP), such as the noncholinergic EPSP of the giant cell R-2 and the presumably cholinergic, monosynaptic RC1 EPSP of the burster cell R-15, are increased up to threefold by PDAc in the range of 0.5 - 5 μM. In the experiments presented here, we concentrated on the RC1 EPSP. At a holding potential of -90 mV (in current or voltage clamp condition) the amplitude of this

EPSP/EPSC follows a specific sequence of changes during repetitive stimulation (1 c/s; Schlapfer et al. 1974). The most prominent alteration is a frequency facilitation, i.e. during a train of stimuli (30 - 35 in our experiments) the amplitude of the EPSP increases by 30 -60% compared to the control. The frequency facilitation is blocked by 5 μM PDAc, so that with respect to the first strongly enhanced EPSP all following EPSP values are smaller than the first one after PDAc has reached its maximal effect (Fig. 1A).

In contrast to the facilitatory influence of PDAc on synaptic transmission, PDAc reduces sodium, chloride and potassium currents evoked by acetylcholine (ACh) in different neurons. The sodium current induced by ACh in the cell R-15 or in neurons of the RB group is reduced by approximately 15% - 40% by 0.2 - 5 μM PDAc. In the medial cells of the pleural ganglia, ACh evokes an early chloride current followed by a potassium current (Kehoe 1972). PDAc blocks the chloride current almost completely and the potassium current by 40 - 60%. These effects of PDAc are reversible within 1 - 2 h of washout.

Effects of H-7 and Staurosporine

To test the effects of H-7 and staurosporine and their interaction with PDAc, we concentrated exclusively on three phenomena: (1) the effects on the action potential (amplitude and slope of the voltage changes), (2) effects on the amplitude and (3) the frequency facilitation (FF) of the RC1 EPSP in cell R-15.

As demonstrated in Fig. 1, the amplitude of the action potential and its rising slope is strongly reduced by H-7 (Figs. 1, 2), while staurosporine has little effect. The amplitude of the RC1 EPSP

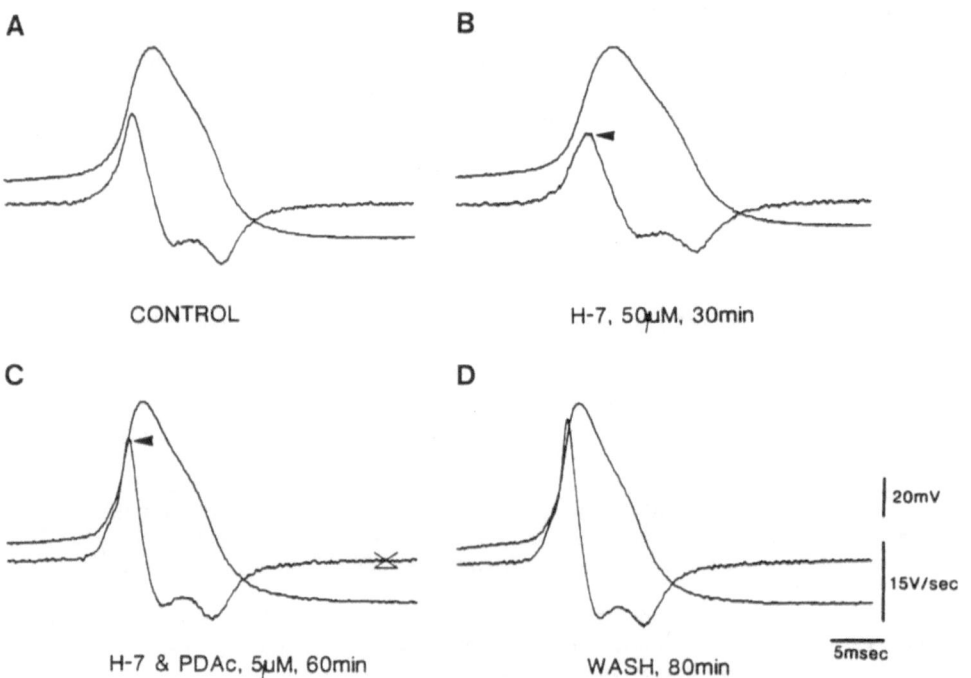

Fig. 1A-D. Effects of H-7 and PDAc on the action potential. **A** Control. **B** Reduced rising slope *(arrow)* and increased duration under H-7. **C** PDAc in the presence of H-7 increases the slope of the rising phase and decreases the AP duration (C versus A). **D** This effect persists after 80 min wash

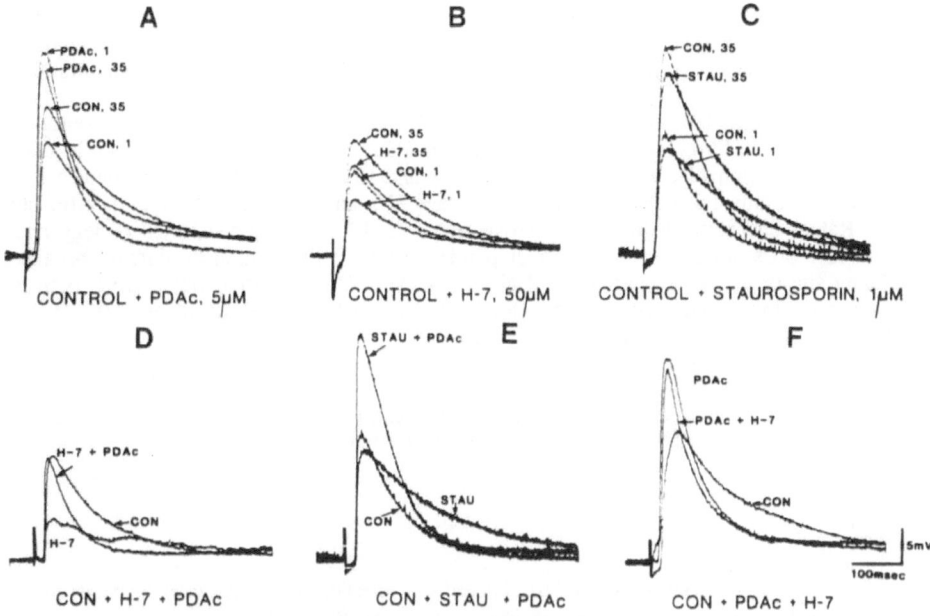

Fig. 2A-F. A Frequency facilitation (FF) in control. EPSP 1 *(CON 1)* is smaller than the last EPSP *(CON 35)*. PDAc increases the EPSP amplitude and blocks FF; the first EPSP *(PDAc 1)* is larger than the following *(PDAc 35)*. **B** H-7 reduces the EPSP amplitudes, but there is still FF *(H-7 1* smaller than *H-7 35)*. **C** Staurosporine depresses the amplitudes of the EPSPs less than H-7. Most prominent is an increased decay time. **D** Strong reduction of a single EPSP by H-7. After adding PDAc the amplitude increases comparable to control. **E** Little effect of staurosporine *(STAU)* on the peak amplitude of the EPSP but strong enhancement following PDAc in the presence of staurosporine. **F** Only slight reduction in the increased EPSP due to the action of PDAc after adding H-7

is reduced by H-7 but with normal FF (Fig. 2B). Adding 5 µM PDAc to the H-7 solution leads to an increase in the amplitude and the rising phase of the action potential as described above as the effect of PDAc alone (Fig. 1C). Staurosporine only slightly reduces the amplitude of the EPSP, however it strongly increases its decay time (Fig. 2C).

The interaction of the two blockers with PDAc is shown in Fig. 2. In Fig. 2D-F, the reduction in amplitude of the EPSP caused by H-7 is completely reversed by PDAc; the amplitude is nearly identical to that in the control. When PDAc is applied following staurosporine, the potentiating effect of PDAc on the amplitude of the EPSP is as strong as under control conditions (Fig. 2E). If H-7 is given after PDAc, its blocking effect is negligible (Fig. 2F). Different actions of PDAc and H-7 can also be deduced from their different time courses. Some effects of H-7 are transient and occur within 10 min, while the effects of PDAc start after about 20 - 40 min and are less reversible.

Discussion

In cat spinal motoneurons and in *Aplysia* bag cells the effects of phorbol esters are identical to those from a direct action of injected PKC, namely an increased AP (DeRiemer et al. 1985; Zhang and Krnjevic 1987). The effects of PKC and phorbol esters on cat motoneuron are similar to those of PDAc on *Aplysia* neurons of the visceral ganglion, which are reported here. They are also similar to the effects of phorbol esters on cat Betz' cells (Baranyi et al. 1987), since the rising slope of the action potential is increased in each type of neuron. Increased synaptic responses caused by phorbol esters have been reported previously (Hvalby et al. 1988; Shapira et al. 1987).

In our experiments, H-7 decreases the slope of the AP and reduces the EPSP amplitude, opposite to the effect of PDAc. However, the action of PDAc in the presence of H-7 is only slightly reduced. The action of staurosporine on the AP and EPSP is much less pronounced than that of H-7. These results raise the question of whether H-7 and staurosporine are specific blockers of PKC. Rather unspecific actions of staurosporine and H-7 have been reported (Rüegg and Burgess 1989). H-7 does not block the action of phorbol esters at the regulatory domain because of their different site of action (Huang 1989; Nakadate et al. 1988); in other preparations both have been shown to exhibit a synergistic effect (Hockberger et al. 1989). In our experiments H-7 is able to reduce but not to block the action of PDAc.

Acknowledgements. We thank Mrs. M. Duesmann for technical assistance, Mrs. M. Ehms-Sommer for the photography, and Mr. N. Steinberg for editing the manuscript.

References

Baranyi A, Szente MB, Woody CD (1987) Intracellular injection of phorbol ester increases the excitability of neurons of the motor cortex of awake cats. Brain Res 424: 396-401

DeRiemer SA, Strong JA, Albert KA, Greengard P, Kaczmarek LK (1985) Enhancementof calcium current in *Aplysia* neurons by phorbol ester and protein kinase C. Science 313: 313-316

Hidaka H, Hagiwara M (1987) Pharmacology of isoquinoline sulfonamide protein kinase C inhibitors. Trends Physiol Sci 8: 162-164

Hidaka H, Inagaki M, Kawamoto S, Sasaki Y (1984) Isoquinolinesulfonamides, novel and potent inhibitors of cyclic nucleotide dependent protein kinase and protein kinase C. Biochemistry 23: 5036-5041

Hockberger P, Toselli M, Swandulla D, Lux HD (1989) A diacylglycerol analogue reduces neuronal calcium currents independently of protein kinase C activation. Nature 338: 340-342

Huang KP (1989) The mechanism of protein kinase C activation. Trends Neurosci 12:425-432

Hvalby O, Reyman K, Andersen P (1988) Intracellular analysis of potentiation of CA1 hippocampal synaptic transmission by phorbol ester application. Exp Brain Res 71: 588-596

Kehoe JS (1972) Ionic mechanisms of a two-component cholinergic inhibition in *Aplysia* neurones. J Physiol 225: 85-114

Klee MR (1988) Effects of phorbol esters on membrane properties of *Aplysia* neurons. 1st Europ Congr Neurol, Prague, pp 198-199

Klee MR, Hoyer J (1989) Different actions of a phorbol ester on cholinergic transmission and ACh-evoked currents. Proc XXXI Int Congr Physiol Sci, p 406

Nakadate T, Jeng AY, Blumberg PM (1988) Comparison of protein kinase C functional assays to clarify mechanisms of inhibitor action. Biochem Pharmacol 37: 1541-1545

Rüegg UT, Burgess GM (1989) Staurosporine, K-252 and UCN-01: potent but nonspecific inhibitors of protein kinases. Trends Physiol Sci 10: 218-220

Schlapfer WT, Woodson PBJ, Tremnlay JP, Barondes, SH (1974) Depression and frequency facilitation at a synapse in *Aplysia californica*: evidence for regulation by availability of transmitter. Brain Res 76: 267-280

Shapira R, Siloberberg SD, Ginsburg S, Rahamimoff R (1987) Activation of protein kinase C augments evoked transmitter release. Nature 325: 58-60

Tamaoki T, Nomoto H, Takahashi I, Kato Y, Morimoto M, Tomita F (1986) Staurosporine, a potent inhibitor of phospholipid/ Ca^{2+} dependent protein kinase. Biochem Biophys Res Comm 135: 397-402

Zhang L, Krnjevic K (1987) Effects of intracellular injections of phorbol ester and protein kinase C on cat spinal motoneurons in vivo. Neurosci Lett 77: 287-292

Genetically Determined Seizure Susceptibility in the Rat

M. Kunz, J. Mareš and R. Brdička

Institutes of Physiology and Biology, Faculty of Medicine, Charles University, Prague, Czechoslovakia

Introduction

In biological sciences the use of animal models is a common practice, and the experiments are usually performed at a technical level that permits one to draw broad conclusions. Frequently, however, it is difficult to compare results of experiments using different animal species. Moreover, this complication may even be encountered within one species. In neurobiology, *Rattus norwegicus* has become a typical example of this situation. Many physiological functions of the CNS are genetically determined and the hereditary character of many seizure disorders has been confirmed in both clinical and experimental research. Sensory epilepsy was one of the very first animal models of spontaneously occurring seizures and new strains of rats spontaneously epileptic or highly sensitive to epileptogenic stimuli, were selected (Naquet and Meldrum 1972; Collins 1972; Seyfried 1982); however, even in the first kindling studies, differences in the development of this phenomenon in various strains of rats were observed (Goddard et al. 1969; Racine et al. 1973). In addition, the relation between the progression of epileptic seizures (kindling) and the course of learning was studied in genetically homogeneous inbred strains of mice (Leech and McIntyre 1976).

The present paper attempts to arrive at a solution of the following questions: (a) Are the differences between responses to epileptogenic stimuli genetically determined? (b) If they are, to what extent do randomly bred animals represent the species *Rattus norwegicus*? (c) Is an increased seizure susceptibility nonspecific or it is restricted to a particular stimulus or agent? (d) Are such differences present even at the level of electrophysiological parameters (EEG, evoked potentials) in naive animals?

Methods

Metrazol-induced seizures (a model of primary generalized epilepsy) were used to compare responses of random-bred and inbred rats. Random-bred male albino Wistar rats (18 animals) and inbred male rats, AVN, $n = 11$; BDV, $n = 14$; BDX, $n = 7$; BN, $n = 6$; BP, $n = 8$; DA, $n = 9$; LEP, $n = 13$; LEW, $n = 10$; SHR, $n = 7$; WAG, $n = 9$, were used. None of the rats suffered from spontaneous seizures. The rats were treated intraperitoneally with 65 mg/kg Metrazol (10% solution, no anesthesia). The animals were then placed into plastic cages and the behavioral manifestations of seizures were followed for the next 30 min. The following parameters were recorded: (a) the incidence and latency of the first jerk, (b) the incidence and latency of clonic seizures, (c) the incidence and latency of tonic-clonic seizures and, (d) mortality. A statistical significance of differences between the incidence of seizures was evaluated using the chi-square test (twofold and fourfold tables); differences between the latencies were tested using one way analysis of variance. The intra-strain variances of responses were compared to that of random-bred rats using the F-test; a stepwise discrimination analysis was also used.

According to the severity of seizures, two of the most (BDX, $n = 11$ and BN, $n = 6$) and two of the least (LEP, $n = 10$ and LEW, $n = 10$) susceptible inbred strains and random-bred Wistar rats were chosen for the electrophysiological part of our study. The chosen parameters were: (a)

124

spontaneous EEG from frontal and occipital cortical regions (power density spectra); (b) somatosensory evoked potentials elicited by stimulation of the sciatic nerve and interhemispheric response to stimulation of the right frontal cortex registered from the left frontal cortex; (c) initial phases of acute kindling, i.e., the duration of four subsequent epileptic afterdischarges evoked by an 8 Hz electrical stimulation of the cortex (duration 20 s) in 10 min intervals (for details see Mareš et al.1981).

Results

The administration of Metrazol induced seizures in all experimental animals; however, interstrain differences between the latencies of the first jerk as well as between the tonic seizures were observed. For example in the LEW strain no tonic seizures and no mortality occurred. By contrast the response of the BDX strain was characterized by a 100% incidence of tonic seizures with the shortest latency observed, by the absence of clonic seizures, and by 100% mortality (Fig. 1). No interstrain differences were observed between the latencies of the clonic seizures. Stepwise discrimination analysis showed that similarly to the individual inbred strains-random-bred rats displayed their own characteristic pattern of responses. The variance in Wistar rat responses did not differ from variances observed within inbred strains. Several slight modifications in the seizure pattern were noticed. In BN rats, the extension part of the tonic seizure was absent, and the flexion part of the tonic phase was immediately followed by the clonic phase. In WAG rats, the clonic seizures were frequently accompanied by abdominal muscle spasms. One of the BDX rats reacted with two successive tonic seizures. In some LEW and LEP animals, occasional motor automatism, i.e., chewing and cleaning movements, occurred before the first jerk.

Our visual evaluation of EEGs did not reveal any differences between the four inbred strains compared; however, some parts of their power density spectra differed significantly. In addition, the differences between individual strains in the amplitudes and latencies of the first waves of the two evoked potentials were significant. The epileptic afterdischarges in all of the animals displayed similar characteristics. They began with spike and wave complexes of 4 - 5 Hz frequency and terminated, in most cases, abruptly and simultaneously in all observed cortical regions. Their

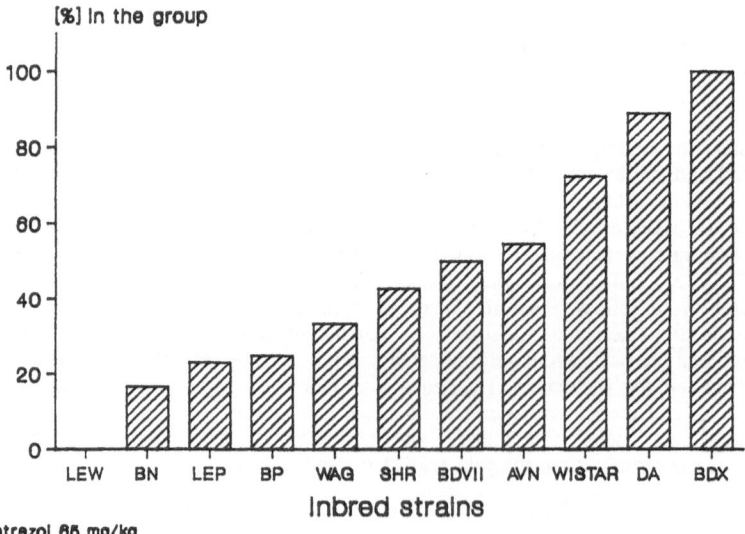

Metrazol 65 mg/kg

Fig. 1.Percent mortality in individual strains after Metrazol application

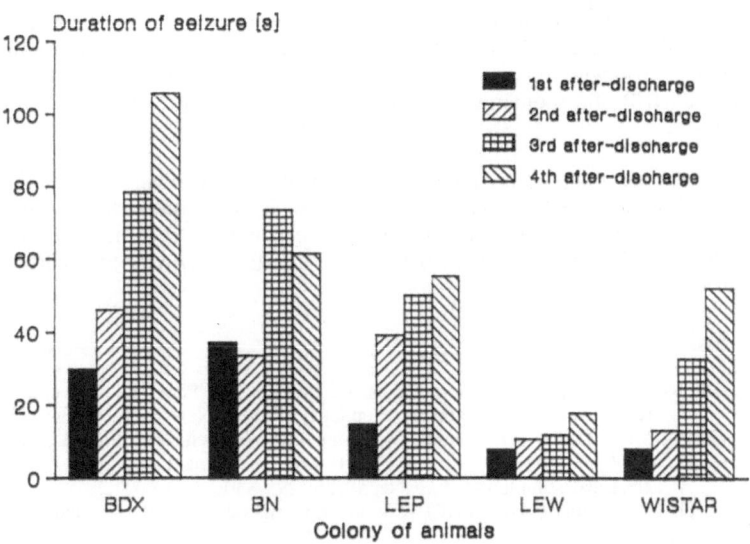

Fig. 2. Duration of repeated electrically-induced self-sustained afterdischarges in five rat strains; Wistar rats are random-bred

character remained unchanged till the end of the seizure, although the frequency decreased. With repetition of the seizures in 10 min intervals, their lengthening was apparent in all groups (Fig. 2).

The duration of seizures differed between inbred strains. The shortest seizures were observed in the LEW strain, whereas the seizures in BDX were the longest. The duration of individual seizures in the BN strain, compared to the other strains, varied considerably. Compared with the results in inbred strains, differences in the duration of individual seizures in the Wistar rats were not obvious.

Discussion

There exist significant inter-strain differences in the susceptibility of rats to Metrazol. Inbred strain BDX is the most susceptible, LEW rats are relatively resistant and do not react with tonic seizures. The BN rats are intermediately susceptible and display slight modifications in the seizure pattern. A comparison of variations of results in the Wistar rats with intra-strain variations of individual inbred strains revealed no differences. This fact may be explained by a presumed partial inbreeding within the random-bred colony. A random-bred strain is considered to ensure genetic variability, comparable to that of a wild outbred population. This presumption was challenged by recent biological research. Festing (1979) points out that in random-bred strains, under conditions of routine breeding regimens, inbreeding with subsequent restriction of the phenotypic spectrum and expression of uncommon recessive features occurs. This fact limits the power of experimental results and prevents generalizations. Instead of common randombred strains of laboratory animals. Festing suggests using a group of randomly chosen inbred strains. Such an approach makes possible to evaluate extreme reactions and to create a model of an outbred population. Analysis of genetic determination of any phenomenon is also possible.

With most physiological parameters, we have observed interstrain differences. As the breeding regimen of all strains as well as the experimental conditions were well controlled, we believe that the observed differences are genetically determined.

In general, the distribution of susceptibility to seizures induced by electrical stimulation is similar to that of Metrazol induced seizures. This fact suggests that a common seizure susceptibility (i.e., nonspecific with regard to the stimulus) may exist. Causes of differences between other parameters (SEP, IHR and spontaneous EEG) remain unclear, but it is obvious that they are also of genetic origin.

All of our findings on inbred strains together with other results, introduce the question of the validity of any experimental result. Extrapolation to research into antiepileptic drugs is probably especially difficult since rats of different genotypes may differ not only in nonspecific susceptibility to epileptogenic stimuli, but also in specific sensitivity to a particular drug.

References

Collins RL (1972) Audiogenic seizures. In: Purpura DP, Penry KJ, Tower DB, Woodbury RM, Walter RD (eds) Experimental models of epilepsy. Raven, New York, pp 347-372
Festing MFW (1979) Inbred strains in biomedical research, McMillan, London
Goddard GV, McIntyre DC, Leech CK (1969) A permanent change in brain function resulting from daily electrical stimulation. Exp Neurol 25: 295-330
Leech CK, McIntyre DC (1976) Kindling rates in inbred mice: an analog to learning? Behav Biol 16: 439-452
Mareš J, Mareš P, Trojan S, Langmeier M (1981) Cortical selfsustained afterdischarges in the rat. Acta Univ Carol (Med Monogr) (Praha) CIV
Naquet R, Meldrum BS (1972) Photogenic seizures in baboon. In: Purpura DP, Penry JK, Tower DB, Woodbury DM, Walter RD (eds) Experimental models of epilepsy. Raven, New York, pp 373-406
Racine R, Burnham W, Gartner J, Levitan D (1973) Rates of motor seizure development in rats subjected to electrical brain stimulation: strain and interstimulus interval effect. Electroencephalogr clin Neurophysiol 35: 553-556
Seyfried TM (1982) Developmental genetics of audiogenic seizures susceptibility in mice. In: Anderson OK, Hauser WA, Penry JK, Sing KO (eds) Genetic basis of epilepsies. Raven, New York, pp 199-221

The Role of the Dentate Gyrus in Transmission of Epileptiform Activity from the Entorhinal Cortex to the Hippocampus

J. D. C. Lambert* and R. S. G. Jones

Division of Neuroscience, John Curtin School of Medical Research, Australian National University, Canberra, ACT 2601, Australia
* Permanent address: Institute of Physiology, University of Aarhus, DK-8000 Århus C, Denmark

Introduction

The entorhinal cortex has extensive reciprocal connections with most major subfields of the hippocampus (Witter 1986). Neurones in layer II of the entorhinal cortex form the origin of the perforant path (PP), the axons of which terminate in a highly organized manner on the dendrites of the granule cells in the fascia dentata. It has been shown that epileptogenesis can be readily induced in the entorhinal cortex by experimental manipulations. We have investigated how discharges which originate in the entorhinal cortex are reflected in the granule cells.

Methods

Experiments were performed on combined slices of the entorhinal cortex and the ventral hippocampus in which the fibre pathways are largely preserved. Conventional electrophysiological techniques were used to record intracellularly from neurones. Simultaneous intracellular recordings were made from pairs of neurones at various locations within the slice to study the temporal relationship between epileptiform events.

Epileptogenesis in the Entorhinal Cortex

Epileptiform events were evoked in the entorhinal cortex by perfusion with $GABA_A$ blockers (picrotoxin or bicuculline) or perfusion with Mg^{2+}-free medium. The events either occurred spontaneously or could easily be evoked by weak stimuli delivered to the appropriate pathway (Fig. 1). Simultaneous recordings from pairs of neurones in the entorhinal cortex showed that epileptiform events occurred first in layer IV-V neurones and appeared to drive synchronous activity in layer II neurones on a one-to-one basis (Jones and Lambert 1990; Jones and Lambert 1990). Characteristics of the discharges were as follows:

1. Events evoked in Mg^{2+}-free medium appeared in all parts of the entorhinal cortex 40 - 80 min after the start of perfusion. They resembled ictal activity, consisting of a large plateau of depolarization with smaller discharges superimposed on it. These events are primarily due to removal of the potential-dependent block of the N-methyl-D-aspartate (NMDA) ionophore by Mg^{2+} since they could be rapidly and reversibly blocked by the NMDA receptor antagonist, 2-amino-5-phosphonovalerate (APV) (Jones and Heinemann 1988). They were only slightly reduced by the non-NMDA (quisqualate/kainate) receptor blocker, 6-cyano-7-nitroquinoxaline-2,3-dione (CNQX 5-10 µM) (Jones and Lambert 1990).

2. Events evoked by disinhibition resembled interictal activity and consisted of an initial paroxysmal depolarizing shift (PDS) surmounted by a few action potentials (AP). This was often followed by a series of afterdischarges (Fig. 1). All of the activity could be readily blocked by CNQX (Jones and Lambert 1990), while APV only reduced the amplitude and duration of the PDS and reduced or abolished the afterdischarges (Jones 1988). The activity probably originates through enhanced

128

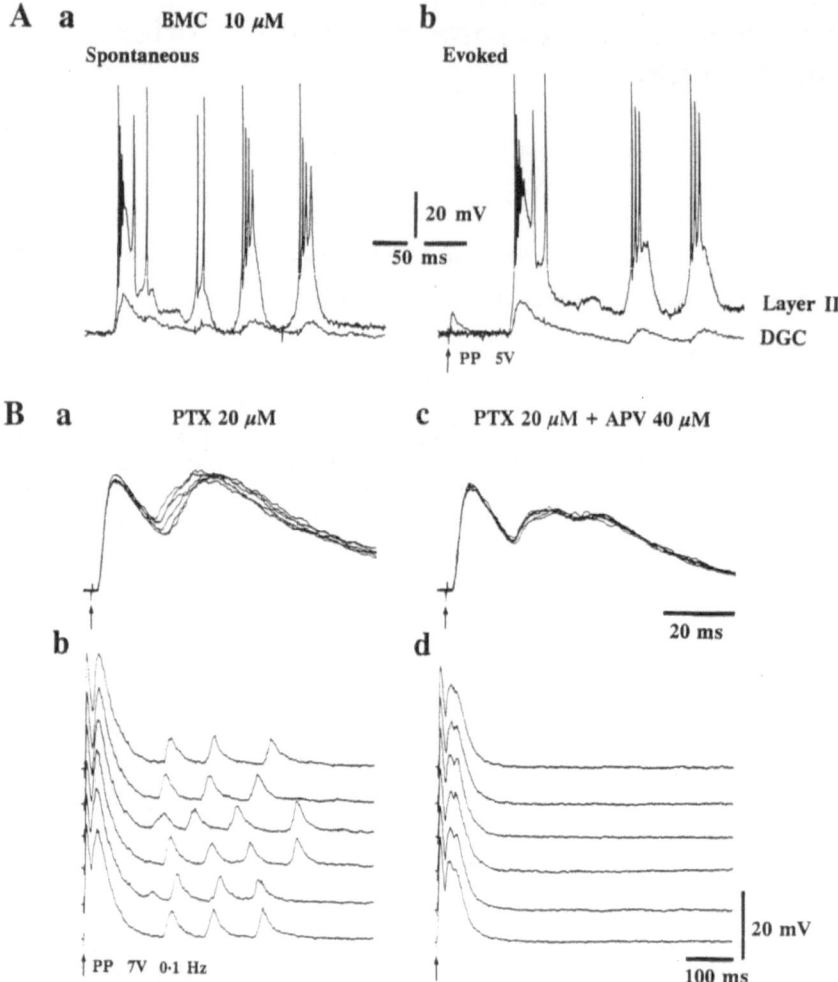

Fig. 1A,B. **A** Simultaneous recordings from a layer II neurone in the entorhinal cortex and a dentate granule cell (*DGC*) to show epileptiform activity induced by bicuculline methochloride (*BMC*, 10 μM). **a** Spontaneously arising events. These consist of a PDS and a series of afterdischarges in the cortical cell which were followed by depolarizing events in the DGC on a one-to-one basis. **b** Stimulation at low intensity in the perforant path (PP) evoked a small monosynaptic epsp in the DGC and initiated a PDS in the cortical cell. The following events were similar to those arising spontaneously. **B** The effect of APV on PTX-induced discharges in a DGC. **a** Superimposed responses (n = 6) evoked by stimulation of the PP 40 min after the addition of PTX (20 μM). A depolarizing potential is present on the falling phase of the monosynaptic epsp. **b** The same 6 sweeps as in **a** at a slower speed showing three or four depolarizing waves. **c, d** 12 min after the addition of ±APV (40 μM) the initial epsp was depressed by about 5% and the wave on its falling phase by about 50%. The later waves were completely abolished

activity in recurrent excitatory pathways which are no longer held in check by GABAergic inhibition.

Reflection of Cortical Epileptiform Events in Granule Cells

Simultaneous intracellular recordings showed that discharges occurring in layer II neurones were followed at a fixed latency (1 - 3 ms) by corresponding events in the granule cells (Fig. 1). A large depolarization (10 - 30 mV) followed the initial PDS in the presence of $GABA_A$ blockers or the rising phase of the ictal event in the case of Mg^{2+}-free medium. This was followed by a series of small (2 - 10 mV) depolarizations in the granule cells which were correlated with the after-discharges in the cortical neurones. Stimulation in the PP evoked epileptiform events in the cortex at a relatively long latency (Fig. 1). The corresponding depolarizing wave was then seen on the falling phase of the monosynaptic epsp in the granule cell. Only in very rare cases did the depolarizations in granule cells reach threshold for AP generation (Jones and Lambert 1990; Jones and Lambert 1990).

The Effect of Mg^{2+}-free Medium and GABA Blockers on the Granule Cell EPSP

Stimulation of the PP on the cortical side of the hippocampal fissure evoked a monosynaptic epsp in the granule cells. Although this was up to 40 mV in amplitude, it rarely reached threshold for AP generation. In isolated hippocampal slices, neither Mg^{2+}-free medium nor $GABA_A$ blockers induced any spontaneous or evoked epileptiform discharges in granule cells. Picrotoxin had only small and inconsistent effects on the psp evoked by stimulation of the PP, which contrasts with the multiple firing seen in CA1 and CA3 pyramidal neurones. In the absence of orthodromically-induced firing, it is not possible to judge the role of recurrent inhibition; however, GABAergic feed-forward inhibition does not appear make a large contribution to the psp. This may be because the interneurones are longitudinally oriented and have been damaged during slicing. It is possible that inhibition in the intact nervous system is more pronounced and more effective at curtailing the excitatory events described here.

On perfusion with Mg^{2+}-free medium, the granule cell epsp became larger and prolonged and occasionally gave rise to an AP. This would be consistent with the participation of NMDA receptors in the epsp. Since the PP-evoked epsp had previously been thought to be mediated entirely by non-NMDA receptors (Crunelli et al. 1983), the possibility of a contribution from NMDA receptors was investigated (Lambert and Jones 1989).

Characteristics of the Granule Cell EPSP

A detailed investigation of the mechanism underlying the granule cell epsp was made in 95 neurones (resting E_M: -85 ± 0.7 mV). The threshold for AP generation was some 40 mV positive to the resting E_M, and the neurones usually did not fire an AP, even with suprathreshold orthodromic stimulation. The receptors involved in the mediation of the epsp have been identified. CNQX (5 - 10 μM) decreased the epsp by 80%-90% (see Fig. 2 for 2 μM CNQX). There was, however, a residual depolarization which was resistant to CNQX and showed characteristics consistent with its mediation by NMDA receptors. Thus, it was reversibly blocked by APV; it was markedly enhanced in Mg^{2+}-free medium; its potential dependency was consistent with operation of the NMDA ionophore in that it increased in size on depolarization and decreased in size with hyperpolarization (Lambert and Jones 1989). This strongly suggests that NMDA receptors do indeed contribute to the normal epsp. This was further confirmed by showing that APV reduced the amplitude and duration of the untreated epsp by 5% - 10% (Fig. 1). APV also removed a small component from the large depolarizing event which follows a spontaneous epileptiform discharge in the entorhinal cortex.

Discussion

Thus, the present experiments show that epileptiform events originating in the entorhinal cortex are transmitted to the dentate gyrus. Here they only evoke subthreshold depolarizations and the activity is therefore not passed on to follower areas in the hippocampus. The excitatory transmission involves the operation of both NMDA and non-NMDA receptors. It is known that death of granule cells accompanies some forms of temporal epilepsy (Dam 1980). It is possible that this is a result of overstimulation of NMDA receptors, which can initiate a cascade of events ending in cell death.

A

PTX 20 μM

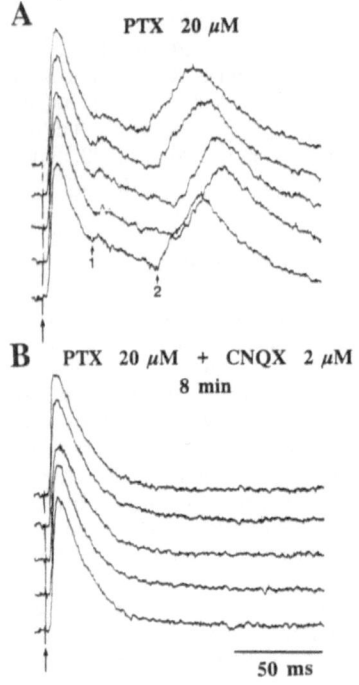

B   PTX 20 μM + CNQX 2 μM
              8 min

50 ms

C   20 min

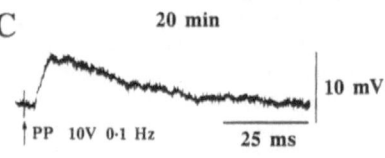

PP 10V 0·1 Hz   25 ms   10 mV

Fig. 2. A-C The action of CNQX on PTX-induced discharges. A and B are rasters showing the responses of a DGC to five successive stimulations of the PP. A Recorded 130 min after the addition of PTX (20 μ*M*) when the epsp is followed by two depolarizing waves on its falling phase. The latency of the second wave was somewhat variable. B 8 min after the addition of CNQX (2 μ*M*), the after discharges were completely abolished at time when CNQX has had little effect on the initial epsp. C After 20 min, the CNQX has reduced the fast epsp by about 60%. Much of the remaining epsp could be blocked by the addition of APV (not shown)

References

Crunelli V, Forda S, Kelly JS (1983) Blockade of amino acid-induced depolarizations and inhibition of excitatory post-synaptic potentials in rat dentate gyrus. J Physiol (Lond) 341: 627-640
Dam AM (1980) Epilepsy and neurone loss in the hippocampus. Epilepsia 21: 617-629
Jones RSG (1988) Epileptiform events induced by GABA-antagonists in entorhinal cortical cells *in vitro* are partly mediated by N-methyl-D-aspartate receptors. Brain Res 457: 113-121
Jones RSG, Heinemann U (1988) Synaptic and intrinsic responses of medial entorhinal cortical cells in normal and magnesium-free medium *in vitro*. J Neurophysiol 59: 1476-1496
Jones RSG, Lambert JDC (1990) Synchronous discharges in the rat entorhinal cortex *in vitro*: site of initiation and role of excitatory amino acids. Neuroscience 34: 657-670
Jones RSG, Lambert JDC (1990) The role of excitatory amino acid receptors in the initiation and propagation of epileptiform discharges from the entorhinal cortex to the dentate gyrus *in vitro*. Exp Brain Res (in press)
Lambert JDC, Jones RSG (1989) Activation of NMDA receptors contributes to the EPSP at perforant path synapses in the rat dentate gyrus *in vitro*. Neurosci Lett 97: 323-328
Witter MP (1986) A survey of the anatomy of the hippocampal formation, with emphasis on the septotemporal organization of its intrinsic and extrinsic connections. In: Schwarcz R, Ben-Ari Y (eds) Excitatory Amino Acids and Epilepsy. Plenum, New York, pp 67-82

Changes in Parvalbumin and Calbindin D28K Immunoreactivity in Rat Brain Following Pilocarpine-Induced Seizures

E. A. Cavalheiro[1] and M. Bentivoglio[2]

[1] Laboratório de Neurologia Experimental, Escola Paulista de Medicina, São Paulo, Brazil
[2] Istituto di Anatomia Umana, Universitá di Verona, Italy

Introduction

Structural damage of the human brain may lead to chronic epilepsy in survivors (Shorvon 1984). Such seizures are usually partial or mixed in type, may develop at any age, and are difficult to control. We have recently shown that rats subjected to structural damage of the brain, induced by means of sustained convulsions, developed spontaneous seizures after a mean latency of 14-15 days (Leite et al. 1990). The mean frequency of these spontaneous recurrent convulsions remained constant for several months. Such a model can be compared to some forms of epilepsy in humans.
Calcium ions (Ca) are involved in many important, and yet only partially understood, neuronal functions including neurotransmitter release and the triggering of intracellular biochemical events (for review see Heinemann et al. 1986). The effects of Ca in cellular reactions are mediated by calcium-binding proteins (CaBP), which may act as cellular regulators and intraneuronal calcium-buffering systems (Cheung 1980).
The present study was undertaken to examine alterations in the immunocytochemical expression of two CaBP, calbindin-D28K (CB) and parvalbumin (PV), during different periods following status epilepticus induced by pilocarpine (Turski et al. 1989). These preliminary results show a significant decrease in PV and CB-IR in the thalamus and hippocampal formation of rats which is more evident during the period of spontaneous recurrent seizures.

Methods

Male adult Wistar rats were used. Sustained seizures were induced by pilocarpine hydrochloride (380 mg/kg i.p.; PILO). Methyl-scopolamine nitrate (1 mg/kg s.c.) was administered 30 min before the injection of PILO to limit peripheral toxic effects. Some rats died in the course of sustained seizures induced by PILO. The survivors were continuously monitored for 120 days after PILO.
For immunocytochemistry of PV and CB in the thalamus and hippocampal formation, rats submitted to PILO-induced seizures were killed in three different periods, i. e, 2-10 h after PILO during status epilepticus; 5-15 days later during the "seizure silent period"; and 110-130 days following PILO during the period of spontaneous recurrent seizures. Rats were perfused under deep barbiturate anesthesia with saline followed by 4% buffered paraformaldehyde. The brains were then immediately removed and stored in fixative solution for a few days and then soaked in 30% buffered sucrose for cryoprotection. Control brains of animals not injected with PILO were always subjected to the same procedures as those of PILO injected ones. Thirty um frozen serial sections were cut in the coronal plane and collected in phosphate-buffered saline (PBS). One out of every fifth section was processed for immunocytochemistry with anti-CB and anti-PV monoclonal antibodies (kindly provided by Dr. M. Celio, University of Kiel, FRG), following the avidin-biotin (ABC) protocol. The following dilutions were used for the overnight incubation: anti-PV 1:10 000 and anti-CB 1:5000. Adjacent unreacted sections were counterstained with cresyl violet.

Results

Behavioral and electrographic features of PILO-induced seizures were similar to those reported previously (Turski et al., 1989). The "acute period" lasted 24-48 h after PILO administration and was followed by a progressive normalization of behavior. First spontaneous recurrent seizures (SRS) were observed 4-44 days after PILO injection. The mean "seizure silent period" lasted for 14.8 q 3.0 days (mean ± SEM). During the "chronic period" SRS were observed with a frequency of 2-4 per animal/week. The duration of SRS rarely exceeded 60 s and were characterized by facial automatisms, head nodding, forelimb clonus, rearing, and falling.

The immunohistochemical study of the brains of the control animals which had not received PILO injections evidenced an inhomogeneous distribution of CB and PV (in agreement with the findings of Celio 1990). In particular, in the hippocampus, CB-positive cells were represented by dentate granule cells, some pyramidal cells, and some interneurons; PV-positive cells by interneurons.

In the thalamus, CB immunopositivity was mainly present in cell bodies of the midline, anterior intralaminar nuclei and in the ventromedial nucleus; thus, it was distributed in neurons of "nonspecific" thalamic structures. PV immunopositivity was mainly confined to cell bodies of the reticular thalamic nucleus (Rt); a dense PV-positive innervation, deriving presumably from Rt cells, was evident throughout most of the dorsal thalamic structures.

Changes in CB and PV immunoreactivity following PILO-induced seizures were especially evident in rats killed during the "seizure silent period" (5-15 days after PILO) and in the "chronic period" of spontaneous recurrent seizures (110-130 days after PILO), both of which displayed basically the same alterations but were much more severe in the latter cases. A severe decrease of both CB- and PV-positive cell populations was evident in the hippocampal formation so that only a very few hippocampal neurons displayed immunoreactivity in the "chronic period" (Fig. 1B). Within the thalamic Rt, zones of complete absence of PV positivity as well as zones displaying shrinkage of cell bodies were evident (Fig. 1D). Marked alterations were evident in PV immunoreactivity of the neuropil of dorsal thalamic structures: The PV-positive innervation appeared very intense (even more than in control normal cases) in the ventral nuclear complex, whereas no PV-positive fibers were evident in more medially located structures, particularly midline and intralaminar ones. In the same structures the CB-positive cell bodies were severely reduced in number and mostly replaced by "ghosts" of cell bodies and some stained cell processes.

The comparison with cresyl violet adjacent sections revealed neuronal degeneration affecting the CA1 subfield of the intermediate hippocampus and the CA3 subfield of the rostral hippocampus. The CA4 area was also consistently affected, whereas the dentate gyrus was rarely injured. Severe cell loss and gliosis could be seen in thalamic structures, such as the Rt, and those embedded in the internal medullary lamina fibers and some of the midline nuclei.

Rats killed during the "acute period", i.e. 2-10 h after PILO injections, showed no significant alterations in CB or PV immunoreactivity both in the thalamus and hippocampal formation.

Discussion

The present report demonstrates a progressive decline of CB and PV immunoreactivity in the thalamus and hippocampal formation following PILO-induced seizures. The decrease in CB- and PV-immunocytochemical expression was very evident during the period of recurrent seizures, a relatively long time (3 or 4 months) after the onset of spontaneous seizures. The time course of the alteration of CB- or PV-containing cell bodies needs to be further investigated. In the present preliminary study, however, neurochemical expression of CB and PV was also found to be slightly decreased during the "seizure silent period", so that this feature cannot be simply ascribed to the ocurrence of spontaneous seizures per se.

Both acute and chronic epilepsies are associated with an enhanced Ca uptake capability into nerve cells through voltage-dependent or NMDA-operated channels (Heinemann and Hamon 1986). Several lines of evidence have stressed that the resulting intracellular load of Ca could represent one of the factors involved in the phenomena of neuronal degeneration as a result of epileptic activity (for review see Cavalheiro et al, 1988).

For the interpretation of the present results, it is important to verify whether the decrease in the neurochemical expression of CB and PV parallels strictly the PILO-induced neuropathological

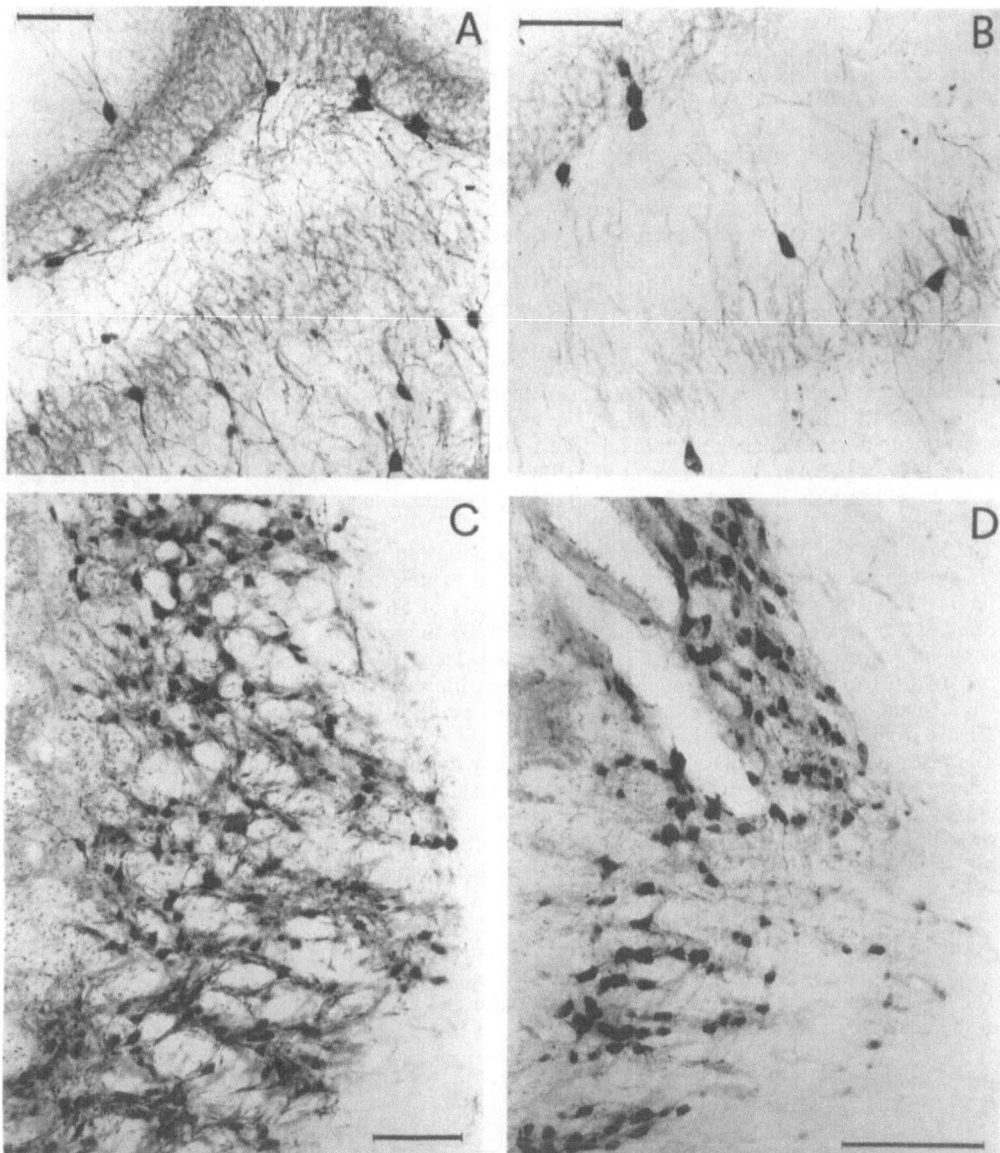

Fig. 1A-D. Parvalbumin-immunoreactivity in the **A, B** hippocampus and **C, D** thalamic reticular nucleus of control rats (**A, C**) and experimental animals (**B, D**) during the period of spontaneous recurrent seizures (120 days after PILO administration). Note in **B** and **D** the severe decrease in parvalbumin-immunoreactive cell bodies and, in **D**, the shrinkage of neurons of the thalamic reticular nucleus. Scale bars correspond to 100 μm in **A** and **B**, and to 200 μm in **C** and **D**

changes (thus reflecting a pattern of neuronal loss), or represents , at least in part, an independent phenomenon. In the present experimental cases the thalamus and hippocampal formation showed considerable shrinkage and neural degeneration, according to the general picture outlined in PILO-induced brain damage (Turski et al. 1989). A systematic investigation and quantitative analysis of the decrease of PV- and CB-positive neurons, as compared to general cell loss, are at present in progress. In any case, it should be stressed that PV- or CB-containing neurons are not spared in PILO-induced brain damage. On the contrary, at least at the thalamic level, cell degeneration was undoubtedly more severe in nuclear structures with pronounced PV or CB immunoreactivity than in other thalamic nuclei, suggesting that PV- or CB-containing thalamic structures could be selectively involved in this experimental model of chronic epilepsy. In the rat, CB-positive neurons are localized to structures which give rise to multiple channels of cortical and subcortical output (see Macchi and Bentivoglio 1986; Bentivoglio and Schiff 1989). In this respect, it should also be mentioned that the neurochemical expression of CB and PV in the thalamus varies in phylogeny, so that the distribution of PV- and CB-immunoreactive thalamic neurons is more widespread in monkey (Jones and Hendry 1989) than in rat. Furthermore, it is worthwhile stressing that the PV-containing Rt nucleus act as a pacemaker of thalamic activity (see Steriade and Llinas 1988). In particular, Rt innervation is, in the rat, the major regulator of the thalamocortical activity, since most of the dorsal thalamic nuclei lack interneurons in this species (e.g., Barbaresi et al. 1986). Rt cells are GABAergic (Houser et al. 1980); thus, in these neurons PV is co-localized with GABA, as has been detected in other neural subpopulations, e.g., rat cerebral cortex (Celio 1986) and hippocampus (Kamphuis et al. 1989).

The literature findings hitherto available on the relationship between CB and PV and epileptogenesis are equivocal, indicating that the neurochemical expression of these proteins may vary in different models of experimental epilepsy. In the study of Sloviter (1989), perforant path stimulations, which induced destruction of hippocampal hilar mossy cells, selectively spared CB- or PV-containing neurons. Thus, Sloviter (1989) concluded that these hippocampal cells were relatively resistant to seizure-induced damage. By contrast, hippocampal PV-immunoreactive cell bodies were found to be increased in the initial stages of kindling and returned to normal after the last seizure (Kamphuis et al. 1989).

The present findings do not allow one to draw any definite conclusion on the functional significance of CaBP in epileptogenesis. Under the present experimental conditions, however, the decrease in CB and PV immunocytochemical expression in the thalamus and hippocampus appeared so striking that the development and maintenance of spontaneous recurrent seizures induced by PILO administration could, at least in part, be ascribed to this phenomenon.

Acknowledgements.

The authors thank Drs. J. P. Leite, Z.A. Bortolotto, and D. Schiff for their help in the preparation of the material and animal observation.

References

Barbaresi P, Spreafico R, Frassoni C, Rustioni A (1986) GABAergic neurons are present in the dorsal column nuclei but not in the ventroposterior complex of rats. Brain Res 382:305-326

Bentivoglio M, Schiff D (1989) Calbinding-immunoreactive neurons give origin to thalamolimbic pathways. Eur J Neurosci (Suppl) 2:270

Cavalheiro EA, Lehmann J, Turski L (eds) (1988) Frontiers in excitatory amino acid research. Alan R Liss, New York

Celio MR (1986) Parvalbumin in most gamma-aminobutyric acid-containig cells of the rat cerebral cortex . Science 231:995-997

Celio MR (1990) Calbindin-D28k and parvalbumin in the rat brain. Neuroscience (in press)

Cheung WY (1980) Calmodulin plays a pivotal role in cellular regulation. Science 207:19-27

Heinemann U, Hamon B (1986) Calcium and epileptogenesis. Exp Brain Res 65:1-10

Heinemann U, Klee M, Neher E, Singer W (eds) (1986) Calcium eletrogenesis and neuronal functioning. Springer, Berlin Heidelberg New York

Houser CR, Vaughn JE, Barber RP, Roberts E (1980) GABA neurons are the major cell type of the nucleus reticularis thalami. Brain Res 200:341-354

Jones EG, Hendry SHC (1989) Differential calcium binding protein immunoreactivity distinguishes classes of relay neurons in monkey thalamic nuclei. Eur J Neurosci 1:222-246

Kamphuis W, Huisman E, Wadman WJ, Heizmann CW, Lopes da Silva FH (1989) Kindling induced changes in parvalbumin immunoreactivity in rat hippocampus and its relation to long-term decrease in GABA-immunoractivity. Brain Res 479:23-34

Leite JP, Turski L, Bortolotto ZA, Cavalheiro EA (1990) Spontaneous recurrent seizures following pilocarpine-induced status epilepticus in rats. Neurosci Biobehav Rev (in press)

Macchi G, Bentivoglio M (1986) The thalamic intralaminar nuclei and the cerebral cortex. In: Jones EG, Peters A (eds) Cerebral cortex, vol 5. Plenum, New York, pp 355-401

Shorvon SD (1984) The temporal aspects of prognosis in epilepsy. J Neurol Neurosurg Psychiatry 47:1157-1165

Sloviter RS (1989) Calcium-binding protein (calbindin-D28k) and parvalbumin immunocytochemistry: Localization in the rat hippocampus with specific reference to the selective vulnerability of hippocampal neurons to seizure activity. J Comp Neurol 280:183-196

Steriade M, Llinas RR (1988) The functional states of the thalamus and the associated neuronal interplay. Physiol Rev 68:649-742

Turski L, Ikonomidou C, Turski WA, Bortolotto ZA, Cavalheiro EA (1989) Cholinergic mechanisms and epileptogenesis. The seizures induced by pilocarpine: a novel experimental model of intractable epilepsy. Synapse 3:154-171

III DEVELOPMENTAL ASPECTS OF EPILEPTOGENESIS

Developmental Alternations in NMDA-Mediated Excitation in Rat Visual and Somatosensory Cortex

H. J. Luhmann[1] and D. A. Prince[2]

[1] Inst. für Normale & Pathologische Physiologie, Universität Köln Robert-Koch-Str. 39 D-5000 Köln 41, FRG
[2] Stanford University School of Medicine; Department Neurology and Neurological Sciences, Stanford, CA 94305, USA

Introduction

The functional status of the brain depends on the balance between excitatory and inhibitory processes. Even minor changes in this equilibrium can lead to the occurrence of pathophysiological phenomena, e.g. a slight reduction in the efficiency of the neocortical $GABA_A$ receptor-mediated inhibition leads to the generation and propagation of epileptiform events over several millimeters in neocortical slices (see Connors and Chagnac-Amitai, this volume). Anatomical and biochemical data indicate that the balance between excitatory and inhibitory actions changes during development in neocortex. At a time when the GABAergic system is still relatively immature (Kvale et al. 1983; Miller 1988), excitatory connections are already well established (Miller 1988) or even transiently expressed in an exuberant manner (Luhmann et al. 1986; Luhmann et al. 1990). These observations may underlie an increased susceptibility to epileptiform phenomena in juvenile neocortex. We therefore studied excitatory events in the neocortical network and its single cellular elements by recording evoked extracellular field potentials and intracellular signals in layers II/III of rat neocortex at different developmental stages under *in vitro* conditions.

Methods

Neocortical slices were prepared and maintained using techniques described previously (McCormick and Prince 1987). In brief, young (postnatal day P4 - 10), juvenile (P11 - 20) and adult ($\geq$ P28) Sprague-Dawley rats were deeply anesthetized, decapitated, and a small block of the brain was rapidly removed and placed in cold Ringer's solution. Coronal slices (400 - 500 μm) of the primary visual and somatosensory cortex were cut on a vibratome and transferred to an interface-type recording chamber. The control bathing solution contained (in mM): NaCl 126, KCl 2.5, NaH_2PO_4 1.25, $MgSO_4$ 2, $CaCl_2$ 2, $NaHCO_3$ 26, dextrose 10, and when saturated with 95% O_2, 5% CO_2 had a pH of 7.4 at the experimental temperature of 35° $\pm$ 1°C. Sulphate and phosphate were omitted from the bath solution when the extracellular Ca^{2+} concentration ($[Ca^{2+}]_o$) was increased to 7 mM. Field potentials (FP) were recorded with 3 - 5 MΩ electrodes containing Ringer's solution. Electrodes for intracellular recording had resistances from 80 - 160 MΩ and contained 4 M K^+ acetate. Synaptic responses in layers II/III were obtained by orthodromic stimulation of the underlying white matter with a bipolar tungsten electrode. Stimulus intensities were adjusted to obtain maximal and submaximal FP responses and subthreshold intracellular responses (threshold to evoke an action potential from a short-latency EPSP = 100%), and were applied at low frequencies ($\leq$ 0.1 Hz, unless otherwise noted).

Results

At all ages, high intensity stimuli evoked an early, biphasic FP response that had a duration of 20 - 30 ms and a peak-to-peak amplitude of > 1 mV. A reduction in the stimulus intensity led to a decrease in the early FP amplitude and, in about 70% of the juvenile slices, to the appearance of a late (onset latency: 30 - 220 ms), long-lasting (duration: 120 - 600 ms) response that had a clear

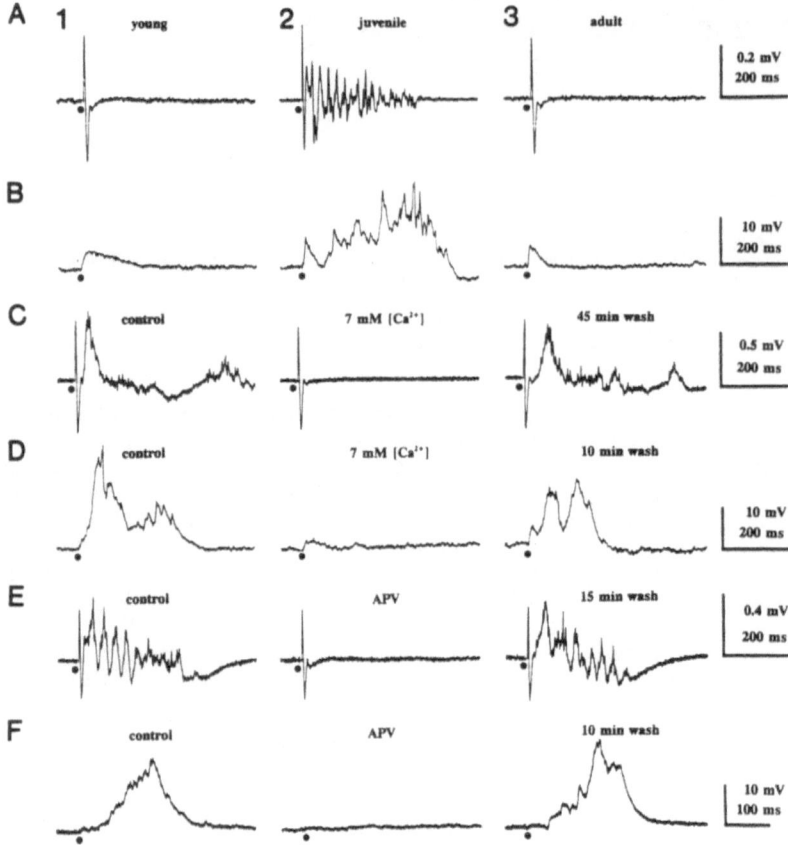

Fig. 1A-F. Age-dependent expression and properties of late field potential (FP) response (A, C, E) and long-latency EPSP (B, D, F). A FP recordings in a young (P7), juvenile (P20) and adult (P29) rat neocortical slice. Stimulus intensities were adjusted to obtain submaximal responses and the peak-to-peak amplitude of the initial evoked component is about the same in all three recordings. Only the juvenile slice expresses an oscillatory, long-lasting response with an average frequency of 44 Hz (A2). B intracellular recordings from young (P7), juvenile (P16) and adult (P29) upper layer regularly spiking cells. A multiphasic l-EPSP could be demonstrated with subthreshold stimulus intensities only in the P16 cell (B2). C, D FP and intracellular recordings from a P16 slice (C) and P15 neuron (D) under control conditions (C1, D1), in 7 mM extracellular [Ca^{2+}] (C2, D2), and after 45 and 10 min wash (C3 and D3, respectively). The late FP response and the l-EPSP are clearly reduced in elevated [Ca^{2+}]$_o$, indicating a polysynaptic origin of these responses. E, F FP and intracellular recordings from a P18 neocortical slice (E) and a P15 regularly spiking cell (F) under control conditions (E1, F1), in D-APV (20 μM bath and 100 μM locally, for E2 and F2, respectively), and after 15 and 10 min washes (E3 and F3, respectively). Note the complete suppression of the late responses in D-APV, indicating a participation of NMDA receptors in mediating these events

oscillatory behavior in the frequency range of 30 - 50 Hz (Fig. 1A). Long-latency FP responses of smaller amplitude (< 0.1 mV) could be also observed in layer IV and infragranular laminae of the juvenile neocortex, but only in 9% of the young and 1.6% of the adult animals. The intracellular correlate of this multiphasic response was a long-lasting (up to 810 ms) excitatory postsynaptic potential (l-EPSP) of variable peak amplitude (10 - 35 mV) and onset latency (10 - 320 ms) (Fig. 1B). Such a l-EPSP could be demonstrated in more than 80% of the juvenile neurons, but in only 3% of the young and adult cells, suggesting an age-dependent manifestation of these events. The multiphasic character, the variability in duration, amplitude and onset latency, and the suppression at higher stimulus frequencies (not shown) of both the late FP response and the l-EPSP suggest that this activity is mediated in a polysynaptic circuit. Evidence supporting this hypothesis was obtained from recordings in high $[Ca^{2+}]_o$. Under these conditions, polysynaptic activity was reduced without a significant effect on monosynaptic events (Berry and Pentreath 1976). In 7 mM $[Ca^{2+}]_o$ both the late FP response (Fig. 1C) and the l-EPSP (Fig. 1D) were clearly reduced or completely blocked, indicating that these excitatory long-lasting events are generated in a local neural network.

In order to test which receptor subtype was involved in mediating these polysynaptic potentials, we added the specific NMDA receptor antagonist D-APV to the bath solution (20 μM) or applied it locally on the slice surface near the recording site (100 μM). In all juvenile slices tested, this manipulation resulted in a reversible reduction or blockade of the late FP response (Fig. 1E) and l-EPSP (Fig. 1F) without having a significant effect on the early FP response or early EPSP.

Discussion

The question arises as to why these polysynaptic NMDA receptor-mediated potentials are transiently expressed in rat neocortex only during an early postnatal phase. In has been shown in mature rat neocortex, that a small reduction of the GABAergic system by the GABA$_A$ antagonist bicuculline uncovers a delayed APV-sensitive EPSP (Artola and Singer 1987), suggesting that the presence of long-lasting, NMDA receptor-mediated events in juvenile cortex might be due in part to an immature GABAergic system at this developmental stage. Conductance measurements in regularly spiking cells of upper layers in three different age groups support this hypothesis (Luhmann and Prince 1988). The peak conductance of the fast, Cl⁻-dependent IPSP and the long-latency, K⁺-dependent IPSP increases significantly (t test, $p < 0.01$ and $p < 0.001$, respectively) during the first postnatal month (Fig. 2A). Further support for this hypothesis comes from observations that the

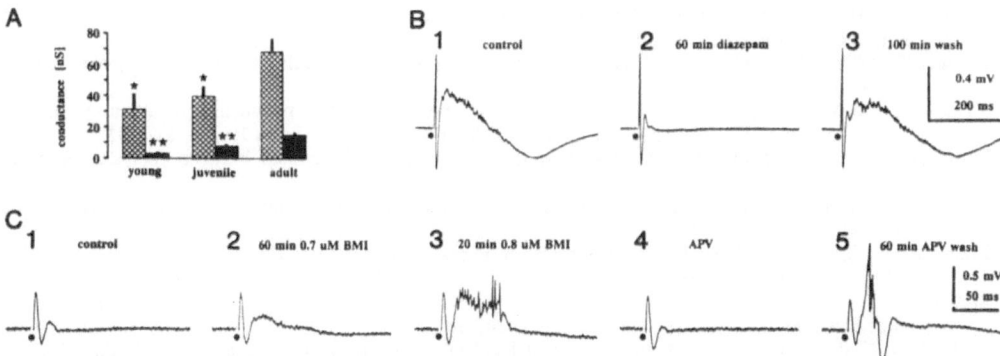

Fig. 2 A-C. Role of GABAergic inhibition in controlling the expression of polysynaptic, NMDA receptor-mediated responses in rat neocortex. **A** Bar histogram showing the average peak conductance ± SE of the fast IPSP (F-IPSP), measured 24 ms after the stimulus (*shaded bars*), and the long IPSP (I-IPSP), measured 140 ms poststimulus (*filled bars*), obtained from 10 young, 24 juvenile and 26 adult upper layer regularly spiking cells. *Arrows* above *bars* indicate significance levels (t test, * $p < 0.01$; ** $p < 0.001$) for differences in f- and l-IPSP peak-conductance between young and juvenile cells, respectively, versus adult cells. **B** FP recordings (average of 10 subsequent sweeps) in a P17 slice obtained under control conditions (**B1**), after 60 min wash-in of 1 μM diazepam (bath) (**B2**), and after 100 min wash-out (**B3**). **C** Single FP recordings in a P30 slice under control conditions (**C1**), after 60 min wash in of 0.7 μM bicuculline methiodide (BMI) (**C2**), after 20 min wash-in of 0.8 μM BMI (**C3**), after local application of 100 μM D-APV (**C4**), and after 60 min wash-out of D-APV (**C5**). Recordings shown in C3-5 were obtained in 0.8 μM BMI (bath)

late FP response was blocked in all juvenile slices tested by bath application of the benzodiazepine diazepam (1 μM) which increased the efficacy of GABAergic inhibition (Fig. 2B). The hypothesis that the strength of the intracortical GABAergic system controls the manifestation of NMDA receptor-mediated activity also implies that a slight suppression of the inhibitory system in adult animals should promote the expression of long-latency, APV-sensitive potentials. A 10% - 20% suppression of the GABAergic inhibition (see Connors and Chagnac-Amitai, this volume), by low doses of bath-applied bicuculline methiodide (0.5 - 1.0 μM), evoked a delayed response in adult cortex, which could be reversibly blocked by D-APV (Fig. 2C). These findings suggest that the powerful, transiently expressed NMDA receptor-mediated potentials in juvenile animals are at least in part a consequence of the relative immaturity of GABAergic inhibition.

The imbalance between excitatory and inhibitory actions during an early phase of postnatal maturation has important implications for pathophysiological processes and normal developmental events. Our results provide one explanation for a pronounced susceptibility of neocortex to epileptiform discharge during a critical period of development (Hablitz 1987). The present findings also may explain why long-term potentiation, which requires activation of NMDA receptors, is most readily induced in juvenile animals (Perkins and Teyler 1988). Both phenomena coincide temporally with the expression of polysynaptic NMDA receptor-mediated activity in juvenile neocortex, indicating a possible functional relationship between these developmental processes.

Acknowledgements. We are grateful to Prof. W. Haefely and Dr. R. Eigenmann of Hoffmann-La Roche (Basel) for kindly providing diazepam. These experiments were supported by NIH grants NS06477, NS12151, the Morris and Pimley research funds (D.A.P.) and a fellowship from the Deutsche Forschungsgemeinschaft (H.J.L.).

References

Artola A, Singer W (1987) Long-term potentiation and NMDA receptors in rat visual cortex. Nature 330: 649-652

Berry MS, Pentreath VW (1976) Criteria for distinguishing between monosynaptic and polysynaptic transmission. Brain Res 105: 1-20

Hablitz JJ (1987) Spontaneous ictal-like discharges and sustained potential shifts in the developing rat neocortex. J Neurophysiol 58: 1052-1065

Kvale I, Fosse VM, Fonnum F (1983) Development of neurotransmitter parameters in lateral geniculate body, superior colliculus and visual cortex of the albino rat. Dev Brain Res 7: 137-145

Luhmann HJ, Martinez Millan L, Singer W (1986) Development of horizontal intrinsic connections in cat striate cortex. Exp Brain Res 63: 443-448

Luhmann HJ, Prince DA (1988) Postnatal development of GABA-ergic inhibition in rat neocortex. Soc Neurosci (abstract) 14: 189

Luhmann HJ, Greuel JM, Singer W (1990) Horizontal interactions in cat striate cortex: III. Ectopic receptive fields and transient exuberance of tangential interactions. Eur J Neurosci, (in press)

McCormick DA, Prince DA (1987) Post-natal development of electrophysiological properties of rat cerebral cortical pyramidal neurones. J Physiol 393: 743-762

Miller MW (1988) Development of projection and local circuit neurons in neocortex. In: Cerebral Cortex, vol. 7, Peters A, Jones EG (eds), Plenum, New York, pp 133-175

Perkins AT, Teyler TJ (1988) A critical period for long-term potentiation in the developing rat visual cortex. Brain Res 439: 222-229

Consequences of Epileptic Seizures in the Developing Brain

G. L. Holmes[1], P. Mareš[2] and S. L. Moshé[3]

[1] Department of Neurology, The Children's Hospital, Harvard Medical School, Boston, Massachusetts, USA
[2] Institute of Physiology, Czechoslovak Academy of Science, Prague, Czechoslovakia
[3] Departments of Neurology, Neuroscience and Pediatrics, Albert Einstein College of Medicine, Bronx, New York

Introduction

Whether seizures in the immature brain adversely influence later intellectual development and behavior is one of the most important, yet controversial, questions in pediatric epilepsy. This question is difficult to study in children since clinical studies cannot control for many of the variables that might effect cognition and behavior. These variables include age of onset of the seizures, type of seizures, frequency and duration of the seizures, and etiological agents responsible for the seizures. Many of these variables can be controlled by using animal models of epilepsy. This paper will review recent studies using the immature animal.

Status Epilepticus in the Immature Brain

Studies of status epilepticus in the immature brain have suggested that status epilepticus can lead to brain cell loss or damage and behavior changes. Three days following severe status epilepticus in 4-day-old rats, Wasterlain and colleagues (Dwyer et al. 1986; Wasterlain and Dwyer 1983; Fando et al. 1979) found reductions in forebrain RNA, DNA, protein, cholesterol, and weight, demonstrating a reduction in number of brain cells without a significant change in mean cell size. However, when the brains were examined 20 days after the seizures only a small reduction in forebrain weight persisted, demonstrating that recovery can occur. Mild flurothyl seizures resulted in smaller deficit in brain growth at age 7 days and complete recovery by age 30 days.

Whether brain damage occurs following status epilepticus in the immature brain may relate to the precipitating agent or duration of the status. Wasterlain and Dwyer (1983) found that bicuculline-induced status also curtailed brain growth, but had milder effects than flurothyl seizures with complete recovery occurring by the age of 30 days. de Feo et al. (1986) reported that status epilepticus induced by kainic acid (KA) resulted in impairment of conditioned avoidance responses later in life while pentylenetetrazol-induced status epilepticus, equivalent in severity and duration to that of KA, did not affect this learned behavior.

Age is another critical factor that appears to be important in determining whether brain damage occurs following status epilepticus. Okada et al. (1984) produced status epilepticus with intraperitoneal injections of KA in 15 day-old rats. Rats that survived the status epilepticus did not demonstrate an increased seizure susceptibility as adults nor were pathological changes seen in their brains. Holmes and Thompson (1988) also found that in 12-and 18-day-old rats intraperitoneal KA administration did not alter seizure susceptibility or cause discernible pathological changes, despite causing prolonged status epilepticus. However, the authors found that rats that received KA at age 27 days kindled faster than control animals.

Changes in behavior, learning and memory have been reported in animals following KA induced status epilepticus by Holmes and colleagues (1988). Status epilepticus was induced in prepubescent male rats with an amygdala electrode in place by administration of KA intraperitoneally while controls received phosphate buffered saline (PBS). The rats were then tested as adults for learning,

144

memory, emotionality, social interaction, and activity level using the T maze, water maze, handling test, home cage intruder test, and open field test. The KA-treated rats learned at a slower rate in the water maze and T maze than the controls. In addition, the KA-treated animals were found to be more submissive and less aggressive than the control animals in the home cage intruder test and significantly more active than the control animals in the open field test. Whether KA administration to younger animals would cause similar impairments in learning, memory, behavior, and activity level is not yet known.

Effects of Brief Seizures on Brain Development

There have been few studies examining the effects of serial seizures on brain development. Wasterlain and Plum (1973) demonstrated permanent reductions in brain weight and cell number in rats subjected to ten daily electroconvulsive seizures between the ages of 2 and 11 days. Daily seizures in older rats (age 19 to 28 days) did not cause any deficits in brain development.

To determine whether frequent seizures could cause deficits in learning and behavior, Holmes et al. (1990) subjected pubescent genetically epilepsy-prone rats (GEPRs) to a series of 66 audiogenic stimulations. The sound-induced seizures in these animals consisted of wild running followed by generalized tonus. Seizures were brief, rarely lasting over 90 s, and were associated with a low mortality rate. Control rats consisted of GEPR littermates who were handled and placed in the sound chamber but were not stimulated. Following the 66 stimulations, the rats were tested for learning, memory, and behavior using the T maze, water maze, open field activity test, home cage intruder test, and handling test. When compared to the control animals, rats subjected to frequent audiogenic seizures reached criteria less frequently in the T maze, required longer times to find the platform in the water maze (Fig. 1 and 2), were less active in the open field activity test, less aggressive in the home cage intruder test, and more irritable and aggressive in the handling test. No pathological lesions were found on routine histology and cell counting in the hippocampus revealed no differences between the experimental animals and controls.

These studies suggest that seizures may have detrimental long-term neurological effects. However, the consequences of seizures in the developing brain are different than those seen in the mature animal. More work is necessary to determine the effects of seizure frequency, type, age at the time of seizure onset, and intensity on subsequent neurological outcome.

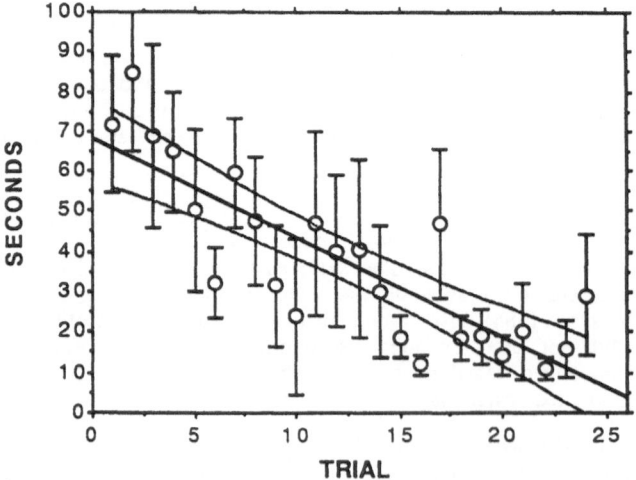

Fig. 1. Simple regression line of time (seconds) versus trial in water maze for the GEPR rats that received 66 audiogenic stimulations

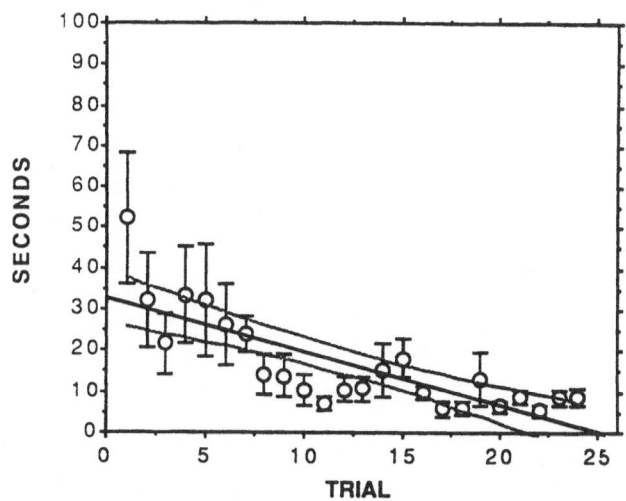

Fig. 2. Simple regression line of time (seconds) versus trial in water maze for the GEPR controls

References

de Feo MR, Mecarelli O, Palladini G, Ricci GF (1986) Long-term effects of early status epilepticus on the acquisition of conditioned avoidance behavior in rats. Epilepsia 27: 476-482

Dwyer BE, Wasterlain CG, Fujikawa DG, Yamada L (1986) Brain protein metabolism in epilepsy. Adv Neurol 44: 913-918

Fando JL, Conn M, Wasterlain CG (1979) Brain protein synthesis during neonatal seizures: an experimental study. Exp Neurol 63: 220-228

Holmes GL, Thompson JL (1988) Effects of kainic acid on seizure susceptibility in the developing brain. Dev Brain Res 39: 51-59

Holmes GL, Thompson JL, Bates T, Feldman DS (1988) Behavioral effects of kainic acid administration on the immature brain. Epilepsia 29: 721-730

Holmes GL, Thompson JL, Marchi TA, Gabriel PS, Hogan MA, Karl FG, Feldman DS (1990) Effects of seizures on learning, memory, and behavior in the genetically epilepsy-prone rat. Ann Neurol 27: 24-32

Okada R, Moshé SL, Albala BJ (1984) Infantile status epilepticus and future seizure susceptibility in the rat. Dev Brain Res 15: 177-183

Wasterlain CG, Dwyer BE (1983) Brain metabolism during prolonged seizures in neonates. Adv Neurol 34: 241-260

Wasterlain CG, Plum F (1973) Vulnerability of developing rat brain to electroconvulsive seizures. Arch Neurol 29: 38-45

Epileptogenesis and the Immature Brain
I. Cortical Mechanisms

P. Mareš[1], S. L. Moshé[2] and G. L. Holmes[3]

[1] Institute of Physiology, Czechoslovak Academy of Sciences, Prague, Czechoslovakia
[2] Department of Neurology, Neuroscience and Pediatrics, Albert Einstein College of Medicine, Bronx, N.Y., USA
[3] Department of Neurology, Harvard Medical School, Boston, MA, USA

Cortical mechanisms are of primary importance in the pathogenesis of simple and complex partial (focal) seizures and generalized seizures of the absence type. Among these seizures absences represent an age-dependent type of epileptic seizure - it cannot appear before a certain stage of maturation is reached. Both types of focal seizures are observed throughout development, but their symptomatology changes with maturation.

Models of Absence Seizures

The most adequate model of human absences is represented by feline generalized penicillin epilepsy (Gloor 1984). Unfortunately, this model was not studied in kittens. Rhythmic electrocorticographic activity, elicited by low doses of pentamethylenetetrazol (Metrazol) in rats, shares many features with generalized penicillin epilepsy and may also be used as a model of human absences (Schickerová et al. 1989). This rhythmic Metrazol activity (RMA) can be recorded starting from the 12th postnatal day but in the immature form (irregularity of rhythm, low frequency). By the age of 18 days all characteristics of RMA are identical to those seen in adult rats (Zouhar et al. 1980; Schickerová et al. 1984). There is also a time coincidence with the development of rhythmic thalamocortical phenomena (for review see Mares et al. 1982) suggesting a thalamocortical originšof RMA. A period of rapid cortical synaptogenesis starting at the end of the second postnatal week in rats (Aghajanian and Bloom 1967; Caley and Maxwell 1968) supports a decisive role for the cortex in this model and is in agreement with Gloor's conclusion about the pathogenesis of generalized penicillin epilepsy: The spike-and-wave rhythm represents a response of the diffusely hyperexcitable cerebral cortex to normal thalamic afferent impulses (Gloor 1984).

Models of Simple Partial Seizures

Neocortical foci induced by local application of penicillin represents a commonly used model of simple partial seizure. Development of penicillin foci was originally described in rabbits (Bishop 1950); the discharges of penicillin foci in immature cortex were of long duration and low amplitude. These findings were confirmed in kittens (Volanschi 1960) and in rats (Mareš 1973a) and two descriptions of feeble projections to the opposite hemisphere was added (Mareš 1973 a,b). Poor interconnections among immature cortical neurons (Caley and Maxwell 1968) probably form a background of a low degree of synchronization of paroxysmal depolarization shifts (PDS) in immature cortex (Prince and Gutnick 1972) expressed as a long duration of focal discharge. Low amplitude might be due to the inability of many developing neocortical neurons to generate PDS (Kriegstein et al. 1987; Hablitz 1987). A higher level of $[K^+]_o$, reached in immature cortex during epileptic activity (Hablitz and Heinemann 1987), and a lack of afterhyperpolarization in immature neurons (Hablitz 1987) might explain the increased epileptogenesis in rat cerebral cortex during the second and third postnatal weeks (our unpublished results).

Experimental Brain Research Series 20
© Springer-Verlag Berlin · Heidelberg 1991

Models of Complex Partial Seizures

The original report of epileptic afterdischarges (ADs) elicited by electrical stimulation of hippocampus starting from the very first postnatal days in kittens (Cadilhac et al. 1960) described particularities of epileptogenic phenomena in immature hippocampus similar to those recorded in immature neocortex : Poor spread of epileptic phenomena into other structures and slow and irregular waves as the predominant EEG pattern. Hippocampal ADs could be regularly recorded in 5-day-old rats, i.e., in the youngest group studied (Marešová et al. 1987). More detailed descriptions of model complex partial seizures in rats aged 12 days and older may be found in the kindling literature (Moshé 1981; Holmes 1983 for amygdala kindling; Lee et al. 1989 for hippocampal kindling). All groups studying kindling in immature animals found a period of increased epileptogenesis due to a failure of post-ictal depression at the end of the second and in the third week of postnatal life in rats (for review see Moshé et al. this volume; Lee et al. 1989). In a detailed study of hippocampal ADs it was demonstrated that at this stage of development even a 1 min interval between the end of an AD and the beginning of the second stimulation did not block the generation of the second AD (Velíšek and Mareš, this volume). This peculiarity of the immature limbic system was also demonstrated in hippocampal slices (Swan and Brady 1984), i.e., increased epileptogenesis is due to the properties of the hippocampus. Swan and coworkers described higher ceiling levels of $[K^+]_o$ and the spontaneous appearance of ictal episodes in slices of immature hippocampus (Swan et al. 1986; Smith and Swan 1987). These findings suggested a basic similarity between the neocortex and older cortical structures with respect to focal epileptogenesis

References

Aghajanian GK, Bloom FE (1967) The formation of synaptic junctions in developing rat brain: A quantitative electron microscopic study. Brain Res 6: 716-727
Bishop EJ (1950) The strychnine spike as a physiological indicator of cortical maturity in the post-natal rabbit. Electroencephalogr clin Neurophysiol 2: 309-315
Cadilhac J, Passouant-Fontaine T, Mihailovic L, Passouant P (1960) L'epilepsie experimentale du chaton en fonction de l'age. Etude corticale et sous-corticale Pathol Biol (Paris) 8: 1571-1581
Caley DW, Maxwell DS (1968) An electron microscopic study of neurons during postnatal development of the rat cerebral cortex. J Comp Neurol 133: 17-43
Gloor P (1984) Electrophysiology of generalized epilepsy. In: Schwartzkroin PA, Wheal HW (eds.) Electrophysiology of epilepsy, Academic Press, London, p 107
Hablitz JJ (1987) Spontaneous ictal/like discharges and sustained potential shifts in the developing rat neocortex. J Neurophysiol 58: 1052-1065
Hablitz JJ, Heinemann U (1988) Extracellular K^+ and Ca^{2+} changes during epileptiform discharges in the immature rat neocortex. Dev Brain Res 36: 299-303
Holmes GL (1983) Effect of serial seizures on subsequent kindling in the immature brain. Dev Brain Res 6: 190-192
Kriegstein AR, Suppes T, Prince DA (1987) Cellular and synaptic physiology and epileptogenesis of developing rat neocortical neurons in vitro. Dev Brain Res 34: 161-171
Lee SS, Murata R, Matsuura S (1989) Developmental study of hippocampal kindling. Epilepsia 30: 266-270
Mareš P (1973a) Ontogenetic development of bioelectrical activity of the epileptogenic focus in rat neocortex. Neuropediatrie 4: 434-445
Mareš P (1973b) Symmetrical epileptogenic foci in cerebral cortex of immature rat. Epilepsia 14: 427-435
Mareš P, Marešová D, Trojan S, Fischer J (1982) Ontogenetic development of rhythmic thalamo-cortical phenomena in the rat. Brain Res Bull 8: 765-770
Marešová D, Mareš P, Velíšek L (1987) Ontogenetic development of a model of human complex partial seizures - hippocampal afterdischarges in rats. Sb.Lek 89: 358-367 (in Czech)
Moshé SL (1981) The kindling phenomenon and its possible relevance to febrile seizures. In: Nelson KB, Ellenberg JH (eds) Febrile seizures, Raven, New York, pp 59-63
Prince DA, Gutnick MJ (1972) Neuronal activities in epileptogenetic foci of immature cortex. Brain Res 45: 455-468
Schickerová R, Mareš P, Trojan S (1984) Correlation between electrographic and motor phenomena induced by metrazol during ontogenesis in rat. Exp Neurol 84: 153-164

Schickerová R, Mareš P, Trojan S (1989) Rhythmic metrazol activity in rats as a model of human absences. Activ Nerv Sup 31: 16-19

Smith KL, Swan JW (1987) Carbamazepine suppresses synchronized afterdischarging in disinhibited immature rat hippocampus in vitro. Brain Res 400: 371-376

Swann JW, Brady RJ (1984) Penicillin-induced epileptogenesis in immature rat CA3 hippocampal pyramidal cells. Dev Brain Res 12: 243-254

Swann JW, Smith KL, Brady RJ (1986) Extracellular K⁺ accumulation during penicillin-induced epileptogenesis in the CA3 region of immature rat hippocampus. Dev Brain Res 30: 243-255

Volanschi D (1960) Cercetari experimentale asupra reactivitatii convulsivante a creierului immatur. Stud Cercet Neurol 11: 505-515

Zouhar A, Mareš P, Brozek G (1980) Electrocorticographic activity elicited by metrazol during ontogenesis in rats. Arch Int Pharmacodyn 248: 280-288

Epileptogenesis and the Immature Brain
II. Subcortical Mechanisms

S. L. Moshé[1], G. L. Holmes[2] and P. Mareš[3]

[1] Departments of Neurology , Neuroscience and Pediatrics, Albert Einstein College of Medicine, Bronx, NY, USA
[2] Department of Neurology, Harvard Medical School, Boston, MA, USA
[3] Czechoslovak Academy of Science, Prague, Czechoslovakia.

Introduction

During development, there is a period during which the immature CNS is more susceptible to the development of seizures than the mature CNS. In rats, this period appears to be between postnatal days 15-21 (Moshé 1987). Areas especially prone to the development of seizures include the amygdala and hippocampus as well as neocortical structures. The increase in seizure susceptibility is not due to decreases in local thresholds. Both cortical mechanisms (see Mareš et al. this volume) and subcortical mechanisms appear to be important. The latter include subcortical circuits which are capable of controlling seizures in adulthood. Two such systems will be discussed: the norepinephrine (NE) system and the system involving the substantia nigra (SN) pars reticulata and its GABA-sensitive efferent projections. Their delayed maturation may account for the inability of the developing CNS to limit the spread of seizures and prevent their recurrences.

The NE System

In adult rats, depletion of brain NE facilitates the development of seizures (Mason and Corcoran 1979). This phenomenon has been extensively studied in the kindling model. In kindling, repetitive low intensity stimulation of specific intracranial structures leads to the development of local afterdischarges (AD), behavioral automatisms and eventually generalized convulsions (Goddard et al. 1969). The behavioral manifestations of kindled seizures were initially classified into five stages. However, if the animals continued to be stimulated for prolonged periods of time (over 60 stimulations), severe seizures, stages 6 and 7, occur (for review, see Sperber et al. in press).

NE depletions decrease the number of stimulations that are necessary for the development of generalized convulsions without altering the AD threshold. Furthermore, NE depletions decrease the refractory periods and kindling can proceed even when the stimuli are delivered at short intervals (McIntyre et al. 1987). If brain NE is not depleted, this type of stimulation fails to induce kindling in adults (Goddard et al. 1969). The facilitation is mainly due to rapid propagation of seizures away from the stimulation site, since generalized convulsions occur quickly (Albala et al. 1986; McIntyre et al. 1987).

NE appears to play an important role in the development of kindling antagonism. This phenomenon is a variation of typical kindling; the stimulations are delivered to two sites on an alternating basis instead of being delivered consistently to the same site (Burchfiel and Applegate 1989). With the alternating pattern, one site kindles while the other does not, despite the occurrence of prolonged ADs. Kindling antagonism is abolished by total brain NE depletions but it can be maintained when forebrain depletions are combined with hypertrophy of hindbrain NE neurons (Burchfiel and Applegate 1989).

Kindling of rat pups reveals that 15 - 18 day old rats behave similarly to adults with total brain NE depletions. They have high local thresholds (Moshé et al. 1981) but can be kindled with stimulations delivered at 15 min intervals. Rat pups spend relative short periods of time in the early kindling stages as they develop bilateral although asymmetrical behavioral convulsions quickly. Within 20-30 stimulations, severe seizures, stages 6 and 7, can be elicited. Furthermore, kindling can be triggered at the same rate from the amygdala or dorsal hippocampus. In contrast, adult animals stimulated from the hippocampus require more stimulations to kindle than amygdala-stimulated rats. Finally, the phenomenon of kindling antagonism does not occur in rat pups. Stimulations alternating between the amygdala and dorsal hippocampus lead to the development of severe stages 6 and 7 from both sites (Sperber et al. in press).

The concentration of NE in the brains of 15 - 18 day old rats is approximately 56% of the adult values (Table 1). Based on the above observations, we propose that the "adult" NE level participates in the containment of seizures in the focus and adjacent areas. Decreased NE levels allow for the propagation of the seizures to structures involved in seizure generalization. Thus, the low NE levels, naturally present in pups, are not capable of containing the seizures in the focus accounting in part for the apparent increased seizure susceptibility of the immature brain. Additional decreases of total NE concentrations, produced by intracerebral 6-hydroxydopamine infusion, further accentuate the development of kindling supporting the proposed role for the NE system (Konkol et al. in press).

Table 1. NE concentration in amygdala-pyriform cortex as a function of age

Age (days)	Water %	Protein (mg/kg)	NE (ng/mg)
15	86.4	0.71	1.37 ± 0.1*
35	84.7	0.84	2.53 ± 0.3
60	84.7	0.83	2.44 ± 0.21

Values are means $\pm$ SEM with concentration expressed relative to tissue dry weight.
*$p < 0.01$ compared to the two other age groups

The SN system

In adults, deoxyglucose autoradiographic studies have disclosed that the SN pars reticulata exhibits increase metabolic activity during generalized seizures (Ackermann et al. 1989). There is also evidence that GABAergic activation of the SN can suppress generalized seizures. The GABAergic effect appears to be mediated by the local $GABA_A$ receptors because nigral infusions of the $GABA_A$ agonist, muscimol, suppress flurothyl seizures while infusions of the $GABA_A$ antagonist, bicuculline, facilitate the development of flurothyl seizures (Sperber et al. 1989b). Infusions of the $GABA_B$ agonist, baclofen, do not have any effects on seizures. The current hypothesis is that the SN pars reticulata is an important site involved in the control of generalized seizures in adult animals.

Deoxyglucose autoradiographic studies in 15 - 16 day old rat pups have revealed that the SN pars reticulata does not exhibit any increases in metabolic activity (Ackermann et al. 1989). Furthermore, the response of the SN to GABAergic stimulation is idiosyncratic perhaps because there is a site-specific paucity of $GABA_A$ high affinity receptors (Wurpel et al. 1988). Thus, both muscimol and bicuculline facilitate the development of flurothyl seizures (Sperber et al. 1989a). The failure of nigral $GABA_A$ergic transmission may account for two important features of epileptogenesis in rat pups: the inability of the immature CNS to suppress recurrent generalized seizures and the early emergence of severe kindled seizures, stages 6 and 7. Consequently, rat pups are more prone to develop severe status epilepticus than adults with an abundance of tonic features among the convulsive manifestations (Albala et al. 1984).

Recent evidence indicates that there is a GABA-sensitive system that can modify seizures in rat pups. It involves the nigral $GABA_B$ system because nigral infusions of the $GABA_B$ agonist, baclofen, suppress the development of flurothyl seizures in this age group (Sperber et al. 1989a). The system appears to require the exogenous administration of baclofen to booster the GABA effect, because of the proconvulsive effects of the $GABA_A$ system. Nevertheless, this finding leads to the investigation of the possible anticonvulsant effects of systemic baclofen infusions in developing animals. Baclofen suppresses the occurrence of both flurothyl and kindled seizures, at doses that do not produce marked drowsiness and ataxia (Sperber et al. in press). In contrast, the efficacy of baclofen in the control of the same type of seizures has not been demonstrated. In fact, baclofen, at higher doses than those used in pups, exacerbates myoclonus in kindled rats (Cottrell and Robertson 1987). These observations emphasize the need to better understand the mechanisms involved in the epileptogenesis as a function of age in order to develop age-appropriate therapeutic approaches.

Acknowledgments. Supported by NIH grant NS-20253 from the NINDS and grant R-36986 from the United Cerebral Palsy Associations.

References

Ackermann RF, Moshé SL, Albala BJ (1989) Restriction of enhanced ^{14}C-2-deoxyglucose utilization to rhinencephalic structures in immature amygdala-kindled rats. Exp Neurol 104: 73-81

Albala BJ, Moshé SL, Cubells JF, Sharpless NS, Makman MH (1986) Unilateral peri-substantia nigra catecholaminergic lesion and amygdala kindling. Brain Res 370: 388-392

Albala BJ, Moshé SL, Okada R (1984) Kainic acid induced seizures: a developmental study. Dev Brain Res 13: 139-148

Burchfiel JL, Applegate CD (1989) Stepwise progression of kindling: perspectives from the kindling antagonism model. In: Cain DP, Teskey, D (eds) Neuroscience and biobehavioral reviews. Pergamon, New York 13: 289-300

Cottrell GA, Robertson HA (1987) Baclofen exacerbates epileptic myoclonus in kindled rats. Neuropharmacology 26: 645-648

Goddard GV, McIntyre DC, Leech CK (1969) A permanent change in brain function resulting from daily electrical stimulation. Exp Neurol 25: 295-330

Konkol RJ, Thompson JL, Holmes GL The effect of regional differences in noradrenergic neuron growth patterns on juvenile kindling. Dev Brain Res (in press)

Mason ST, Corcoran ME (1979) Catecholamines and convulsions. Brain Res 170: 497-507

McIntyre DC, Rajalla J, Edson N (1987) Suppression of amygdala kindling with short interstimulus intervals: effect of norepinephrine depletion. Exp Neurol 95: 391-402

Moshé SL (1987) Epileptogenesis and the immature brain. Epilepsia 28[Suppl]: S3-S15

Moshé SL, Sharpless NS, Kaplan J (1981) Kindling in developing rats: Afterdischarge thresholds. Brain Res 211: 190-195

Sperber EF, Haas K, Moshé SL Mechanisms of kindling in developing animals. In: Wada JA (ed) Kindling 4, Plenum, New York (in press)

Sperber EF, Wurpel JND, Moshé SL (1989a) Evidence for the involvement of nigral $GABA_B$ receptors in seizures of rat pups. Dev Brain Res 47: 143-146

Sperber EF, Wurpel JND, Zhao DY, Moshé SL (1989b) Evidence for the involvement of nigral $GABA_A$ receptors in seizures of adult rats. Brain Res 480: 378-382

Wurpel JND, Tempel A, Sperber EF, Moshé SL (1988) Age-related changes of muscimol binding in the substantia nigra. Dev Brain Res 43: 305-307

Involvement of Excitatory Amino Acid Receptors in Epileptiform Activity in the Immature Neocortex

W. L. Lee and J. J. Hablitz

Neurobiology Research Center and Department of Physiology and Biophysics, University of Alabama at Birmingham, Birmingham AL 35294, USA

Previous work from this laboratory has shown that slices of immature rat neocortex display prolonged ictal-like discharges after exposure to the convulsant drug picrotoxin (Hablitz 1987; Sutor and Hablitz 1989a). Initial observations that the amplitude of epileptiform discharges in 9- to 14-day-old animals increased with membrane hyperpolarization and decreased with depolarization suggested the presence of an underlying synaptic component (Hablitz 1987). However, it was not clear if a synaptic input persisted for the 10- to 30-sec period such discharges encompassed. Since several factors are thought to be involved in the generation of epileptiform discharges in the neocortex, including intrinsic membrane currents (Gutnick et al. 1982), synaptic interactions (Connors 1984; Hablitz 1987), and alterations in interstitial ion concentrations (Hablitz and Heinemann 1987), we wished to determine the extent of synaptic involvement in generation of ictal-like discharges and evaluate the contribution of various subtypes of excitatory amino acids receptors.

Excitatory amino acids are prominent neurotransmitter candidates in a variety of neocortical areas (Streit 1984). Although the evidence pointing towards an excitatory amino acid as a neurotransmitter in the neocortex is strong, the nature of the receptor mediating various types of synaptic responses is less clear. Synaptic responses mediated by both NMDA and non-NMDA receptors have been observed (Sutor and Hablitz 1989bc; Thomsen 1986). Until recently, specific and potent antagonists have been available only for NMDA receptors, but the advent of the quinoxalinedione group of glutamate receptor antagonists (Honore et al. 1988) provides a pharmacological tool for the study of non-NMDA receptor mediated responses. In the present experiments, specific NMDA and non-NMDA receptor antagonists were used to identify the role of these receptors in elaboration of epileptiform discharges. We also performed voltage-clamp studies of picrotoxin-induced, ictal-like, paroxysmal discharges to determine if a sustained synaptic input was present.

Neocortical brain slices were prepared from immature rats 9 - 14 days of age, as described previously (Hablitz 1987). After a preincubation period of at least 1 h, the slices were transferred to an interface-type chamber where they were continuously perfused with a saline solution containing (in mM): NaCl 125, KCl 5, NaH_2PO_4 1.25, $CaCl_2$ 2, $MgSO_4$ 2, $NaHCO_3$ 25, and glucose 10. Intracellular recordings were obtained from layer II/III pyramidal neurons using microelectrodes (60 - 80 MΩ) filled with 4 M potassium acetate. A specially constructed single-electrode current- and voltage-clamp amplifier (Sutor and Hablitz 1989b) was used for recording. Bath application of 50 µM picrotoxin was used to induce epileptiform activity. Paroxysmal discharges were evoked by intracortical stimulation via a bipolar electrode.

A typical ictal-like epileptiform discharge recorded in a slice from a 10 - day - old rats is shown in Fig. 1A. As described previously (Hablitz 1987), paroxysmal activity consisted of an initial sharply rising membrane depolarization (paroxysmal depolarizing shift or PDS) followed by a second PDS prior to complete repolarization. Multiple PDSs accompanied by a sustained membrane depolarization followed. When this neuron was voltage clamped at a holding potential of -62 mV (Fig. 1B), an inward current was observed to underlie the initial PDS. Subsequently there was another transient inward current followed by development of a sustained inward current with

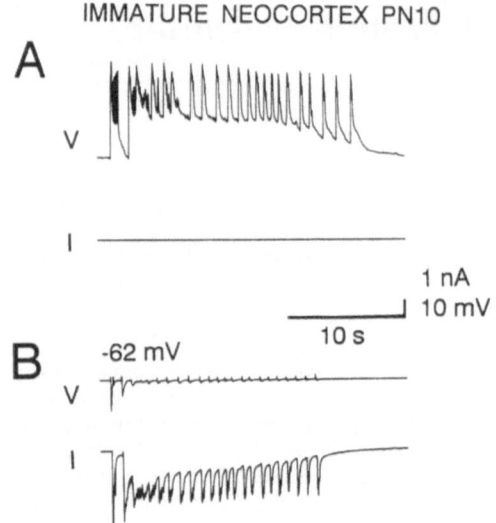

IMMATURE NEOCORTEX PN10

Fig. 1A,B. Specimen records of intracellular recordings of evoked ictal - like discharges in a picrotoxin-treated rat neocortex. Slice prepared from a 10 - day - old rat. **A** Current clamp recording obtained at the resting potential of this neuron (-65 mV). **B** Voltage-clamp recordings obtained from the same neuron. Holding potential was -62 mV. Data taken from chart recorder tracings

superimposed transients. The duration of the ictal-like discharge was similar under current and voltage clamp conditions suggesting that the long duration of the event was not due to activation of voltage-dependent intrinsic membrane currents.

To assess the role of non-NMDA receptors, the antagonist CNQX was bath applied after control responses to orthodromic stimulation had been obtained. In all cases, each neuron served as its own control. Chart records of evoked ictal-like events are shown in the upper trace of Fig. 2A and faster time records of the initial PDS are shown in Fig. 2B. Bath application of 5 μM CNQX suppressed the sustained depolarization and late PDSs of the ictal-like event (Fig. 2A, lower portion) while reducing the duration of the initial PDS (Fig. 2B, lower portion). It is apparent that after exposure to CNQX the latency to onset of the evoked initial PDS had increased and late PDSs were no longer triggered. These effects were reversible upon washing (not shown) and the results indicate that quisqualate and/or kainate receptors contribute significantly to ictal-like epileptiform responses in the immature neocortex.

The effects of the NMDA receptor antagonist D-APV on evoked paroxysmal activity are shown in the lower portion of Fig. 2. D-APV decreased the duration of the ictal-like discharge (Fig. 2C). The NMDA receptor antagonist also reduced the amplitude of the initial PDS (Fig. 2D). Spontaneous activity, when present, was blocked or markedly reduced by D-APV. These effects were reversible upon washing and indicate that NMDA receptors contribute significantly to epileptiform activity in the developing nervous system.

The present results using the single-electrode voltage clamp technique indicate that a sustained synaptic current underlies the paroxysmal discharges seen in picrotoxin-treated slices of immature rat neocortex. This findings argues against the possibility that the ictal-like discharges characteristic of the neonatal period are due to prolonged intrinsic responses triggered by a short synaptic input. The exact form of the epileptiform discharge under unclamped conditions is likely to be governed by this synaptic current and the intrinsic membrane properties of neocortical neurons.Voltage-dependent currents are known to influence synaptic responses in the neocortex (Sutor and Hablitz

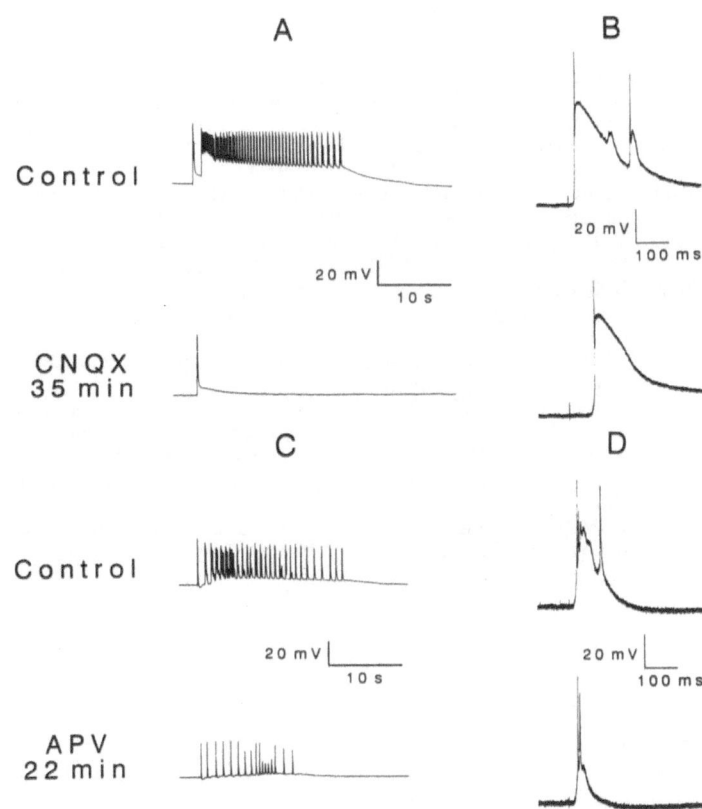

Fig. 2A-D. Effects of NMDA and non-NMDA receptor antagonists on evoked paroxysmal events. **A** Records obtained before (*upper trace*) and after (*lower trace*) exposure to 5 μM CNQX. **B** Same data as in A but on a faster time scale. **C** Recordings in another neuron before (*upper trace*) and after (*lower trace*) bath application of D - APV (10 μM). **D** Same records as in C but on a faster time scale

1989bc) and this is likely to also be true for epileptiform discharges (Gutnick et al. 1982). The present results suggest, however, the presence of a strong synaptic component and the factors responsible for this sustained input need to be determined.

The studies with NMDA and non-NMDA antagonists suggest that both types of excitatory amino acids receptors contribute to epileptiform discharges in the immature brain. Epileptiform activity was reduced by D-APV and CNQX. However, complete suppression of evoked activity was not seen with NMDA antagonists. CNQX increased the latency to onset of the initial PDS and decreased or abolished the later components of the ictal-like event. It remains to be seen whether this residual response in the presence of CNQX is sensitive to NMDA antagonists. This will allow determinatiEon of the role of different receptor types in the initiation versus maintenance of epileptiform discharges.

Acknowledgements. This work was supported by National Institutes of Health grant NS22373.

References

Connors BW (1984) Initiation of synchronized neuronal bursting in neocortex. Nature 310: 685-685.

Gutnick MJ, Connors BW, Prince DA (1982) Mechanisms of neocortical epileptogenesis in vitro. J Neurophysiol 48: 1321-1335

Hablitz JJ (1987) Spontaneous ictal-like discharges and sustained potential shifts in the developing rat neocortex. J Neurophysiol 58: 1052-1065

Hablitz JJ, Heinemann U (1987) Extracellular K^+ and Ca^{2+} changes during epileptiform discharges in the immature rat neocortex. Dev Brain Res 36: 299-303

Honore T, Davies SN, Drejer J, Fletcher EJ, Jacobsen P, Lodge D, Nielsen FE (1988) Quinoxalinediones: Potent competitive non-NMDA glutamate receptor antagonists. Science 241: 701-704

Streit P (1984) Glutamate and aspartate as transmitter candidates for system of the cerebral cortex. In: Jones EG, Peters A (eds) Cerebral Cortex. Functional Properties of Cortical Cells vol. 2 Plenum, New York pp 119-143

Sutor B, Hablitz JJ (1989a) Cholinergic modulation of epileptiform activity in the developing rat neocortex. Dev Brain Res 46: 155-160

Sutor B, Hablitz JJ (1989b) Excitatory postsynaptic potentials in rat neocortical neurons in vitro. I. Electrophysiological evidence for two distinct EPSPs. J Neurophysiol 61: 607-620

Sutor B, Hablitz JJ (1989c) Excitatory postsynaptic potentials in rat neocortical neurons in vitro. II. Involvement of N-methyl-D-aspartate receptors in the generation of EPSPs. J Neurophysiol 61: 621-634

Thomson AM (1986) A magnesium-sensitive post-synaptic potential in rat cerebral cortex resembles neuronal responses to N-methylaspartate. J Physiol (Lond) 370: 531-549

Susceptibility to Bicuculline-Induced Seizures after Neonatal Hypoxia

M. R. de Feo, O. Mecarelli, G. Palladini, G. Cusimano, M. F. Rina and G. F. Ricci

I Neurofisiopatologia, Dip. Scienze Neurologiche, Universita
"La Sapienza" Roma, Viale Regina Elena, 336, Roma, Italy

Introduction

Neonatal hypoxia, whether or not associated with other neuropsychiatric disorders, is considered to be one of the most important etiological factors in epilepsy. The results of a recent study (Bergamasco et al. 1984) have shown that the epileptic risk in hypoxic newborns is 5.1 times higher than in a control group of normal babies. Moreover, although the incidence of epilepsy is greater in the first year after birth, perinatal hypoanoxia related seizures may appear in early childhood or even later.

In spite of the relevance of the argument to human pathology, experimental research on this subject is scarce and results are often discordant (Chiba 1985; Moshe et al. 1985; Shimomura et al. 1988). As to anatomical and biochemical effects of hypoxia on the immature central system, it has been recently demonstrated that hypoxia preferentially destroys GABAergic neurons in developing rat neocortex explants in culture (Romijn et al. 1988). Due to controversies regarding the "epileptogenic effect" of early postnatal hypoxia and taking into consideration the particular sensitivity of GABAergic neurons to this condition, the present experiment was designed to test the susceptibility to bicuculline-induced seizures of neonatal rats exposed to prior cerebral anoxia.

Methods

Wistar albino rats were randomly divided into three groups (A, B and C). Animals from groups A and B were exposed to three sessions of anoxia at postnatal days 2, 4 and 6. During each session, which lasted 5 min in group A and 10 min in group B, the behaviour and the EKG were continuously monitored. Anoxia was achieved by placing animals in a plexiglass box in which 99.9% nitrogen was introduced at a rate of 5 l/min. The temperature in the box was kept constant at 37 °C. At the end of each anoxic session, surviving animals were taken back to their mothers and their behavior was observed daily up to the age of 28 days. The third group of rats (group C) was not exposed to anoxia and was used as a control to test the susceptibility to bicuculline seizures in the absence of anoxia. At 28 days of age, bicuculline (2.5 mg/kg, i.p.) was injected. The latency to the appearance of the first epileptic feature, the incidence of convulsive manifestations (isolated myoclonic jerks and/or generalized tonic-clonic seizures) and the incidence of status epilepticus were observed. At the end of the experiments, surviving rats were killed and their brains removed for histological examination. The morphometric and morphological evaluations were carried out on dorsal, lateral and ventral cortical areas (122.5 μm) corresponding to those shown in the A17, A23 and A33 sections of Craigie's atlas (Zeiman et al. 1963). Equivalent areas of CA2, CA3 and CA4 hippocampal subfields and of the basal and cortical amygdaloid nuclei were also studied.

Results

A - Behavioral changes
About 1 min after nitrogen administration animals showed hyperactivity with incoordinated movements of all limbs followed by a marked reduction of motility up to complete immobility and cyanosis. A progressive reduction in cardiac and respiratory rate was also noted.

B - Mortality
During anoxic sessions and the successive 3 weeks, the mortality rate observed in group A was similar to that found in the control group C (5/36 and 4/38, respectively). On the other hand, a significantly higher mortality rate was noted in group B (20/36; $p < 0.001$ versus A and C groups).

C - Body weight changes
A significant decrease in body weight was observed during all observation period in groups A and B when compared to control rats (Table 1).

Table 1. Body weight changes in control and experimental rats

	controls	Group A	Group B
	34	31	16
14 days	35.17 ± 0.71	$29.51 \pm 0.74^{**}$	$26.50 \pm 1.20^{*}$
21 days	42.29 ± 1.00	39.54 ± 1.46	$37.75 \pm 1.40^{*}$
28 days	71.70 ± 2.06	$64.45 \pm 2.03^{*}$	$63.63 \pm 2.48^{*}$

Body weight is expressed in grams (mean $\pm$ SE).

 * p < 0.01 versus control
** p < 0.001 versus control

Table 2. Latency to epileptic features induced by bicuculline

	controls	Group A	Group B
	34	31	16
Myoclonic jerks	156.19 ± 6.42	138.05 ± 6.85	$129.15 \pm 9.62^{*}$
First genera-lized seizure	175.35 ± 10.92	157.21 ± 9.93	150.70 ± 13.85

Latency times are expressed as seconds (mean $\pm$ SE)

* p < 0.02 versus control

D - Convulsive disorders

The i.p. injection of bicuculline induced epileptic manifestations which consisted of myoclonic jerks. These were isolated at first and then became gradually more numerous and massive, leading up to typical tonic-clonic seizures which could evolve to status epilepticus, as previously reported by us (de Feo et al. 1985). The latency of appearance of both myoclonic jerks and tonic/clonic fits are shown in Table 2. The incidence of diverse kinds of epileptic manifestations (myoclonic jerks, tonic-clonic seizures and status epilepticus) is reported in Table 3. As shown in Tables 2 and 3 no significant changes could be observed in anoxic rats when compared to controls, although a tendency toward an increased incidence of bicuculline-induced seizures was evident in group B.

Table 3. Incidence of epileptic manifestation

	controls	Group A	Group B
Myoclonic jerks	23.5%	16.1%	18.8%
Single seizure	20.6%	25.8%	43.7%
Status epilepticus	17.6%	19.4%	18.8%
No seizures	38.3%	38.7%	18.7%

Discussion

Our data show that in developing rats repeated periods of postnatal hypoxia are ineffective in modifying the late susceptibility to bicuculline seizures; however, some differences may be noted in animals exposed to longer anoxic sessions.   The results obtained in group A, in which anoxia lasted 5 min per session, were in fact very similar to those observed in control animals, since no remarkable changes in the latency to initial epileptic manifestations, in their incidence and in their evolution toward status epilepticus were found.   These data agree with the observations of Moshe and Albala (Moshe et al. 1985) who did not find modifications in the susceptibility to kindling or flurothyl seizures in young rats which had been exposed to hypoxia (6% O_2) at 1 or 10 days of life.   On the other hand, group B rats, in which anoxia was more prolonged, showed a trend toward an increased susceptibility to bicuculline seizures, as demonstrated by the shorter latency to initial epileptic signs and by the greater incidence of generalized seizures as compared to controls. Since this trend, observed in group B, cannot be explained by neuronal and/or glial lesions (no substantial differences between controls, group A and B animals have been found in morphometric and morphological examinations), it is possible that these changes may be ascribed to modifications at a biochemical level, as already suggested by the works of Chiba (Chiba 1985) and Shimomura and Ohta (Shimomura et al. 1988). Both groups have shown an increased susceptibility to pentylenetetrazol (PTZ) seizures in rats exposed to severe cerebral anoxia during the postnatal period without any related neuroanatomical lesion. According to these authors, it is possible that various biochemical processes are involved in the greater epileptic susceptibility of anoxic rats, which also includes changes in the GABAergic system. In conclusion, only when anoxia is so severe as to cause a high mortality rate, is it possible to observe in surviving rats an increased susceptibility to bicuculline seizures; minor degrees of anoxia are very well tolerated by the immature brain. If the GABAergic system is preferentially involved in hypoxia (Romijn et al. 1988; Sloper et al. 1980), it may be that compensatory reactions take place during successive development in order to stabilize inhibitory processes in their original functional mode.

References

Bergamasco B, Benna P, Ferrero P, Gavinelli R (1984) Neonatal hypoxia and epilepsy risk: clinical prospective study. Epilepsia 25: 131-136

Chiba S (1985) Long-term effects of postnatal hypoxia on the seizure susceptibility in rats. Life Sci 37: 1597-1604

de Feo MR, Mecarelli O, Ricci GF (1985) Bicuculline- and allylglycine-induced epilepsy in developing rats. Exp Neurol 90: 411-421

Moshé SL, Albala BJ (1985) Perinatal hypoxia and subsequent development of seizures. Physiol Behav 35: 819-823

Romijn HJ, Ruijter JM, Wolters PS (1988) Hypoxia preferentially destroys GABAergic neurons in developing rat neocortex explants in culture. Exp Neurol 100: 332-340

Shimomura C, Ohta H (1988) Behavioral abnormalities and seizure susceptibility in rat after neonatal anoxia. Brain Dev 10: 160-163

Sloper JJ, Johnson P, Powell TPS (1980) Selective degeneration of interneurons in the motor cortex of infant monkeys following controlled hypoxia: possible cause of epilepsy. Brain Res 198: 204-209

Zeiman W, Maitland-Innes JR (1963) Craigie's neuroanatomy of the rat. Academic, New York

Cortico-Cortical Transfer During Initial Phases of Kindling in the Rat: A Developmental Study

J. Mareš and S. Trojan

Institute of Physiology, Faculty of Medicine, Charles University, Prague, Czechoslovakia

Introduction

Epileptic seizure generates short and long lasting changes in functions of the nervous system (for review see Racine 1978). These changes occur during each seizure as well and throughout the entire process repetition of epileptic seizures (kindling). We have suggested that the long-lasting remodeling of CNS function is a continuous process and that the short interval after the end of a seizure plays an extremely important role in later epileptogenesis (Mareš et al. 1981).

Developmental literature suggests that various processes thought to depend on neuroplasticity proceed at different rates and reach different end points in immature and adult animals. In our previous studies we have developed a model of the initial phases of the acute kindling and described its ontogeny (Mareš et al. 1981). We have also examined the influence of interstimulation intervals upon the duration of the next seizure (Mareš et al. 1981). On the basis of our experiments we know that the second seizure is longer than the first one when the interval between the end of the first seizure and the following stimulation exceeds 7 min. This is in very good correlation with morphological findings in cortical synapses. In comparison to controls, the number of synaptic vesicles is extremely low within the first minute after the end of a seizure and becomes two times higher than the normal control values within next 10 min (Mareš et al. 1981).

The morphological picture is almost normal after 1 h. These changes in the number of synaptic vesicles were observed in the cortex of the opposite, unstimulated hemisphere. In chronic kindling experiments a phenomenon reflecting the spread of plastic changes throughout the central nervous system has been described and called transfer (for review see Racine 1978). Transfer is responsible for the shifts of increased excitability between synaptically connected areas in subcortical structures.

The present paper deals with studies on transfer in the neocortex during the beginning of acute kindling, ontogenetic development of transfer, and the significance of relations between frontal and occipital neocortex.

Methods

Experiments were performed on male albino rats (Wistar) bred in our institute. Adult and young rats aged 12, 15, 18, 21, 25, 30 and 45 days post partum were used. Five experiments were performed. The minimum number of animals in each age group of each experiment was eight, the total number of animals was 429.

Surgical preparation was the same in all experiments. All wounds and pressure points were infiltrated by procaine, the animals were restrained and artificially ventilated. Bipolar stimulation electrodes were placed on the cortex surface. The intensity of stimulation eliciting epileptic self-sustained after-discharges (SSAD) was fivefold the threshold for eliciting an interhemispheric response. The frequency of pulses in the train was 8 Hz and the duration of each train was 20 s.

Experimental Brain Research Series 20
© Springer-Verlag Berlin · Heidelberg 1991

The interval between the end of the previous epileptic SSAD and the next stimulation was always 10 min. Stimulation was repeated four times in each animal. Frontal and/or occipital neocortex was stimulated; in three experiments the location of stimulation was changed after the second seizure and in two experiments all four stimulations were delivered at one cortical region (for details see Table 1). An EEG was recorded from frontal and occipital cortical regions of both hemispheres and the duration of elicited SSADs were evaluated.

Table 1. Localization of stimulation on the cerebral cortex

Stimulation / Experiment					
	1	2	3	4	5
1	RF	RF	RO	RO	RO
2	RF	RF	RO	RO	RO
3	RF	LF	RO	LO	LF
4	RF	LF	RO	LO	LF

RF, right frontal; LF, left frontal; RO, right occipital; LO, left occipital

Results

The response of 12 - day - old rats depended on the site of the first stimulation. When the frontal cortex was stimulated first (experiments 1 and 2), the incidence of SSADs was lower than if the occipital cortex was stimulated first (experiments 3 - 5). SSADs could be elicited in frontal cortex regularly from the 15th day of life and from the 12th day in occipital cortex (more than 70% of the initial stimulations evoked afterdischarges).

In the course of ontogeny there are two maxima in the duration of SSADs. The second one is reached in adulthood.The first maximum was observed in the occipital cortex in the age 18 days group and in the frontal cortex in the age 25 days group (Fig. 1). With most age groups the duration of SSAD increased with repeated stimulations in all five experiments (Figs. 1,2). Thus, we conclude that the initial phases of acute kindling can be elicited throughout ontogeny. This phenomenon occurs regardless of the stimulated cortical area, but the prolongation of SSADs was slower in occipital cortex than in frontal cortex (Fig. 1). SSADs elicited in identically aged rats by the first stimulation of the occipital cortex were always longer than those elicited by the first stimulation of the frontal cortex. The same was true for the second, third and fourth stimulations. Despite the translocation of stimulations (experiments 2, 4, and 5) the first SSADs were regularly shorter than the fourth ones. Some exceptions occurred in the groups of 30- (experiment 5) and 45 - day - old (experiment 2) animals (Fig. 2). Prolongation of SSADs after translocation of the stimulation between the occipital cortical areas of both hemispheres was much less prominent than after the same experimental manipulation in the frontal cortex.

Discussion

The age at which regular formation of neocortical SSADs was first observed is in good correlation with results on developing neocortical neurons in vitro (Kriegstein et al. 1987) and with the first occurrence of audiogenic seizures in genetically epilepsy-prone rats (Savage et al 1986). This age is also a phase of rapid maturation of long-term post-tetanic potentiation in visual cortex (Perkins and Teyler 1988).

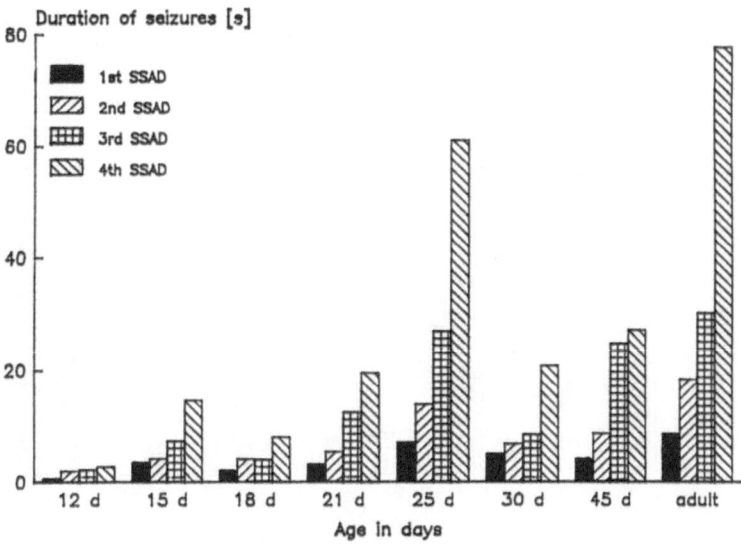

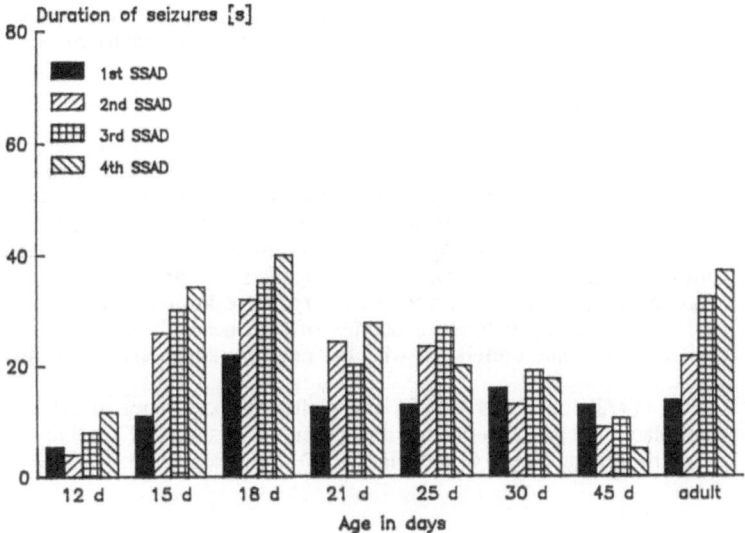

Fig. 1. Duration of the first four SSADs during the start of acute kindling in the frontal cortex in individual age groups (*top*); Kindling in occipital cortex (*bottom*)

The divergence in the development of SSADs in frontal and occipital cortical regions is probably due to different rates of ontogeny of sensorimotor and visual cortex. The observed different durations of SSADs show that the reaction of both cortical areas to epileptogenic stimuli is also different. Racine (1975) also described different courses of classical kindling in various cortical regions.

TRANSFER (occipito—frontal)

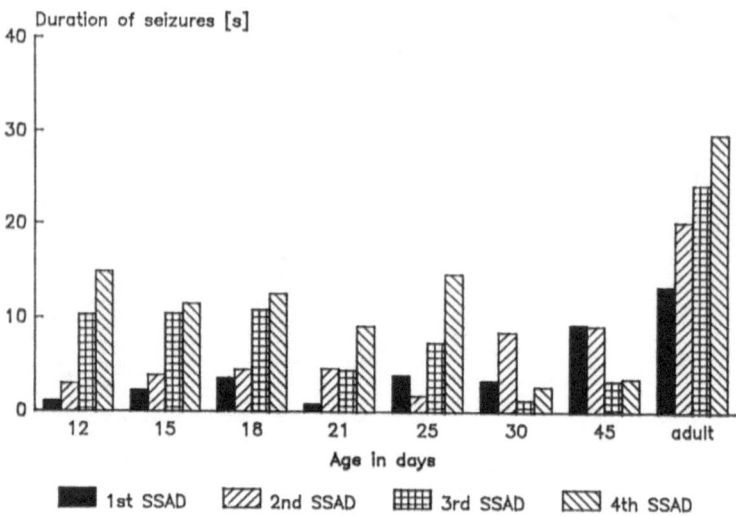

Fig. 2. The first two columns represent duration of the first two SSADs elicited from occipital neocortex and the next two columns show duration of SSADs elicited from frontal neocortex after the transfer of stimulation

The transfer between the frontal cortical areas of both hemispheres is possible starting from the 12th day of life. The neuronal pathway for this shift of increased excitability is at least monosynaptic. The multisynaptic pathway for transfer was also active (experiment 5) during the followed stage of ontogeny. According to the results of transfer between occipital and frontal cortex, the pathway composed by at least two synapses starts its action approximately parallel with the monosynaptic one. It is apparent that the corpus callosum plays a decisive role in the transfer of increased excitability between neocortex of both hemispheres. Our results on the development of transfer shows some correlation with the rate of ontogenesis of myelination of corpus callosum.

Racine (1972) found that transfer arises after the occurrence of the epileptic afterdischarge at the target place. We can also assume that during early stages of kindling the occurrence of transfer is caused by generalization of the first SSAD throughout the neocortex.

References

Kriegstein AR, Suppes T, Prince DA (1987) Cellular and synaptic physiology and epileptogenesis of developing rat neocortical neurons. Dev Brain Res 34: 161-171
Mareš J, Mareš P, Trojan S, Langmeier M (1981) Cortical selfsustained after-discharge in the rat. Acta Univ Carol Med Monogr (Praha) 54
Perkins AT,IV, Teyler TJ, (1988) A critical period for long-term potentiation in the developing rat visual cortex. Brain Res 439: 222-229
Racine RJ (1972) Modification of seizure activity by electrical stimulation. II. Motor seizures. Electroencephalogr Clin Neurophysiol 32: 281-294
Racine RJ (1975) Modification of seizure activity by electrical stimulation: cortical areas. Electroencephalogr Clin Neurophysiol 38: 1-12
Racine R (1978) Kindling the first decade. Neurosurgery 3: 234-252
Savage DD, Riegel CE, Jobe PC (1986) Angular bundle kindling in rats with a genetic predisposition to acoustic stimulus induced seizures. Brain Res 376: 412-415

Plasma Levels and Anticonvulsant Action
of Phenytoin and Carbamazepine in Immature Rats

S. Marešová[1], P. Mareš[1], D. Marešová[2], M. Pometlová[2] and S. Trojan[2]

[1] Institute of Physiology, Czechoslovak Academy of Sciences, Prague
[2] Institute of Physiology, Faculty of Medicine, Charles University, Prague, Czechoslovakia

Introduction

Carbamazepine (CBZ) and phenytoin (PHT) are used in the treatment of human partial focal seizures in adults and in children. An experimental study published by Vernadakis and Woodbury (1969) provided evidence for qualitative changes in the action of PHT during ontogenetic development. Since one of the reasons for such a change may be the different pharmacokinetics of PHT in immature rodents, we studied the plasma levels and anticonvulsant action of PHT and CBZ (for comparison) in adult and 12-day-old rats.

Methods

Male Wistar rats (adult and 12-day-old) were used for the experiments. Plasma PHT and CBZ analysis and testing of anticonvulsant effects were performed on different animal groups.

Plasma PHT and CBZ Analysis

PHT (60 mg/kg) and CBZ (50 mg/kg) were administered to animals in both age groups. Decapitation was performed either 30 min or 60 min later and blood samples were collected. The plasma drug levels were determined using gas chromatography. PHT was analyzed on a packed column (Pometlová et al. 1981) and CBZ on a capillary column (CP SIL 19CB, 10 m x 0.32 mm). Column temperature was programmed in both types of the analysis.

Tests of Anticonvulsant Action

Metrazol-Induced Seizures. Animals of both age groups were pretreated with PHT (60 mg/kg, i.p.) or CBZ (50 mg/kg, i.p.) and Metrazol was administered 15 min later in doses of 70, 80, 90 and/or 100 mg/kg, s.c. Animals were observed for 30 min following the Metrazol injection and the quality of the seizures was evaluated.
Hippocampal Electrical Stimulation. The methods of operation and stimulation were described elsewhere (Marešová and Mareš, this volume). Following the end of the first self-sustained afterdischarge (SSAD) the drugs (PHT and CBZ) were administered and the stimulations were performed 15, 25 and 35 min following the first stimulation. The lengths of the ADs were compared, and the localization of stimulating electrodes was verified by means of histological serial sections.

Results

Plasma levels measured 30 and 60 min after the administration of antiepileptic drugs were within the human therapeutic range. The only difference found was a slow increase in plasma levels of

168

PHT in 12-day-old pups such that the value at 30 min was significantly lower than the corresponding value in adult rats. The results are shown in Table 1.

Both CBZ and PHT acted against major Metrazol-induced seizures. A selective action against the tonic phase was observed following treatment with both drugs.

Table 1. Plasma levels of antiepileptic drugs in 12-day-old and adult rats

Drug		30 min (μM)	60 min (μM)
CBZ	12-day-old	63.2	59.8
	adult	76.0	84.1
PHT	12-day-old	28.4	44.2
	adult	61.0	51.7

Hippocampal ADs were reliably suppressed by CBZ in both adult and young animals. PHT, by contrast, did not exhibit any significant action in adult rats, but the duration of ADs was increased in immature rats (Fig.1).

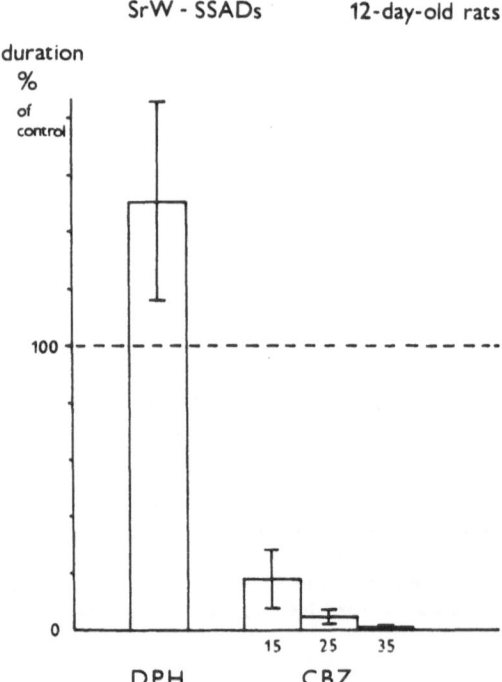

Fig. 1. Lengths of hippocampal ADs. Ordinate, length of ADs as a percent of the length of the first AD; *PHT*, phenytoin group; *CBZ*, carbamazepine group. *Dashed line* shows the 100% length of the first AD. The shortening of ADs following CBZ administration and prolongation of ADs following PHT administration are demonstrated

169

Discussion

The results of the tests of anticonvulsant action are controversial. Both drugs suppressed major Metrazol-induced seizures; however, the hippocampal afterdischarges were suppressed by CBZ only. Moreover, PHT administration led to an increase in the duration of the hippocampal AD in 12-day-old animals. We assume that the reasons for this prolongation of AD in young animals are qualitative rather than quantitative, since the plasma levels of both drugs remained in human therapeutic ranges. This result is in agreement with data of Vernadakis and Woodbury (1969) and with the results of our previous experiments. The results also indicate that CBZ is the drug of choice in cases of focal seizures with complex symptomatology.

References

Pometlová M, Marešová D, Mareš P, Trojan S (1981) Motor behavior and plasma levels of DPH in the rat. Physiol Bohemoslov 30: 181-182
Vernadakis A, Woodbury DM (1969) The developing animal as a model. Epilepsia 10: 163-179

Effect of Hypoxia on Hippocampal Afterdischarges in Young Rats

D. Marešová[1] and P. Mareš[2]

[1] Department of Physiology, Medical Faculty, Charles University, Prague, Czechoslovakia
[2] Institute of Physiology, Czechoslovak Academy of Sciences, Prague, Czechoslovakia

Introduction

All experimental models of epileptic seizures and/or epileptic processes include one important positive prognostic factor - they are elicited in a healthy brain (Löscher and Schmidt 1988, Mareš and Zouhar 1987). In human epileptology, patients with poor outcome are usually those in which the epileptic process has progressed in a damaged brain. Since hypoxic-ischemic injury seems to be of primary importance (Hill and Volpe 1989), we decided to study epileptic afterdischarges (ADs) in experimental animals exposed to hypoxia.

Methods

Experiments were performed in 12- and 18 - day - old albino rats of the Wistar strain. Animals were implanted with hippocampal stimulating electrodes (dorsal hippocampus, mostly CA3 area) and with cortical recording electrodes cemented to the skull by a dental acrylic. An indifferent electrode was attached to the nasal bone. After surgical preparation, the animals were allowed to recover from ether anesthesia for at least 1 h, fed with a 5% solution of glucose, and then exposed to hypobaric hypoxia at a simulated altitude of 7000 or 9000 m for 1 h. Control rat pups were left in atmospheric pressure. Immediately after exposure to hypoxia the stimulation series started. All animals were stimulated with rectangular pulses of 0.3 ms duration and 8 Hz frequency. One stimulation series lasted 15 s, the intensity was two times higher than that necessary for eliciting a clear cortical response when single stimuli were used. The interval between the end of the AD and the start of the next stimulation series was 15 min; an additional group of 18-day-old rats was stimulated with 3 min intervals. The duration of epileptic AD was measured and the results from control and experimental groups were evaluated by means of analysis of variance. The level of statistical significance was set at 5%. After the end of experiment the localization of stimulation electrodes was confirmed histologically in serial sections.

Results

Control Animals

Twelve-day-old rats (n = 52): An epileptic AD was elicited in all animals. Electrographically it was formed by fast spikes with low amplitude which transgressed into large delta waves (0.5 - 2 Hz) with superimposed fast activity (i.e., serrated waves). The average duration of the first AD was 47 $\pm$ 4 s (M $\pm$ SEM). Out of 52 rats, 24 were stimulated three or four times with 15 min intervals. Repeated stimulations led to a progressive increase in duration of ADs. The most marked behavioral correlate was represented by wet dog shakes (WDS), which appeared toward the end of ADs and after them.

Eighteen-day-old rats (n = 41): The same type of epileptic activity (electrocorticographic and behavioral) was recorded in this age group. The mean duration of the first AD was 23.5 $\pm$ 2.5 s, i.e., significantly shorter than in 12 - day - old rat pups. Repeated stimulations performed in 23 animals resulted again in a progressive lengthening of ADs. A group of 8 rats was stimulated with 3 min intervals. The second AD was significantly longer than the first one, this difference was maintained in the third and fourth ADs.

Animals Exposed to Hypoxia

Twelve-day-old rats (n = 35): ADs were nearly the same as in the control rats, only the amplitude tended to be lower. All animals exhibited WDS. The mean duration of the first AD was 79 ± 10 s (n = 27) and 98 ± 18.5 s (n = 8) for rats exposed to simulated altitudes of 7000 and 9000 m, respectively. Both values were significantly longer than control one (Fig. 1).

Repeated stimulations in 14 (7000 m) and 8 (9000 m) animals failed to induce progressive prolongation of ADs. On the contrary, the duration of the fourth AD in the 9000 m group was significantly shorter than that of the first AD.

Eighteen-day-old rats(n = 39): The results were similar to those in 12 - day - old animals. The mean duration of the first AD was again increased (51.5 ± 13.5 s, Fig 1); progressive prolongation of ADs with repeated stimulations in 14 out of 26 rats (7000 m) and 13 out of 13 rats (9000 m) also failed to appear in this age group.

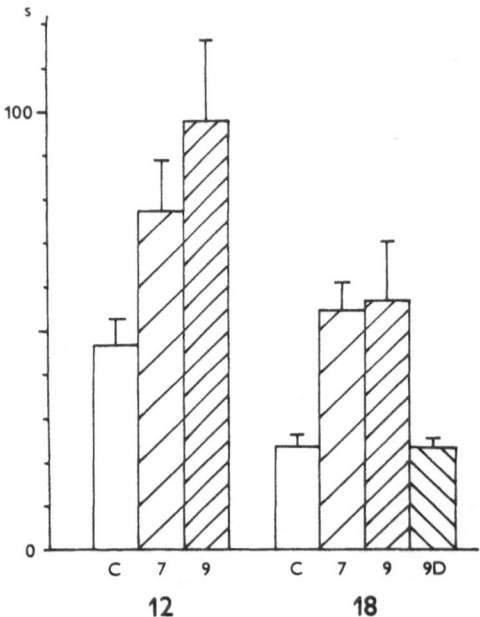

Fig. 1. Duration (M + SEM) of the first afterdischarge (AD) induced by hippocampal stimulation in rat pups 12- (*left*) or 18-day-old (*right*). C control animals; 7 and 9, two groups exposed to hypobaric hypoxia at simulated altitudes of 7000 and 9000 m, respectively; *9D*, animals exposed to 9000 m at the age of 12 days and tested at 18 days

An additional group of 7 rats exposed to 9000 m altitude was stimulated with 3 min intervals. Again the first AD was longer than in control rats; the second AD was significantly longer than the first one, but the significance of the difference was lost with the third and fourth stimulations.

To establish if the prolongation of the first AD by an exposure to hypoxia was permanent, a group of 12 animals was exposed to 9000 m at the age of 12 days and electrophysiologically examined 6 days later, at the age of 18 days. The duration of the first AD (23.5 $\pm$ 2 s) was the same as in 18 - day - old controls; the progressive lengthening of ADs with repeated stimulation at 15 min intervals was also present.

Discussion

Results of control groups were identical with those published recently (Marešová et al. 1987). Exposure of both 12- and 18 - day - old rats to hypoxia led to a significant increase in the first AD, i.e., to damage of a system responsible for termination of the epileptic seizure - probably an active inhibition. A failure of progressive changes following repeated stimulations suggests that an injury of excitatory systems using NMDA type receptors might be involved. In agreement with the sparse literary data on the influence of hypoxia on epileptic phenomena in developing animals (Chiba 1985; Holmes and Weber 1985; Moshé and Albala 1985), we may conclude that epileptic phenomena are generally aggravated by hypoxic insult, but the situation is rather complicated. An analysis of the underlying mechanisms is necessary.

References

Chiba S (1985) Long-term effect of postnatal hypoxia on the seizure susceptibility. Life Sci 37: 1597-1604
Hill A, Volpe JJ (1989) Hypoxic-ischemic encephalopathy of the newborn. In: Swaiman KF (ed) Pediatric neurology. Moshby, St Louis, pp 372-392
Holmes GL, Weber DA (1985) Effects of hypoxic-ischemic encephalopathies on kindling in the immature brain. Exp Neurol 90: 194-203
Löscher W, Schmidt D (1988) Which animal models should be used in the search for new antiepileptic drug? A proposal based on experimental and clinical considerations. Epilepsy Res 2: 145-181
Mareš P, Zouhar A (1987) Ontogenetic models of epileptic seizures. II. Primary generalized seizures. Electroencephalogr Clin Neurophysiol 67: 17P
Marešová D, Mareš P, Velíšek L (1987) Ontogenetic development of a model of human complex partial seizures - hippocampal afterdischarges in the rat (in Czech). Sb Lek 89: 358-367
Moshé SL, Albala BJ (1985) Perinatal hypoxia and subsequent development of seizures. Physiol Behav 35: 819-823

Epileptiform Activity as Dependent and Independent Variables in Brain Maturation: A Comparison of the Effects of Chronic Tetrodotoxin and Picrotoxin Treatment on Developing Rat Neocortex Neurons Cultured in Vitro

M. A. Corner and G. J. A. Ramakers

Netherlands Institute for Brain Research, Meibergdreef 33, 1105 AZ Amsterdam, The Netherlands

Introduction

Our laboratory has been using dissociated cell cultures derived from late fetal rat occipital cortex for morphophysiological investigations into the role of spontaneous bioelectric activity (SBA) in neural network formation. Chronic deprivation experiments have employed tetrodotoxin (TTX) in order to suppress action potentials; picrotoxin (PTX) has been used for blocking inhibitory synaptic transmission. The former treatment leads to the disappearance of extracellular electrical activity, whereas the latter treatment induces large amplitude "epileptiform" field potentials together with polyneuronal spike discharges (Van Huizen et al. 1987; Van Huizen et al. 1987).

Results

Neuronal spike trains show a characteristic development in vitro which, despite large variations (Fig. 1), can be demonstrated using multivariate statistical analysis (Habets et l. 1987; Ramakers et al. 1990). The major trends observed were: (i) a progressive decline in the incidence of strong interspike interval dependencies (expressed as the Markov Value); (ii) a persisting early decrease in the regularity of grouped spike discharges (expressed by the coefficient of variation (CV) for the "Burst Period", being the time between the onset of successive spike clusters); and (iii) a large

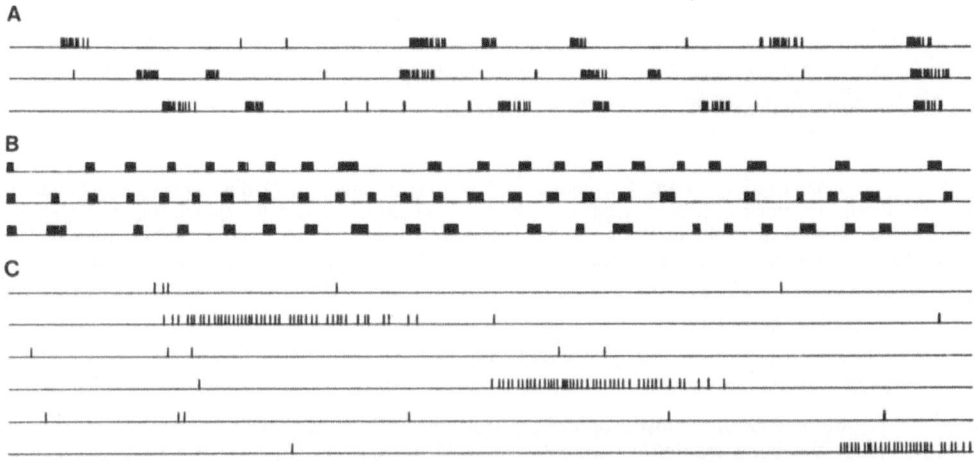

Figure 1. Neurons showing strong interspike interval dependencies (A) and high regularity of clustered discharges (B) seen most often at 14 div, and large minute-to-minute fluctuations in mean firing rate (C) common after 21 div. Each line represents 1 min (continuous recording)

increase in the proportion of units showing prominent, often highly regular, slow fluctuations of their mean firing level (as reflected in the CV for the number of spikes falling into successive 1 min time bins). Modal intervals calculated from interval histograms fell into categories which we have defined as being predominantly BURST (< 10 ms), PHASIC (10 - 25 ms), or TONIC (> 25 ms). The last of these constituted the majority of units at 2 weeks in vitro but were eventually replaced almost completely by "bursters".

Chronic exposure to TTX led to an apparent developmental arrest: interspike interval dependencies, burst regularities and minute-order fluctuations remained indefinitely at the 2 week time point (Ramakers et al. 1990). This retardation was visible even in raw spike-train plots, in which clustered discharges (Fig. 2) are encountered more commonly in TTX-treated than in control cultures. In addition, the proportion of spontaneously silent or very slow firing units was low, and the incidence of stationary spike trains high at all ages in TTX vs control groups (2). Abnormally short modal intervals and a dearth of "tonic" firing neurons already at 14 days in vitro (div) (p < .05) indicate that TTX induced pathophysiology starts even earlier than revealed by other parameters. The permanence of these sequelae is suggested not only by the occasional presence of visibly "epileptiform" activity (Fig. 2), but also by the high incidence of bimodal interval distributions in TTXtreated cultures, with "phasic" being favored over "tonic" firing (p < .025).

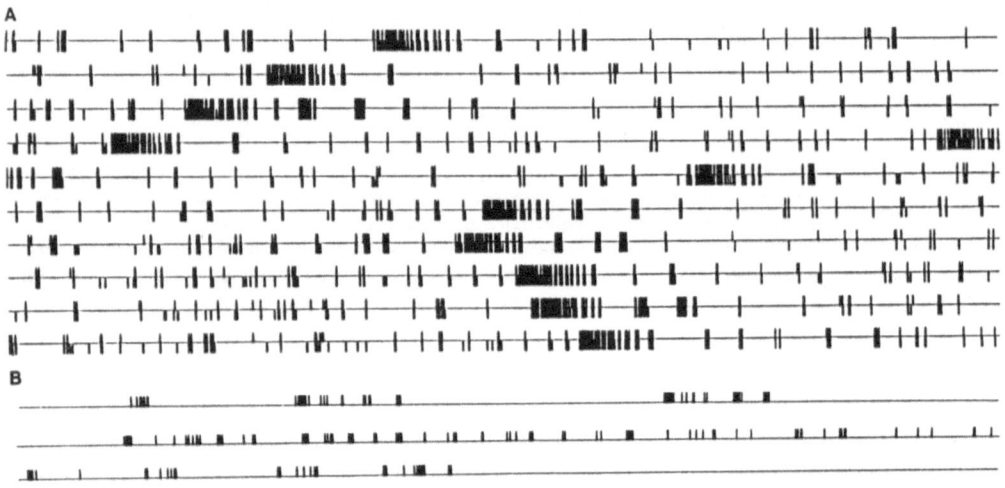

Fig. 2. A Two simultaneously recorded neurons in a TTX-treated culture at 42 div, showing synchronized "interictal" discharges along with short "ictal" episodes every minute or so. B A 9-s stretch from one of these units; interictal as well as ictal episodes are composed of high frequency bursts of action potentials (4 ms peak in the interspike interval histogram) at a rate of 10 - 15/s, giving a secondary peak at 50 ms

Chronic exposure to PTX had mostly opposite effects from TTX: interspike interval dependencies (Markov Value) and burst regularities were lower at 21 div than in controls (p < .05), whereas a higher proportion of units was devoid of stationary epochs (p < .01). Spike-train plots from PTX-treated neurons nevertheless gave the appearance of relatively continuous firing, an impression confirmed by a lower overall incidence of interspike intervals lasting more than 1 s (p < .01). In addition, predominantly "tonic" firing as well as secondary "tonic" (at the expense of "phasic") peaks in the interval histograms occurred more often in the PTX group (p < .05).

Discussion

The complementary nature of the consequences of TTX and PTX treatment argues in favor of SBA itself being the critical factor. Since early SBA in neocortex cell cultures takes the form of synchronous spike barrages (Habets et al. 1987), a certain degree of paroxysmal activity may be not only a normal but also a necessary factor in cortical development. Although its relative infrequence later on may indicate a reduction in the ontogenetic importance of SBA per se, phasic polyneuronal firing could nevertheless participate later in afferent induced maturation. In its absence (PTX treatment) cortical "plasticity" would presumably be curtailed, whereas a delay in its disappearance (TTX treatment) might prolong the "critical" period for stimulus dependent epigenesis.

It is of special interest that the physiological sequelae reported here tend to be in the opposite direction from the acute treatment effects. Thus, neurons become abnormally active and sometimes even show epileptiform firing patterns following long-term deprivation of SBA, whereas chronic disinhibition of the network eventually produces spike trains which lack stereotyped bursting. Although such "compensatory" growth responses suggest a process of, respectively, up- and down-regulation of neurotransmitter receptor sensitivities, part of the explanation may also lie in quantitative synaptic abnormalities (Van Huizen et al. 1987; Van Huizen et al. 1987). Thus, excessive excitatory interactions (TTX treatment) would lead to an increase of synchronized spike barrages, with a deficiency (PTX treatment) giving the opposite result.

References

Habets A M, Van Dongen A, Van Huizen F, Corner M A (1987) Spontaneous neuronal firing patterns i n fetal rat cortical networks during development in vitro: a quantitative analysis. Exp Brain Res 69: 43-52

Ramakers G J A, Corner M A, Habets A M (1990) Development in the absence of spontaneous bioelectric activity results in increased stereotyped burst firing in cultures of dissociated cerebral cortex. Exp Brain Res 79: 157-166

Van Huizen F, Romijn H J, Corner M A (1987) Indications for a critical period for synapse elimination n developing rat cerebral cortex cultures. Dev Brain Res 31: 1-6

Van Huizen F, Romijn H J, Van den Hooff P, Habets A M (1987) Picrotoxin-induced disinhibition of spontaneous bioelectric activity accelerates synaptogenesis in rat cerebral cortex cultures. Exp Neurol 97: 280-288

Motor Seizures in Immature Rats

M. Pohl, P. Mareš, L. Velíšek and L. Stanková

Institute of Physiology, Czechoslovak Academy of Sciences, Prague, Czechoslovakia

Motor seizures induced by Metrazol (pentamethylenetetrazol) in adult rodents are routinely used for testing of antiepileptic drugs (Swinyard and Woodhead 1982). The seizures also can be elicited by numerous other convulsants with different mechanism of action (e.g., isonicotinehydrazide, INH; 3-mercaptopropionic acid, 3-MPA; convulsant benzodiazepine, Ro 5-3663; picrotoxin; kainic acid; N-methyl-D-aspartate, NMDA). Metrazol is able to induce two basic types of seizures characterized by their motor patterns:

1. Minimal Metrazol seizures (mMs) with a dominant clonic component and preserved righting reflexes.
2. Major Metrazol seizures (MS) starting with wild running and followed by the clonic and tonic phases. The righting reflexes are lost.

Criteria to distinguish between these two types of seizures, elicited by convulsant drugs in adults and in immature rats, are shown in Table 1.

Table 1. Criteria to distinguish between minimal (mMs) and major (MS) Metrazol-induced seizures

Component	mMs	MS
Onset	Usually after several myoclonic jerks of the whole body; at early stages of development progressive intensification of chewing or head nodding.	Wild running starting abruptly, in young animals slower "swimming".
Tonic	Faintly expressed in tail and neck muscles during the whole seizure.	generalized - involving whole body, lasting several seconds, extremities exhibit tonic flexion and eventually extension as the most severe part of seizure.
Clonic	Dominant, not generalized, involving mainly head and forelimbs, sometimes asymmetrical especially in rat pups.	Generalized, often symmetrical lasting tens of seconds, in immature rats slower movements which are poorly coordinated.
Righting reflexes	Preserved	Lost

180

In addition to the increased velocity and coordination of movements, the following changes could be observed in the course of ontogenetic development: 1. mMs-like seizures were elicited starting from the first postnatal week by picrotoxin, Ro 5-3663, 3-MPA, INH and kainate (Mareš et al.1983b,1986; Stanková et al. 1988). Bicuculline induced this type of seizures starting from the 12th postnatal day (Zouhar 1989), Metrazol (Fig. 1) and kainate usually from the third postnatal week (Velíšková et al. 1988). However, the combination of Metrazol and some anticonvulsant drugs (e.g., phenytoin, carbamazepine, ethosuximide flunarizine and R 57720) paradoxically induced this type of seizures even in the first and second postnatal weeks (Mareš et al. 1981, 1983a; Pohl and Mareš 1987; Pohl et al. 1988). NMDA administration never led to mMs-like seizures (Mareš and Velíšek 1989). Latencies of mMs exhibited a U-shaped curve during maturation; statistically significant differences were found between adult and both 25- and 18-day-old rats (Fig.2).
2. MS-like seizures could be elicited in all age groups studied by all convulsant drugs tested (Fig. 1 shows Metrazol-induced seizures only). Latencies of MS increased starting from the third postnatal week (Fig. 2).

Comparing the mMs and MS as concerns their motor performance, incidence and latency development, we can conclude that they represent two different phenomena which exhibit a high degree of stereotypy of motor patterns even when elicited by various convulsants. Therefore, two generators probably with different courses of maturation, may be suggested. The disappearance of mMs in some age groups might be explained by the different maturation of latencies of mMs and MS (e.g., a U-shaped curve for mMs) so that MS latency might be shorter and mMs generation might be hidden. The possibility of different generators is supported by the finding of Browning and co-workers who studied electroshock-induced seizures (Browning and Nelson 1981). The exact anatomical localization of the generators of the two seizure types as well as ways of triggering their function should be objects of future research.

INCIDENCE OF SEIZURES

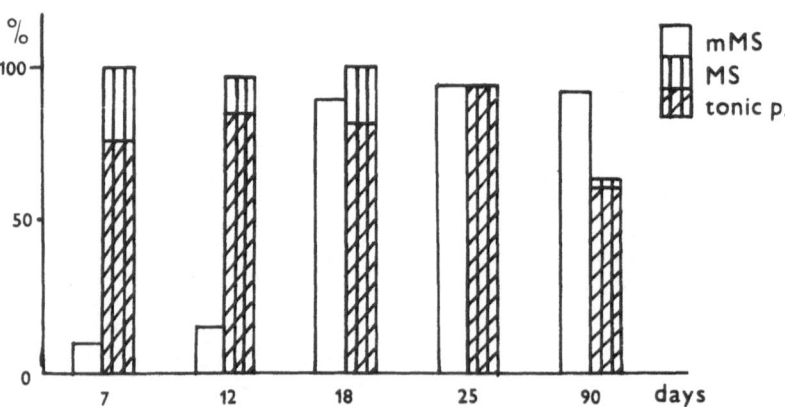

Fig. 1. Incidence of metrazol seizures. *Ordinate* incidence of seizures as a percent of all animals in the age group; *abscissa* 7, 12, 18, 25, and 90 indicate the age groups; *mMs* minimal metrazol seizures; *MS* major metrazol seizures; *tonic p* tonic phase of the major seizure

LATENCY OF SEIZURES

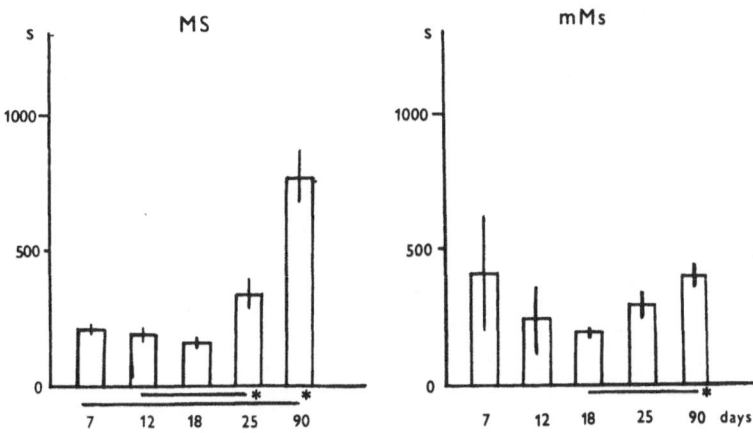

Fig. 2. Latency of metrazol seizures. *Ordinate* mean latencies of seizures in seconds; *abscissa* as in Fig. 1. *Left graph* major seizures; *Right graph* minimal seizures; statistical significance (5%) assigned by *asterisks*

References

Browning RA (1985) Role of the brain-stem reticular formation in tonic-clonic seizures: lesion and pharmacological studies. Fed Proc 44: 2425-2431

Browning RA, Nelson DK (1986) Modification of electroshock and pentylenetetrazol-induced seizures in rats after precollicular transsections. Exp Neurol 93: 546-556

Mareš P, Velíšek L (1989) Seizures induced by N-Methyl-D-Aspartate (NMDA) in developing rats. Physiol Bohemoslov 38: 377

Mareš P, Marešová D, Schickerová R (1981) Effect of antiepileptic drugs on metrazol convulsions during ontogenesis in rats. Physiol Bohemoslov 30: 113-121

Mareš P, Hlavatá J, Líšková K, Mudrochová M (1983a) Effects of carbamazepine and diphenylhydantoin on metrazol seizures during ontogenesis in rats. Physiol Bohemoslov 32: 92-96

Mareš P, Rokyta R, Trojan S (1983b) Epileptic seizures during ontogenesis in the rat. J Physiol (Paris) 78: 862-864

Mareš P, Líšková-Bernášková K, Mudrochová M (1986) Seizures induced by Ro 5-3663 during ontogenesis. Physiol Bohemoslov 35: 546

Pohl M, Mareš P (1987) Effects of Flunarizine on Metrazol-induced seizures in developing rats. Epilepsy Res 1: 302-305

Pohl M, Mareš P, Lanstiaková M (1988) Antimetrazol action of R 57720 during postnatal development in rats. Physiol Bohemoslov 37: 115-121

Stanková L, Mareš P, Zouhar A, Hirsová M, Muchová K (1988) Ontogenetic development of convulsant action of picrotoxin in the rat. Physiol bohemoslov 37: 571

Swinyard EA, Woodhead JH (1982) General principles. Experimental detection, quantification and evaluation of anticonvulsants. In: Woodbury DM, Penry JK, Pippenger CE (eds) Antiepileptic drugs. Raven, New York, pp 111-126

Velíšková J, Velíšek L, Mareš P (1988) Epileptic phenomena produced by kainic acid in laboratory rats during ontogenesis. Physiol Bohemoslov 37: 395-405

Zouhar A, Mareš P, Líšková-Bernášková K, Mudrochová M (1989) Motor and electrographic epileptic activity induced by bicuculline in developing rats. Epilepsia 30: 501-510

Increased Epileptogenesis in the Immature Hippocampus

L. Velíšek and P. Mareš

Institute of Physiology, Czechoslovak Academy of Sciences, Videnska 1083, Prague, Czechoslovakia

Introduction

There is a considerable evidence that the period of increased susceptibility to seizures is during childhood (Kellaway, 1985). A similar period has been searched for in experimental models of epileptic seizures. In the 70s, Mucha and Pinel (1977) described the time course of post-ictal (post-seizure) depression in adult rats following amygdaloid stimulation. Some years later, Holmes and Thompson (1987) revealed that this period is extremely shortened in rat pups, allowing the rapid development of amygdaloid kindling (using short interstimulation intervals which blocked the kindling in adult animals). In order to determine whether increased epileptogenesis could also be observed with hippocampal stimulations in vivo, the present study was performed.

Material and Methods

Male Wistar albino rats (n = 250) in five age groups (7, 12, 18, 25, and 90 days) were used. In adults the electrodes were placed in both hippocampi and on the cortex under pentobarbital anesthesia. The experiments started one week later. In pups the electrodes were inserted into the aforementioned structures under ether anesthesia early in the morning and the experiments were performed in the afternoon.

Stimulation protocol: The rats were stimulated twice by a series of rectangular, constant-voltage pulses of 1 ms duration, 8 Hz; the length of the train was 15 s. The interval between the trains varied from 30 s to 60 min (12 - 18-day-old rats), from 1 min to 60 min (25-day-old pups) and from 3 min to 60 min (adults). The voltage used for stimulation was twofold the threshold value. The duration of the evoked afterdischarges (AD) and number of "wet dog shakes" (WDS) were measured following the first and second stimulation trains.

The results were expressed as the "effective interval 100%" (EI 100), i.e., the calculated interstimulation interval following which the values after the second stimulation were equal to those seen after the first one. For the incidence of the phenomena (i.e., AD and WDS) we computed the 50% incidence interval value (50 - ii) by means of regression analysis.

Results

Figure 1 shows the duration of the interstimulation interval interval necessary for AD or WDS to be induced for the same duration (AD) or number (WDS) as following the first stimulation. A marked shortening of this interval was noted in 12-day-old rats (AD), whereas a constant increase occurred in 18-, 25-, and 90-day-old rats. Regarding WDS, following an initial decrease in the duration of the interstimulation interval a "plateau" was observed in 12-, 18-, and 25-day-old pups with a moderate increase towards adulthood.

Fig. 2 shows the duration of interstimulation intervals in which a 50% incidence of both phenomena was registered. The AD curve has nearly the same course as in Fig.1.:After a decrease

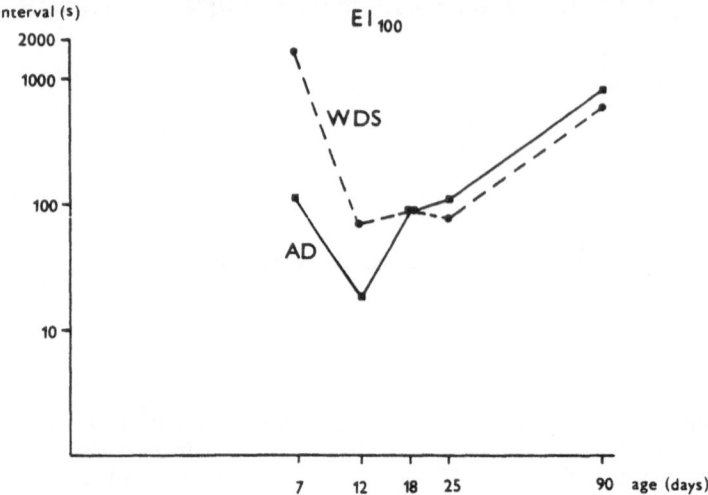

Fig. 1. Effective interval 100% (EI 100, see text). *Abscissa*, age in days (logarithmic scale); *ordinate*, interstimulation interval in seconds (logarithmic scale); *full line* and *black squares* represent values of the afterdischarge duration; *dotted line* and *black circles* denote the values for the number of wet dog shakes

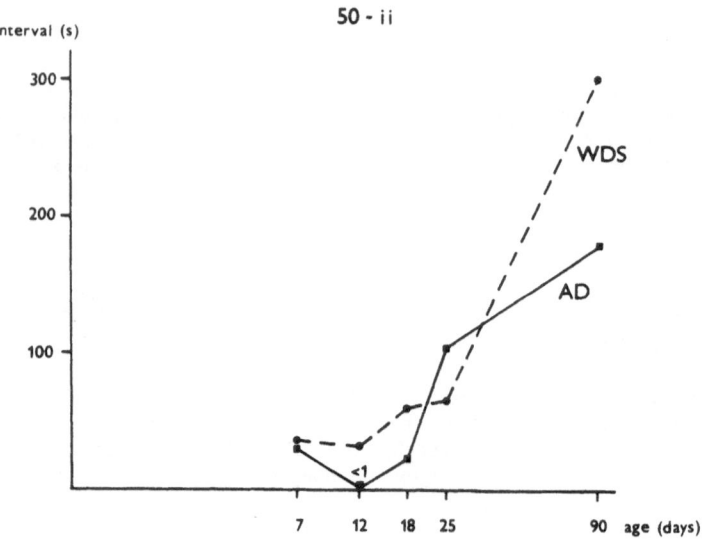

Fig. 2. 50% incidence interval (50-ii, see text). *Abscissa*, same as in Fig.1; *ordinate*, interstimulation interval in seconds (linear scale); *full line* and *black squares* represent the values of the afterdischarge duration; *dotted line* and *black circles* denote the values for the number of wet dog shakes. Note that the value for the 50-ii in 12-day-old pups is less than 1 s

in duration in 12-day-old pups, a progressive increase until adulthood followed. The differences in the 50%-ii for WDS were minor between 7-, 12-, 18-, and 25-day-old rats but then a rapid increase in the interval duration followed.

Discussion

Our results clearly demonstrate that there exists a period of dramatically shortened post-ictal inhibitory events in rat pups. The most pronounced effects were seen in 12-day-old pups and then in 18-day-old pups. The results seen in 7-day-old pups were similar to those found at postnatal day 25.

The existence of shifted boundaries between post-ictal inhibition and normal excitability of hippocampal tissue in pups may explain the rapid development of kindled seizures elicited from the other limbic system site,the of amygdala (Moshé et al. 1983; Holmes and Thompson 1987). As concerns the hippocampus, an extensive developmental study has not been performed yet and the reason for such suppressed post-ictal inhibition remains unclear (for review see Moshé 1987). One possible explanation might be based on the different maturation rates the of various transmitter systems (acetylcholinergic vs serotonergic). Another possibility was described in an ultrastructural study by Schwartzkroin et al.(1982) in rabbits, the excitatory synapses develop earlier than the inhibitory ones. Another explanation is only of speculative nature - the late development of neuronal circuit(s) which exert their inhibitory effects in the hippocampus. A failure of post-ictal depression in immature animals may be taken as a characteristic feature, at least in limbic structures. For other structures there are no data available.

This phenomenon of increased progressive epileptogenesis might serve as a simple model of status epilepticus and can be used for testing of antiepileptic drugs in rat pups.

References

Holmes GL, Thompson JT (1987) Rapid kindling in the prepubescent rat. Dev Brain Res 36: 281-284
Kellaway P (1985) Childhood seizures. In: Gotman J, Ives JR, Gloor P (eds.): Long term monitoring in epilepsy. EEG Suppl 37: 267-283
Moshé SL (1987) Epileptogenesis and the immature brain. Epilepsia 28 (Suppl.1) S3-S15
Moshé SL, Albala BJ, Ackermann RF, Engel J Jr (1983) Increased susceptibility of the immature brain. Dev Brain Res 7: 81-85
Mucha RF, Pinel JPJ (1977) Postseizure inhibition of kindled seizures. Exp Neurol 54: 266-282
Schwartzkroin PA, Kunkel DD, Mathers LH (1982) Development of rabbit hippocampus: anatomy. Dev Brain Res 2: 453-468

Changes in Lipid Metabolism in Epileptic Mirror Foci

R. Rokyta, S. Marešová, J. Velíšková, V. Kubík and K. Bernášková.

Department of Physiology, Medical Faculty of Hygiene, Charles University, Ke Karlovu 4, 120 00 Praha 2, Czechoslovakia

Introduction

Augmented production of free radicals (partially reduced oxygen compounds) can lead to very serious metabolic consequences. Presumably these compounds also play an important role in different pathophysiological processes in the brain. Free radicals start a chain reaction which induces the degradation of polyenoic fatty acids that form essential components of membrane lipids. Increased peroxidation of membrane lipids produces a decrease in essential fatty acids. These chemical reactions could thus produce changes in membrane permeability (Tappel 1973; Fridowich 1987).

Lipid peroxidation during formation of epileptic mirror foci throughout the ontogeny of the rat was studied. We hypothesized, that changes introduced by epileptic phenomena could also provoke degradation of cell membrane polyenoic fatty acids (PFA).

Methods

Symmetrical trephine openings were performed over the sensorimotor cortex in male laboratory Wistar rats ages of 7, 9, 12, 15, 18, and 25 days and in adults. The epileptic focus was induced by local application of Na-penicillin onto the right frontal region. Peroxidation was determined in animals with penicillin foci and in control animals who received an application of physiological saline. Cortical samples were excised from both the primary and mirror foci either 5 or 15 min after the appearance of the mirror focus. Peroxidation was evaluated by Snell's method (Snell and Mullock 1987). In this method oxidative products of PFA react with the 2-thiobarbituric acid. During the incubation oxidative cleavage of three-carbon residues, which is the primary reaction, produces the dialdehydes of malonic acid (MDA). The results are expressed as MDA equivalents per gram of wet tissue weight. Free fatty acids derived from brain lipids were estimated in samples excised 15 min after the appearance of mirror foci and in control specimens by capillary gas chromatography. After homogenization, the crude lipids were extracted from the brain tissue (chloroform: methanol 2:1 v/v). The free fatty acids fraction was obtained by separation on TLC (Silica gel $HF_{254+366}$). After esterification, the methylesters of the fatty acids were analyzed. CHROMPACK 438 A, a gas chromatograph equipped with a flame ionization detector, a split/splitless injector and a capillary column, CP SIL 19 CB (10 m x 0.25 mm) was used. Column temperature was programmed from 160° to 230° C at 5°/min.

Results

The content of MDA equivalents increased both in the epileptic and mirror foci (Fig. 1), whereas the relative content of arachidonic and docosohexaenoic acids diminished (Fig. 2). In the early stages of development (7 and 9 days), peroxidation increased the most in mirror epileptic foci. In animals of 12, 15, 18, and 25 days and in adults, peroxidation increased but not to such a high level (Fig. 3). Total peroxidation increased up to the 18th day and slightly decreased in 25-day-

188

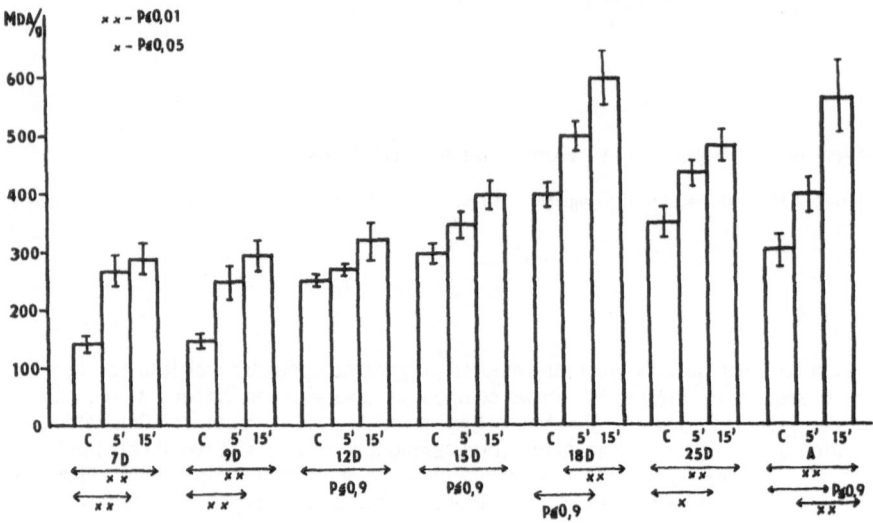

THE DEVELOPMENT OF MDA ACTIVITY IN CONTROL RATS /C/, AND IN RATS AFTER 5 AND 15 MINUTES IN MIRROR EPILEPTIC FOCUS.

Fig. 1. Development of MDA activity in control and experimental rats in different age groups. D age in days; A adult animals; xx, p<0.01, x - p<0.05

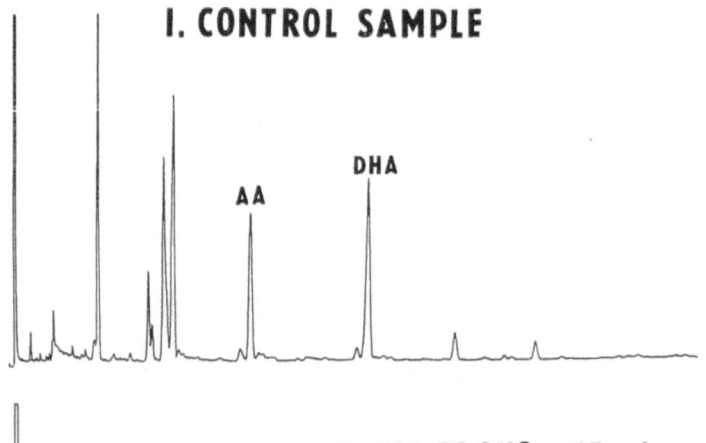

I. CONTROL SAMPLE

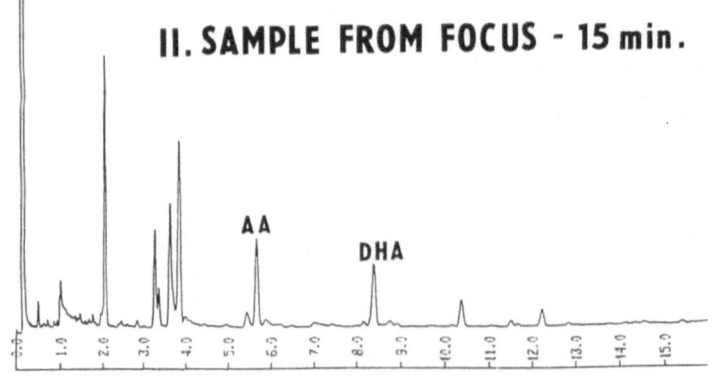

II. SAMPLE FROM FOCUS - 15 min.

Fig. 2.
Analysis of free fatty acid fractions in brain tissue using capillary gas chromatography in samples from adult control and experimental animals with epileptic mirror foci.
AA arachidonic acid;
DHA docosahexaenoic acid

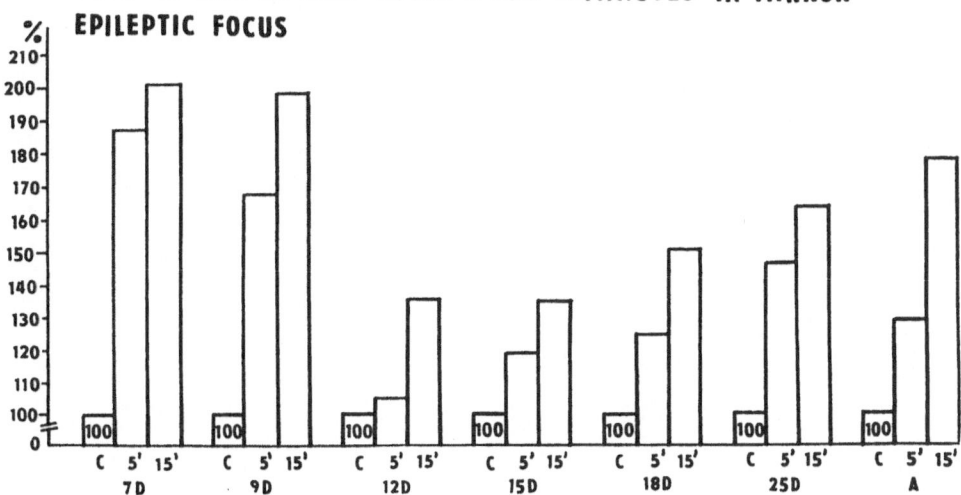

Fig. 3. Relative level of MDA activity during development in animals with epileptic mirror foci. Control level, 100%; **D** age in days; **A** adult animals

old and adult rats. Samples taken from adult animals with mirror epileptic foci and analyzed by capillary gas chromatography did not differ in free fatty acid composition when compared with controls; however the amount of arachidonic acid (20:4) decreased from 9.6% to 6.9%, and the amount of docosahexaenoic acid (22:6) decreased from 13.3% to 6.3% relative to the controls (Fig. 2).

Discussion

The uncontrolled amount of highly reactive forms of oxygen might explained the toxicity of reactive oxygen. Normally, these highly reactive forms of oxygen can be eliminated from the organism by several mechanisms: superoxide dismutase, catalase, peroxidase, and a redox system of glutathione, ceruloplasmin and so called low-molecular scavengers (Lippman 1980, Del Maestro 1980). We suggest that epileptic neurons have either higher levels of reactive oxygen or decreased removal of reactive oxygen by the above mentioned elimination systems.

In summary, it may be concluded that peroxidation of unsaturated fatty acids in an epileptic mirror focus is increased during some stages of development. The relative content of arachidonic acid and docosohexaenoic acid was decreased relative to controls in animals with mirror foci. These findings may explain subsequent changes in epileptic neurons.

References

DelMaestro RF (1980) An approach to free radicals in medicine and biology. Acta Physiol Scand Suppl 492: 153-168

Fridowich I (1978) The biology of oxygen radicals, Science 201: 875-880

Lippman HD (1980) Chemiluminescent measurement of free radicals and autooxidant molecular protection inside living rat mitochondria. Exp Geront 15: 339-351

Snell K, Mullock B (1987) Biochemical toxicology - a practical approach. IRL Press, Oxford

Tappel AL (1969) Inactivation of ribonuclease and other enzymes by peroxidizing lipids and by malonyl aldehyde. Biochemistry 8: 2827-2832

IV ANTICONVULSANT DRUG ACTIONS

Anticonvulsant Drug Mechanisms

W. Löscher

Department of Pharmacology, Toxicology and Pharmacy, School of Veterinary Medicine, Bünteweg 17, 3000 Hannover 71, FRG

Introduction

There are two general ways in which drugs might abolish or attenuate epileptic seizures: (a) through effects on pathologically altered neurons of seizure foci to prevent or reduce their excessive discharge, and (b) through effects that would reduce the spread of excitation from seizure foci and prevent firing and functional disruption of normal neuron aggregates. Most, if not all, antiepileptic drugs in current clinical use act, at least in part, by the second mechanism. A complete discussion and description of all aspects of the mechanisms by which antiepileptic drugs exert their effects on epileptic processes goes beyond the scope of this paper and has been the subject of several recent monographs and articles (e.g., Glaser et al. 1980; Frey and Janz 1985; Macdonald and McLean 1986; Löscher 1987; DeLorenzo 1988). The purpose of this paper is to summarize some of the currently accepted mechanisms of various classes of therapeutic agents. Furthermore, some newly developed anticonvulsant drugs, which may partially act by new mechanisms, will be discussed.

Mechanism of Action of Clinically Used Antiepileptic Drugs

Since initial choice of an antiepileptic drug is usually governed by seizure diagnosis, antiepileptic drugs are generally classified according to their effectiveness against specific seizure types (Table 1). Thus, phenytoin, carbamazepine, phenobarbital, and primidone are drugs of first choice against partial seizures and generalized tonic-clonic seizures. Valproic acid and ethosuximide are primarily used to treat patients with absence and myoclonic seizures, although valproate is also active against other seizure types. Similar to valproate, benzodiazepines possess a wide spectrum of antiepileptic activity but, due to development of tolerance and dependence during long-term administration, these drugs are primarily used to interrupt convulsive status epilepticus. The fact that antiepileptic drugs differ in their spectrum of clinical activity strongly indicates major differences in their mechanisms of action. Furthermore, drugs with a broad spectrum of antiepileptic activity, such as valproate or benzodiazepines, might combine two or more complementary mechanisms of action. It should be noted in this respect, however, that patients with the same seizure type may differ in their response to an antiepileptic drug - especially patients with complex partial seizures who are often drug resistant - possibly indicating that different molecular and cellular events may induce the same seizure type. Thus, there is a need for antiepileptic drugs with diverse mechanisms of action.

With respect to experimental studies on drug mechanisms, several antiepileptic drugs, e.g., barbiturates and benzodiazepines, exert sedative/hypnotic and muscle relaxant effects at doses somewhat higher than those used for antiepileptic treatment. It is therefore important to relate drug concentrations effective in neurophysiological, neurochemical, or pharmacological models to concentrations occurring in the human CNS during chronic antiepileptic therapy. Interestingly, for most antiepileptic drugs, "active" brain concentrations in epileptic patients (determined in biopsy samples from brain surgery) are similar to brain concentrations determined in experimental animals at anticon-

Table 1. Spectrum of clinical activity of common antiepileptic drugs. (From Löscher and Schmidt 1988)

Drug	Partial seizures		Partial seizures evolving to generalized seizures (GTCS, GTS or GCS)	Generalized seizures				
	Simple	Complex		Tonic and/or clonic (grand mal)	Absence (petit mal)	Myoclonic (impulsive petit mal)	Lennox-Gastaut syndrome (myoclonic astatic epilepsy)	West syndrome (infantile spasms)
Phenytoin	+++	+++	+++	+++	NE	NE	NE	NE
Carbamazepine	+++	+++	+++	+++	NE	NE	NE	NE
Phenobarbital	+++	+++	+++	+++	NE	++	NE	NE
Primidone	+++	+++	+++	+++	NE	++	NE	NE
Valproate	+	+	+	+++	+++	+++	++	+
Ethosuximide	NE	NE	NE	NE	+++	+++	NE	NE
Benzodiazepines*	++	++	++	++ (Status +++)	+	+	++	+
Relative frequency (%) of seizure type**	10	30-40	10	10-40	5-10	4	1-6	1-2

+, ++, +++, antiepileptic efficacy; NE, not effective; GTCS, generalized tonic-clonic seizures; GTS, generalized tonic seizures; GCS, generalized clonic seizures; *development of tolerance during long-term treatment; ** range of different investigations

vulsant doses during seizure tests, such as the maximal electroshock seizure (MES) test or models using chemoconvulsants for seizure induction (Tables 2-5). Only in the cases of valproate and ethosuximide are active brain concentrations in rodent models markedly higher than concentrations in the brains of patients undergoing long-termtreatment (Tables 6 and 7). Several investigations have proposed that experimental drug actions should be studied at free plasma or CSF levels occurring in humans during therapy (e.g., Macdonald and McLean 1986). Except for valproate, however, drug concentrations in human brain tissue are markedly higher than free plasma or CSF levels (Tables 2-7) so that correlations between experimental and clinical data should be undertaken on the basis of brain tissue rather than CSF levels. Furthermore, the sensitivity of non-mammalian preparations or cultured neurons, which are often used for investigation of drug mechanisms, may be different from the sensitivity of neurons in the intact human brain.

The potentially most useful classification of antiepileptic drugs is one based on anticonvulsant mechanisms of action. Numerous biochemical and electrophysiological processes have been shown to be modulated by antiepileptic drugs (Tables 2-7), but clear correlations of these effects with anticonvulsant activity are often difficult to determine. Macdonald and McLean (1986) have proposed that the ability of phenytoin, carbamazepine, phenobarbital, valproate, and benzodiazepines to limit sustained, high-frequency, repetitive firing (SRF) of neurons may be an important mechanism involved in the action of these drugs against generalized tonic-clonic seizures and convulsive status epilepticus. Ethosuximide, which is not effective against generalized tonic-clonic seizures, is not able to modify SRF (Macdonald and McLean 1986). SRF is a nonsynaptic property of neurons and reduction of SRF by antiepileptic drugs may be mediated by effects of these drugs on ion currents (Tables 2-6). Indeed, all drugs which are active against SRF have been shown to reduce inward sodium and/or calcium currents. The effect of these drugs on the major voltage-dependent calcium channels is not shared by ethosuximide (DeLorenzo 1988), but the latter has recently been shown to reduce thalamic low threshold activated (T-type) calcium currents. This effect could be important for the anti-absence action of ethosuximide (Coulter et al. 1989).

Table 2. Pharmacological, neurophysiological, and neurochemical effects of phenytoin

---

Clinical activity   Partial seizures, generalized tonic-clonic seizures,
                        inactive against generalized "minor" seizures

Experimental activity   MES, electrical kindling, tonic chemoconvulsions,
                        no activity against clonic chemoconvulsions

Active brain concentrations 0.05-0.1 mM (clinically and experimentally)

Mechanisms   - Reduction of maximal rate of action potential depolarization
                        - Reduction of posttetanic potentiation
                        - Limitation of sustained, high frequency, repetitive firing of
                          sodium-dependent action potentials
                        - Use- or frequency-dependent *reduction of inward sodium current*
                        - Blockade of voltage-dependent calcium entry
                        - Inhibition of calmodulin activation of calmodulin kinase II and
                          thereby of calcium-dependent protein phosphorylation
                        - Effects on potassium conductance?
                        - Potentiation of postsynaptic effects of GABA
                        - Effects on the micromolar benzodiazepine receptor
                        - Effects on DHP but not on TBPS binding at the GABA receptor
                          complex (ligands for the barbiturate/picrotoxinin site)
                        - Reduction of excitatory amino acid release

---

For references see Macdonald and McLean (1986); Löscher (1987),
DeLorenzo (1988) and Löscher and Schmidt (1988)

196

Table 3. Pharmacological, neurophysiological, and neurochemical effects of carbamazepine

Clinical activity	Partial seizures, generalized tonic-clonic seizures, inactive against "minor" generalized seizures; antipsychotic activity
Experimental activity	MES, electrical kindling, tonic chemoconvulsions, no activity against clonic chemoconvulsions
Active brain concentrations	0.02-0.05 mM (clinically and experimentally)
Mechanisms	- Depression of posttetanic potentiation - Limitation of sustained, high-frequency, repetitive firing of action potentials - *Reduction of inward sodium and calcium currents* - Reduction of outward potassium currents - Inhibition of calmodulin activation of calmodulin kinase II and thereby of calcium-dependent protein phosphorylation - No potentiation of postsynaptic effects of GABA - Effects on benzodiazepine receptors (?) - Interactions with central adenosine receptors (?)

For references see Macdonald and McLean (1986), Löscher (1987), DeLorenzo (1988) and Löscher and Schmidt (1988)

Table 4. Pharmacological, neurophysiological, and neurochemical effects of phenobarbital

Clinical activity	Partial seizures, generalized tonic-clonic seizures, myoclonic seizures; ineffective against absence seizures; also used as sedative/hypnotic drug
Experimental activity	All types of experimental seizures
Active brain concentrations	0.05-0.2 mM (clinically and experimentally; (these concentrations also occur during long-term treatment with primidone)
Mechanisms	- General reduction of membrane excitability - Limitation of sustained, high frequency, repetitive firing of action potentials - Reduction of voltage-dependent sodium, calcium, and potassium conductances (at high concentrations) - Enhancement of slow outward potassium currents (at low concentration) - *Increase of chloride conductance* (by direct and indirect effects) - Reduction of glutaminergic and cholinergic excitation - Reduction of transmitter release - *Increase of GABA binding and function* - Inhibition of the enhancement of benzodiazepine binding by pentobarbital - Inhibition of DHP- and TBPS-binding (*barbiturate/picrotoxinin-site* of the GABA receptor-chloride ionophore complex)

For references see Macdonald and McLean (1986), Löscher (1987), DeLorenzo (1988) and Löscher and Schmidt (1988)

Table 5. Pharmacological, neurophysiological, and neurochemical effects of benzodiazepines

Clinical activity	Partial *and* generalized seizures, but development of tolerance during long-term treatment; primarily used as sedative/hypnotic and anxiolytic drugs
Experimental activity	All types of experimental seizures
Active brain concentrations	About 0.05 - 5 μM (clinically and experimentally)
Mechanisms	- Limitation of high-frequency, repetitive firing of action potentials - Reduction of inward sodium and calcium currents - Reduction of outward potassium currents - Inhibition of calmodulin activation of calmodulin kinase II and thereby of calcium-dependent protein phosphorylation - *Enhancement of GABA-produced chloride channel openings* via selective binding sites ("nanomolar benzodiazepine receptors") at the GABA receptor-chloride ionophore complex - In contrast to barbiturates no direct effect on chloride conductance - Decrease of voltage sensitive calcium conductance via "micromolar benzodiazepine receptors"?

For references see Macdonald and McLean (1986), Löscher (1987),
DeLorenzo (1988) and Löscher and Schmidt (1988)

Table 6. Pharmacological, neurophysiological, and neurochemical effects of valproic acid

Clinical activity	Partial *and* generalized seizures; antipsychotic effects
Experimental activity	All types of experimental seizures
Active brain concentrations	About 0.05-0.2 mM in humans, but 0.5-2 mM in rodents
Mechanisms	- Limitation of sustained, high frequency, repetitive firing of sodium dependent action potentials - Reduction of sodium and potassium channel gating (high concentrations) - Increase of potassium conductance (at high concentrations) - Potentiation of the inhibitory effects of GABA on neuronal firing - No direct effect on GABA, benzodiazepine or TBPS binding - *Increase of presynaptic GABA levels* (by enhancement of GABA synthesis or inhibition of GABA degradation or reduction of GABA turnover and release?) - Enhancement of GABA turnover in substantia nigra - Enhancement of GABA release from cultured neurons at 0.3 mM - Increase of CSF GABA levels in animals and humans - Effects on serotoninergic mechanisms?

For references see Macdonald and McLean (1986), Löscher (1987), DeLorenzo (1988),
Gram et al. (1988), Löscher and Schmidt (1988) and Löscher (1989a,b)

Table 7. Pharmacological, neurophysiological, and neurochemical effects of ethosuximide

---

Clinical activity	Absence and myoclonic seizures; ineffective against partial seizures and generalized tonic-clonic seizures
Experimental activity	Clonic chemoconvulsions; not active against MES or kindling
Active brain concentrations	About 0.3-0.7 mM in humans; higher in rodents
Mechanisms	- No effect on sustained, high frequency, repetitive firing of sodium-dependent action potentials - Blockade of sodium and potassium conductance (at high concentrations) - Reduction of low threshold (T-type) calcium currents in thalamic neurons - Reduction of GABA responses (at high concentrations) - absences associated with an excess of GABA-mediated activity ?

For references see Macdonald and McLean (1986), Löscher (1987), DeLorenzo (1988) and Coulter et al. (1989)

In addition to direct effects on voltage-dependent ion channels, another effect common to several antiepileptic drugs is potentiation of GABAergic inhibitory neurotransmission. This has been reported for phenytoin, phenobarbital, valproate, and the benzodiazepines, whereas carbamazepine was not active in this regard and ethosuximide, at high concentrations, even tended to reduce GABA responses (Tables 2-7). Thus, the suggestion of Macdonald and McLean (1986) that the ability of drugs to enhance GABAergic synaptic transmission may correlate with their efficacy against generalized absence seizure is not substantiated by the clinical activity of the respective drugs, since, e.g., phenobarbital enhances GABAergic transmission but aggravates absence seizures. Similarly, new drugs, such as vigabatrin, which selectively potentiate GABAergic function (see below), are effective against various seizure types, including complex partial seizures and generalized tonic-clonic seizures, but, at least in animal models, aggravate absence seizures (Löscher and Schmidt 1988). Moreover, it has been suggested that absences may be associated with an excess rather than a deficit of GABA-mediated activity (Mirsky et al. 1986). Thus, potentiation of GABAergic function appears to be an important mechanism for prevention or attenuation of generalized tonic-clonic and partial seizures but obviously not of absence seizures.

Besides effects on ion conductances and GABAergic transmission, other cellular and synaptic effects of antiepileptic drugs have been described (Tables 2-7). These effects may be related to adverse effects rather than to the antiepileptic action of the respective agents. For example, carbamazepine has been reported to act as an antagonist of central adenosine receptors, an effect which is shared by CNS stimulants, such as caffeine (Daval et al. 1989). This effect might be responsible for the proconvulsant action of carbamazepine observed at high doses both experimentally and clinically (Lerman 1986). Similarly, the effect of valproate on serotoninergic mechanisms (Table 6) is certainly not involved in valproate's mechanisms of anticonvulsant action, but may explain certain side-effects, e.g., "wet-dog shakes", observed in rodents.

In conclusion, the clearest pharmacological and neurophysiological evidence of antiepileptic drug mechanisms relates to direct effects on neuronal membrane ion conductances, notably voltage-dependent sodium and calcium inward currents, and to enhancement of GABAergic inhibitory neurotransmission.

It is by no means certain however that these are the major effects of antiepileptic drugs. For example, the mechanisms involved in effects of drugs on absence seizures are still unsettled.

New Mechanisms of Anticonvulsant Drug Action

A rational approach to the pharmacological control of epilepsy must be based on knowledge of the molecular and cellular events responsible for epilepsy. Based on experimental and clinical evidence that impairment of GABAergic transmission may be involved in the pathogenesis of certain types of epileptic seizures (Löscher 1989a), several drugs that selectively enhance GABAergic function have been developed; two of these drugs, vigabatrin and progabide, have already been evaluated clinically (Table 8). Both drugs are effective against different seizure types in humans (Löscher and Schmidt 1988), but experimental data indicate that, similar to benzodiazepines, there may be problems with tolerance and dependence (Löscher et al. 1989). In order to attenuate or abolish these problems, ligands with partial agonistic activity toward benzodiazepine receptors have been developed (Table 8). Indeed, the benzodiazepine receptor "antagonist" flumazenil (Ro 15-1788), which is known to exert partial agonistic activity in various animal models and neurophysiological preparations, seems to display antiepileptic properties in humans. Another direction in drug development was initiated by the discovery of receptor subtypes for excitatory amino acids, i.e. glutamate and aspartate. Antagonists of the N-methyl-D-aspartate (NMDA) receptor subtype (Table 8) have been shown to exert anticonvulsant effects in various animal models; however, MK-801, the first compound to be evaluated in humans, induced psychotic behavior. Whether the newly developed competitive NMDA antagonists listed in Table 8 are devoid of psychotomimetic effects must be ascertained. A third major strategy in drug development came from the increasing knowledge regarding calcium regulation of neuronal function. Centrally active calcium channel blockers, such as flunarizine, were shown to exert anticonvulsant effects against tonic seizures in animal models, but their clinical efficacy in epileptic patients is limited (Löscher and Schmidt 1988). The various other agents with diverse mechanisms of action listed in Table 8 have been discussed in detail elsewhere (Löscher and Schmidt 1988).

The problems associated with newly developed drugs illustrate the limitations of any rational drug design. For development of new drugs it is certainly necessary to use simple neurophysiological or pharmacological models; however, in contrast to a model in which "epileptic" discharges are induced by known electrical or chemical means, the mechanisms underlying the diverse seizure types in humans are mainly unknown. For example, SRF is an important model for studying the excitability of isolated neurons and may be used in the search of new anticonvulsant agents, but no direct evidence has demonstrated the link between SRF and epilepsy. Furthermore, one of the newly developed compounds, memantine (Table 8), blocks SRF in cultured neurons but induces

Table 8. Newly developed drugs with anticonvulsant action

Already in clinical evaluation:

- GABAmimetic drugs: vigabatrin, progabide
- Partial agonists at benzodiazepine receptors: e.g., ZK 95962 (a ß-carboline)
                                                          flumazenil (a benzodiazepine)
- Calcium channel blockers: e.g., flunarizine
- Noncompetitive NMDA-antagonists: MK-801
- Miscellaneous: gabapentin, stiripentol, denzimol, oxcarbazepine, milacemide,
                              zonisamide, felbamate, flupirtine, lamotrigine, nafimidone

In preclinical evaluation

- Orally active competitive NMDA antagonists: CGP 37849, CGP 39551
- Selective GABA-uptake inhibitors: SKF 100330-A and SKF 89976-A
- Miscellaneous: e.g., trans-2-en-valproate, memantine

For references see Meldrum and Porter (1986) and Löscher and Schmidt (1988)

convulsions in kindled animals, a model of human complex partial seizures (Hönack and Löscher 1989). Thus, it may be misleading to predict antiepileptic activity from effects determined in single models. Instead, a battery of seizure models, including chronic models such as kindling, should be used (Löscher and Schmidt 1988). In view of the possibility that the most clinically useful antiepileptic drugs are those that possess two or more mechanisms of action (see above), the current strategy to develop drugs with a selective effect on a single neurotransmitter system or ion channel function may yield drugs with a narrow spectrum of anticonvulsant activity.

References

Coulter DA, Huguenard JR, Prince DA (1989) Characterization of ethosuximide reduction of low-threshold calcium current in thalamic neurons. Ann Neurol 25: 582-593

Daval J-L, Deckert J, Weiss SRB, Post RM, Marangos PJ (1989) Upregulation of adenosine Al receptors and forskolin binding sites following chronic treatment with caffeine or carbamazepine: a quantitative autoradiographic study. Epilepsia 30: 26-33

DeLorenzo RJ (1988) Mechanisms of action of anticonvulsant drugs. Epilepsia [Suppl 2]: S35-S47

Frey H-H, Janz D (1985) Antiepileptic drugs. Springer, Berlin, Heidelberg, New York (Handbook of experimental pharmacology, vol 74)

Glaser GH, Penry JK, Woodbury DM (1980) Antiepileptic drugs: mechanisms of action. Raven, New York

Gram L, Larsson OM, Johnsen AH, Schousboe A (1988) Effects of valproate, vigabatrin and aminooxyacetic acid on release of endogenous and exogenous GABA from cultured neurons. Epilepsy Res 2: 87-95

Hönack D, Löscher W (1989) Memantine induces seizures in kindled but not in non-kindled rats. Naunyn-Schmiedeberg's Arch Pharmacol 340 [Suppl. II]: R75

Lerman P (1986) Seizures induced or aggravated by anticonvulsants. Epilepsia 27: 706-710

Löscher W (1987) Neurophysiologische und neurochemische Grundlagen der Wirkung von Antiepileptika. Fortschr. Neurol. Psychiat. 55: 145-157

Löscher W (1989a) GABA and the epilepsies. Experimental and clinical considerations. In: Bowery NG, Nistico G (eds) GABA - basic research and clinical applications. Pythagora, Rome, pp 260-300

Löscher W (1989 b) Valproate enhances GABA turnover in the substantia nigra. Brain Res 501: 198-203

Löscher W, Schmidt D (1988) Which animal models should be used in the search for new antiepileptic drugs? A proposal based on experimental and clinical considerations. Epilepsy Res 2: 145-181

Löscher W, Jäckel R, Müller F (1989) Anticonvulsant and proconvulsant effects of inhibitors of GABA degradation in the amygdala-kindling model. Eur J Pharmacol 163: 1-14

Macdonald RL, McLean MJ (1986) Anticonvulsant drugs: mechanisms of action. In: Delgado-Escueta AV, Ward AA, Woodbury DM, Porter RJ (eds) Basic mechanisms of the epilepsies. Molecular and cellular approaches. Raven, New York, pp 713-736

Meldrum BS, Porter JRJ (1986) New anticonvulsant drugs. Libbey, London

Mirsky AF, Duncan CC, Myslobodsky MS (1986) Petit mal epilepsy: a review and integration of recent information. J Clin Neurophysiol 3: 179-208

Mechanism of Block of Thalamic T-Type Ca^{2+} Channels by Petit Mal Anticonvulsants

D. A. Coulter, J. R. Huguenard and D. A. Prince

Department of Neurology and Neurological Sciences Stanford University Medical Center Stanford, CA 94305, USA

Introduction

The low frequency spike-and-wave discharge which is the electroencephalographic signature of typical absence or petit mal seizures is generated by an underlying thalamocortical oscillatory interaction (Gloor and Fariello 1988). A low-threshold Ca^{2+} conductance, found in virtually all thalamic neurons, has been shown to be of primary importance in the generation of thalamocortical oscillatory behavior (Steriade and Llinás 1988). We have previously reported that the petit mal anticonvulsants ethosuximide (Zarontin, Parke Davis), dimethadione (the active metabolite of Tridione, Abbott), and α-methyl-α-phenyl succinimide (MPS; the active metabolite of methsuximide, Celontin, Parke Davis), in clinically relevant concentrations, all blocked the low-threshold (T-type) Ca^{2+} current (Coulter et al. 1989; Coulter et al. 1989; Coulter et al. 1989) which generates the low-threshold Ca^{2+} spike in thalamic relay neurons (Coulter et al. 1989). All three anticonvulsants blocked the current by a similar mechanism, without affecting the time course of activation or inactivation, and without affecting steady-state inactivation. The reduction of T current by these anticonvulsants was voltage-dependent, with maximal block occurring at threshold to elicit the current (approximately -65 mV). Two possible mechanisms could explain this block of T current. Since neither the voltage dependence of probability of opening nor the voltage and time dependence of inactivation of the T-type channels were affected by anticonvulsant exposure (as assessed through recordings of properties of block of the whole-cell current), we hypothesized that these anticonvulsants could be acting by either reducing the single channel conductance or by reducing the number of T-type channels available to be activated.

Methods and Results

In order to differentiate between these two possibilities, we employed ensemble fluctuation analysis, a form of noise analysis developed by Sigworth (Singworth 1980), and applied to Ca^{2+} currents by Bean and colleagues (Bean et al. 1984), to study the trace to trace variance in Ca^{2+} current records under control and drug-exposed conditions. These fluctuations are due to the opening and closing of single channels, and properties of these channels can be studied by analyzing this noise. Comparing sequential traces activated by identical depolarizing responses, one can quantify the fluctuations evident in the records by digitally subtracting the two traces (to remove the DC component of whole-cell current), squaring the resulting difference trace (to compute the variance), and dividing by two (to account for the combination of variance from two traces). Averaging many of these difference traces together gives an estimate of the mean change in variance with time as the whole-cell Ca^{2+} current activates and inactivates. By plotting this variance against the whole-cell Ca^{2+} current amplitude, a graph is generated which can be fitted by the function:

$$\sigma^2 = iI - I^2/N$$

where σ^2 is the variance, i is the single channel current, I is the mean current, and N is the number of channels (assuming that there is one non-zero conductance state for the channel and that the

Experimental Brain Research Series 20
© Springer-Verlag Berlin · Heidelberg 1991

channels are independent). This analysis provides information about the single-channel current and the number of channels involved in generation of the whole-cell current, since the variance and the mean current are known.

Whole-cell voltage-clamp techniques were used to assess anticonvulsant actions on Ca^{2+} currents of enzymatically isolated (Coulter et al. 1989) thalamic cells of young (1 - 15 day old) rat pups. Intracellular and extracellular solutions were designed to ionically and pharmacologically isolate Ca^{2+} currents (Coulter et al. 1989). An ATP reconstitution system was included in the electrode solution to improve cell viability (Forscher and Oxford (1985). Drugs were applied by a concentration clamp technique, which allowed solution changes in <<1 s. MPS was chosen as the anticonvulsant to test using ensemble fluctuation analysis, since the concentration-dependence of its actions on T current has been fully characterized (Coulter et al. 1989) and it has greater maximal efficacy than ethosuximide (100% vs 40%, (Coulter et al. 1989) and (Coulter et al. 1989)). All fluctuation analyses were performed on currents elicited by a command step to -40 mV from a holding potential of -100 mV. This step activates only T current since: (1) both the N and L type Ca^{2+} currents have been reported to have a more depolarized threshold than -40 mV (Carbone and Lux 1987; Coulter et al. 1989; Fox et al. 1987), and (2) the Ca^{2+} current activated by this step command steady-state inactivates as a single component (Coulter et al. 1989). This satisfies the fluctuation analysis requirement for activation of an independent population of T channels with a single open state (Singworth 1980).

Figure 1 illustrates the results of ensemble fluctuation analysis of MPS block of T current. Application of 750 M MPS resulted in a 37% block of T current (Fig. 1A) without affecting the time course of the current, as has been reported previously (Coulter et al. 1989). From analysis of the fluctuations during the rising phase of the current (between the arrows in Fig. 1A), it is evident that the variance of these fluctuations increases as the amplitude of the current increases, as expected, since these fluctuations are due to the opening and closing of individual T-type Ca^{2+} channels, the frequency of which increases as the whole-cell current increases (Fig. 1B). Plotting the mean variance vs the mean whole-cell current under control conditions, the curve can be fitted with the equation described above, using a single channel current (i) of 0.15 pA, and a number of channels estimate (N) of 4750 (Fig. 1C). Following drug application, the mean current was reduced by 37%. Assuming this reduction was accomplished by either blocking 37% of the channels (reducing the N from 4750 to 3000) or by reducing the single channel conductance 37% (reducing the i from 0.15 to 0.09 pA), the plot of mean current vs variance during drug exposure was fitted with these two models. It is evident that a model with a reduced number of T channels (solid line in Fig. 1D) fits the data much better than one with a reduction in the single channel current (dashed line in Fig. 1D). Thus, MPS appears to block T current by reducing the number of channels available to be activated.

Discussion
We have previously shown that the structurally similar petit mal anticonvulsants ethosuximide, MPS, and dimethadione all block T current in thalamic neurons in a similar voltage-dependent manner (Coulter et al. 1989; Coulter et al. 1989; Coulter et al. 1989). On the basis of the above fluctuation analysis data, and assuming that all three anticonvulsants are acting by molecular mechanisms similar to that of MPS, we conclude that the mechanism of block of T current by these anticonvulsants is by reducing the number of channels available to be activated without affecting the voltage-dependence of activation or inactivation or the single channel conductance. This type of block has been termed closed channel block.

Acknowledgements. We thank Edward Brooks for technical assistance. This work was supported by NIH grants NS06477 and NS12151 and the Morris and Pimley research funds.

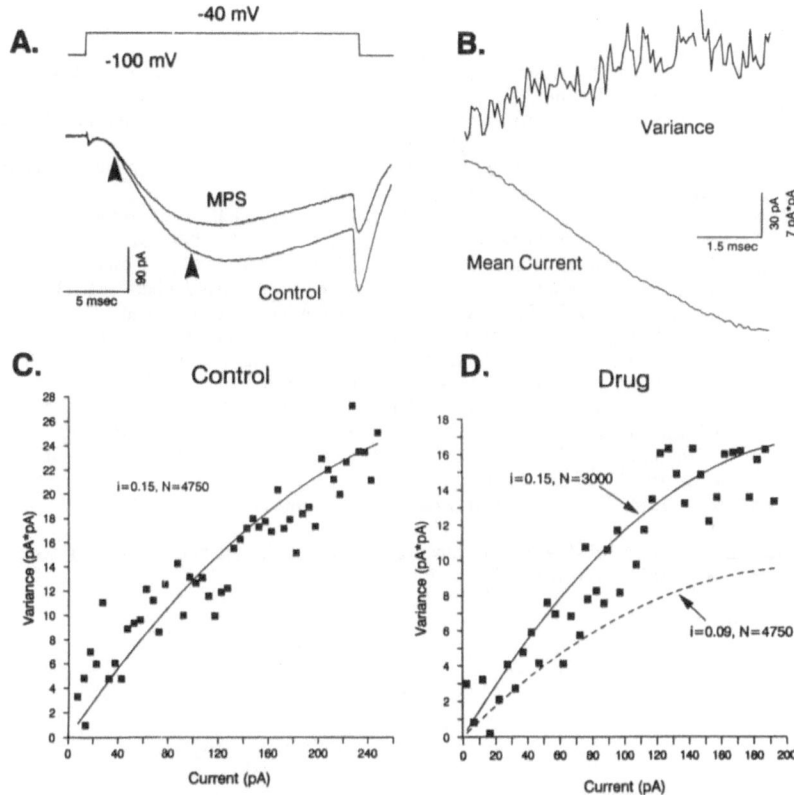

Fig. 1A-D. Ensemble fluctuation analysis of α-methyl-α-phenyl succinimide (MPS) block of T current in thalamic neurons. **A** Application of 750 *M* MPS resulted in a reduction of the T current of 37%. The T current was elicited by a step command to -40 mV from a holding potential of -100 mV (*top*). Ensemble fluctuation analysis was performed on the rising phase of the current (portion between the *arrows*). **B** Average variance (in pA², mean of 94 difference traces, *top sweep*) during the rising phase of the whole-cell Ca²⁺ current (in pA, *bottom trace*). Note that as the whole-cell Ca²⁺ current increases in amplitude, the fluctuation variance increases. **C** Plot of mean variance vs mean whole-cell current under control (wash from MPS) conditions. Background variance (due to leak and thermal noise) was subtracted from the curve by fitting the initial, linear portion of the trace by regression and using the y intercept (variance with 0 whole-cell Ca²⁺ current) as the background fluctuation variance of the cell at rest. The plot was then fitted to the ensemble fluctuation equation (see text) using a single channel current estimate (*i*) of 0.15 pA, and a channel number estimate (*N*) of 4750. The variance is the mean of 94 difference traces, averaged in 5 pA current bins in this plot. The mean whole-cell current is the average of 20 leak subtracted traces, 10 from immediately before the fluctuation analysis, and 10 from immediately subsequent to the analysis. **D** Plot of mean variance versus mean whole-cell Ca²⁺ current under MPS-exposed (750 μM) conditions (calculated as in C). The plot was fitted assuming that the 37% reduction in whole-cell Ca²⁺ current was caused either by a 37% reduction in the single channel current (*i*, from 0.15 to 0.09 pA, *dashed line*), or in the number of channels (*N*, from 4750 to 3000, *solid line*). Note that the data is best fitted by assuming a reduced number of channels

References

Bean BP, Nowycky MC, Tsien RW (1984) β-Adrenergic modulation of calcium channels in frog ventricular heart cells. Nature 307: 371-375

Carbone E, Lux HD (1987) Kinetics and selectivity of a low-voltage-activated calcium current in chick and rat sensory neurones. J Physiol 386: 547-570

Coulter DA, Huguenard JR, Prince DA (1989) Specific petit mal anticonvulsants reduce calcium currents in thalamic neurons. Neurosci Lett 98: 74-78

Coulter DA, Huguenard JR, Prince DA (1989) Characterization of ethosuximide reduction of low-threshold calcium current in thalamic neurons. Ann Neurol 25: 582-593

Coulter DA, Huguenard JR, Prince DA (1989) Differential effects of petit mal anticonvulsants and convulsants on thalamic neurones. I. Calcium current reduction. Brit J Pharmacol (1990) In Press

Coulter DA, Huguenard JR, Prince DA (1989) Calcium currents in rat thalamocortical relay neurones: kinetic properties of the transient, low-threshold current. J Physiol 414: 587-604

Forscher, P, Oxford GS (1985) Modulation of calcium channels by norepinephrine in internally dialyzed avian sensory neurons. J Gen Physiol 85: 743-763

Fox AP, Nowycky MC, Tsien, RW (1987) Kinetic and pharmacological properties distinguishing three types of calcium currents in chick sensory neurones. J Physiol 394: 149-172

Gloor P, Fariello RG (1988) Generalized epilepsy: some of its cellular mechanisms differ from those of focal epilepsy. TINS 11: 63-68

Sigworth FJ (1980) The variance of sodium current fluctuations at the node of Ranvier. J Physiol 307: 97-129

Steriade M, Llinás RR (1988) The functional states of the thalamus and the associated neuronal interplay. Physiol Rev 68: 649-742

CGP 37849 / CGP 39551:
The First Competitive NMDA Receptor Antagonists Suitable for Oral Antiepileptic Therapy

M. Schmutz, C. Portet, H. R. Olpe, K. Klebs, A. Jeker, R. Heckendorn, G. E. Fagg and H. Allgeier

Research and Development Department, Pharmaceuticals Division, CIBA-GEIGY Ltd, CH-4002 Basel, Switzerland

Introduction

N-methyl-D-aspartate (NMDA) receptors are believed to be implicated in the pathophysiology of neurological disorders, such as brain ischemia and epilepsy (see Meldrum 1985; Cavalheiro et al. 1988). For example, they are involved in burst firing and the generation of seizure activity (see Flatman et al. 1983; Herrling et al. 1983; Herron et al. 1985; Slater et al. 1985; Dingledine et al. 1986). Among the various sites of the NMDA receptor complex, the transmitter recognition site and the channel site have been well-characterized. Competitive and noncompetitive ligands can suppress NMDA-induced spiking in vitro and prevent convulsions in animal models of epilepsy;however, the therapeutics use of non competitive receptor antagonists seems limited by their unwanted side effects (Meldrum 1985; Lehmann et al. 1988; France et al. 1989), and orally active competitive antagonists have been unavailable to date.

Here we describe the anticonvulsant properties of a novel, competitive, NMDA receptor antagonist, CGP 37849 (DL-(E)-2-amino-4-mcthyl-5-phosphono-3-pentenoic acid; 4-methyl-APPA), and its carboxy-ethylester, CGP 39551 (Fig. 1). Both compounds are the first competitive antagonists with potent anticonvulsant activity following oral administration and are therefore promising candidates for antiepileptic therapy. Preliminary results from this study have been presented in abstract from (Fagg et al. 1988; Schmutz et al. 1988b).

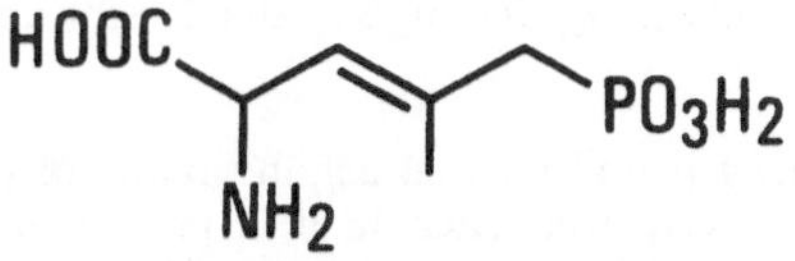

CGP 37849
DL-(E)-2-amino-4-methyl-5-phosphono-3-pentenoic acid
(DL-4-methyl- APPA)

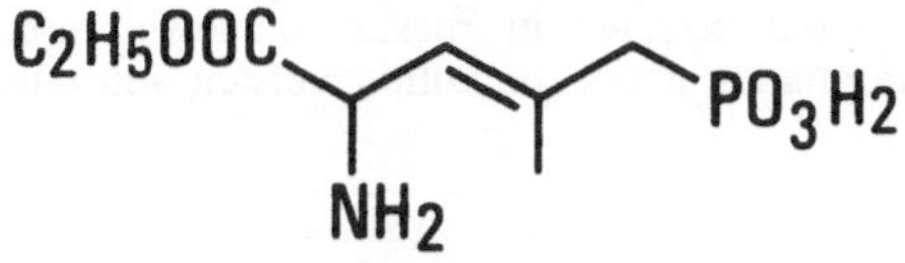

CGP 39551
DL-4-methyl-APPA-ethylester

Fig. 1. Chemical structure of CGP37849 and CGP 39551

The experiment were performed on male mice (Tif: MAGf (SPF)) and male albino rats (Tif: RAI). The animals were caged in groups of 5 - 10 (mice) or 1 - 5 (rats) and allowed water and foot ad libitum. They were kept in an air-conditioned room at 21° - 22°C, with a 12 h light/dark cycle.

Results

In radioligand binding assays CGP 37849 and CGP 39551 inhibited the binding of the specific NMDA receptor antagonist, [³H] CPP, with Ki values of 35 and 310 nM, respectively. Both compounds were weak or inactivate in assays for other neurotransmitters or modulators, including the quisqualate-, kainate- and strychnine-insensitive glycine binding site.

Concerning animal models of convulsions, CGP 37849 was effective in protecting against electroshock-induced tonic hind limb seizures in mice (stimulation parameters: 50 Hz, 10 mA, 0.2 s, alternating electrical current; corneal electrodes; 5 - 10 animals per dose; statistics: Spearman-Kaerber, see: Cavalli-Sforza, 1972), following pretreatment periods 0f 1 - 8 h (Table 1). After an optimum pretreatment time for oral administration, i.e., 4 h, the ED_{50}s ranged from 12.6 to 21,0 mg/kg po (median: 17 mg/kg; four separate experiments). CGP 37849 was more potent by factor of about 15 in this test when given intraperitoneally or intravenously: The ED_{50}s amounted to about 1 mg/kg i.p. or i.v.. The maximal effect was observed after pretreatment periods of 2 h (0.4 - 1.8 mg/kg i.p.; median 1.0 mg/kg; four separate experiments) and 1 h (0.6 - 0.9 mg/kg i.v.; median: 0.8 mg/kg; three separate experiments), respectively.

CGP 39551 was more potent than CGP 37849 following oral administration to mice: The ED_{50}s ranged between 4 and 15.8 mg/kg po after a pretreatment periods of 1 - 24 h. Thereafter, its potency decreased steadily; whereas the ED_{50} after 48 h of pretreatment time amounted to 41 mg/kg po, no protection at 100 mg/kg po was registered following a pretreatment period of 72 h. When given intraperitoneally or intravenously, the anticonvulsive potency of CGP 39551 was comparable to that following oral administration (Table 1). Although lowest ED_{50} values were obtained after 2 h (2.7 mg/kg i.v.) and 4 h (5.2 and 8.7 mg/kg i.p.) pretreatment, the compound still was active following pretreatment periods of 8 and 24 h.

In the electroshock test in rats (50 Hz, 36 mA, 0.63 s; other parameters: see mouse data) similar results were obtained with both compounds (Table 1). After a pretreatment period of 4 h the ED_{50} of GCP 37849 ranged between 8 and 22 mg/kg po (median: 10 mg/kg; five separate experiments). The optimum pretreatment times were 1 h for i.v. (ED_{50}: 0.8 mg/kg) and 2 h for i.p. administration (ED_{50}s of 1.7 and 2.4 mg/kg).

As with CGP 37849, the optimum pretreatment period for oral administration of CGP 39551 was also 4 h (ED_{50}s of 3.7 and 8.1 mg/kg); however, even after 48 h, a protective activity against electroshock-induced seizures in rats was present (ED_{50}: 54 mg/kg po). The highest potencies for i.v. (ED_{50} 4.3 mg/kg) and i.p. administration (ED_{50}s 2.8 and 2.9 mg/kg) were measured following pretreatment times of 2 and 4 h.

CGP 37849 and CGP 39551 suppressed seizures induced by electric shock more potently than those induced by pentetrazole or strychnine. CGP 39551 was also tested at dosages of 5 - 40 mg/kg/day po and proved effective against chronic partial seizures in rhesus monkeys implanted with aluminium hydroxide (for more detailed information about the anticonvulsant and unwanted side effect profile see Schmutz et al. 1990).

With regard to rat amygdaloid kindling evolution (for methodological details see Schmutz et al. 1988a), CGP 37849 at 60 mg/kg po and 6 mg/kg i.p. was hardly effective at all. By contrast, CGP 39551 showed clear effects: It moderately delayed afterdischarge development at 30 mg/kg po daily, and markedly diminished the severity of behavioral symptoms at doses as low as 10 mg/kg po. The suppression seen during the 13 day administration period was highly significant and comparable to that of, e.g., phenobarbital and clonazepam.

Table 1. Anticonvulsant activity of CGP 37849 and CGP 39551 against electroshock-induced seizures in mice and rats

			CGP 37849				CGP 39511	
mouse	1 h	po	30.0	-42.0	(3)		10.0	
		i.p.		1.1			n.t.	
		i.v.	0.6	-6.9	(3)		3.5	
	4 h	po	12.6	-21.1	(4)	4.0	-5.5	(2)
		i.p.		2.0		5.2	-8.7	(2)
		i.v.		2.7			7.4	
	8 h	po		18.0			15.8	
		i.p.	0	15.0			13.2	
		i.v.	ca.	6.0		ca.	12.0	
rat	1 h	po		60.0		0	20.0	
		i.p.		1.9			>10.0	
		i.v.		0.8	(2)		4.8	
	4 h	po	8.0	-22.0	(6)	3.7	-8.1	(2)
		i.p.	2.5	-3.0	(2)	2.8	-2.9	(2)
		i.v.		3.0			4.3	
	8 h	po	25.0	-28.0	(2)		8.0	
		i.p.	ca.	30.0			4.4	
		i.v.	0	8.0			7.0	

Values are ED_{50}s in mg/kg (respectively lowest-highest ED_{50} when more than 1 experiment was done). Numbers in brackets: 2-6 experiments were done under identical conditions.
1h, 4h, 6h, pretreatment period in hours; po, i.p, i.v., oral, intraperitoneal, intravenous administration.
>, ED_{50} not reached at indicated dose; 0, no protection at indicated dose; n.t., not tested

In in vivo electrophysiological studies on anesthetized rats, iontophoretic administration of either NMDA, quisqualate, or kainate to rat CA1 hippocampal pyramidal neurons induced spiking with a frequency of 5 - 8 spikes per second. CGP 37849 (100 mg/kg po) and CGP 39551 (30 and 100 mg/kg po), dissolved in physiological saline solution and administered by means of cannula into the esophagus, selectively suppressed the response evoked by NMDA, leaving the quisqualate- and kainate-induced spiking unaffected (Fig. 2).

Concerning unwanted side effects, the most prominent one was ataxia. In two- to fourfold above the ED_{50} values for the protection against electroshock-induced seizures. The therapeutic ratios based on the rotarod and electroshock tests in mice and rats were about 3 for CGP 37849 and for CGP 39551.

Discussion

The result reported here show that the specific and competitive NMDA antagonists, CGP 37849 and CGP 39551 exhibited potent and long-lasting anticonvulsant activity following oral administration. They are thus the first competitive NMDA antagonists suitable for oral antiepileptic therapy.

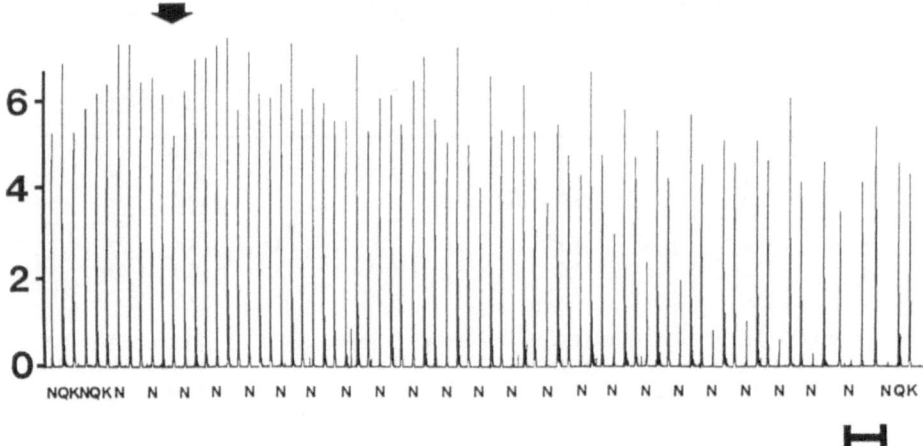

Fig. 2. CGP 39551 at 100 mg/kg po selectively blocks NMDA-induced spiking in the rat hippocampus. Ratemeter record showing increases in the firing rate of CA1 hippocampal neurons in the anesthetized rat evoked by iontophoretically applied NMDA (*N*), quisqualate (*Q*) or kainate (*K*). CGP 39551 was administered orally at the time indicated (*arrow*). *Ordinate*: Number of spikes per second. *Abscissa*: Time in min. (⊢--⊣ = 10 minutes)

Both compounds bind potently a specifically to NMDA receptors (see also Fagg et al. 1990). Specificity for the NMDA-subtype of glutamate receptor was also demonstrated electrophysiologically: CGP 37849 and CGP 39551 suppressed, under in vivo and in vitro conditions, only spiking induced by NMDA and not that evoked by kainate and quisqualate (see also {Pozza et al. 1990).

The potency and duration of anticonvulsants effect of both compounds in preclinical models compare well with that of standard antiepileptic drugs, such as carbamazepine or phenytoin. Their effect profile, i.e., activity against electroshock-induced tonic-clonic seizures and against chronic partial seizures in rhesus monkeys, indicates therapeutic efficacy against generalized tonic-clonic partial seizures. Suppression of rat kindling evolution by CGP 39551 suggest that the compound may have an anti-absence and/or anti-myoclonic component (see Schmutz et al. 1988a). This assumption is strengthened by the findings of Chapman et al. (1990) demonstrating that both compounds are highly effective against myoclonus in DBA/2 mice sensitive to audiogenic stimuli and in photosensitive papio papio-baboons.

The therapeutic window and quality of the unwanted side effects do not differ substantially from standard antiepileptic drugs do not seem to be prohibitive for the use of both compounds as antiepileptics (see Schmutz et al. 1990). The most prominent unwanted effects were ataxia and muscle relaxation; no generalization to MK-801 and ketamine cues was seen in rhesus monkeys (France and Woods, personal communication). Furthermore, no direct detrimental effects on learning and memory have been observed (Mondadori et al. personal communication).

In conclusion, CGP 37849 and CGP 39551 are the first selective, competitive NMDA antagonists with potent and long-lasting oral anticonvulsant activity. They are thus promising candidates for novel antiepileptic therapy in humans.

209

References

Cavalheiro EA, Lehmann J, Turski L (1988) Frontiers in excitatory amino acid research. Liss, New York
Cavalli-Sforza L (1972) Biometrie (Grundzuege biologisch-medizinischer Statistik). Fischer, Stuttgart
Chapman AG Graham JL, Patel S, Meldrum BS (1990) Anticonvulsant activity of two orally active competitive NMDA antagonists, CGP 37849 and CGP 39551 against sound-induced in DBA/2 mice and photically-induced myoclonus in papio papio. Epilepsia (to be published)
Dingledine R, Hynes MA, King GL (1986) Involvement of N-methyl-D-aspartate receptors in epileptiform bursting in the rat hippocampal slice. J Physiol 380: 175-189
Fagg GE, Baud J, Pozza MF, Olpe HR, Schmutz M, Baumann P, Bittiger H, Allgeier H, Heckendorn R, Angst C, Brundish D, Dingwall JG (1990) CGP 37894 and CGP39551: novel and potent competitive N-methyl-D-aspartate receptor antagonists with oral activity. Br J Pharmacol (in press)
Fagg GE, Olpe HR, Bittiger H, Schmutz M, Angst C, Brundish D, Allgeier H, Heckendorn R, Dingwall JG (1988) CGP 37894/CGP 39551: potent and selective competitive NMDA receptor antagonists with oral activity. Soc Neurosci Abstr 14, part 2: 941
Flatman JA, Schwindt PC, Crill WE, Strafstrom CE (1983) Multiple actions of NMDA on cat neocortical neurones in vitro. Brain Res 266: 169-173
France CP, Woods JH, Ornstein P (1989) The competitive NMDA antagonist CGS 19755 attenuates the rate-decreasing effects of NMDA in rhesus monkeys without producing ketamine-like discriminative stimulus effects. Eur J Pharmacol 159: 133-139
Herrling PL, Morris R, Salt TE (1983) Effects of excitatory amino acids and their antagonists on membrane and action potentials of cat caudate neurones. J Physiol 339: 207-222
Herron CE, Williamson B, Collingridge GL (1985) A selective N-methyl-D-aspartate antagonist depresses epileptiform activity in rat hippocampal slices. Neurosci Lett 61: 255-166
Lehmann J, Hutchinson AJ, McPherson SE, Mondadori C, Schmutz M, Sinton CM, Tsai C, Murphy DE, Steel DJ, Williams M, Cheney DL, Wood PL (1988) CGS 19755, a selective and competitive N-methyl-D-aspartate-type excitatory amino acid receptor antagonist. J Pharmacol Exp Ther 246: 65-75
Meldrum BS (1985) Possible therapeutic applications of antagonists of excitatory amino acid neurotransmitters. Clin Sci 68: 113-122
Pozza MF, Olpe HR, Brugger F, Fagg GE (1990) Electrophysiological characterization of a novel potent and orally active NMDA-receptor antagonist: CGP 37849 and its ethylester CGP 39551. Eur J Pharmacol (in press)
Schmutz M, Klebs K, Baltzer V (1988a) Inhibition or enhancement of kindling evolution by antiepileptics. J Neural Transm 72: 245-257
Schmutz M, Klebs K, Olpe HR, Fagg GE, Allgeier H, Heckendorn R, Angst C, Brundish D, Dingwall JG (1988b) CGP 37849/CGP 39551: Competitive NMDA receptor antagonists with potent oral anticonvulsant activity. Soc Neurosci Abstr 14, part 2: 864
Schmutz M, Portet C, Jeker A, Klebs K, Vassout A, Allgeier H, Heckendorn R, Fagg GE, Olpe HR, van Riezen H (1990) The competitive NMDA receptor antagonists CGP 37849 and CGP 39551 are potent, orally-active anticonvulsants in rodents. Naunyn-Schmideberg's Arch Pharmacol (in press)
Slater NT, Stelzer A, Galvan M (1985) Kindling-like stimulus pattern induce epileptiform discharges in the guinea pig in vitro hippocampus. Neurosci Lett 60: 25-31

Effects of Anticonvulsant Compounds on Voltage-and Neurotransmitter-Activated Sodium Conductances of Central Neurons in Cell Culture

M. J. McLean and A. W. Wamil

Department of Neurology, Vanderbilt University Medical Center, 2100 Pierce Ave., Nashville, Tennessee 37212, USA

Introduction

Multiple cellular mechanisms of action may contribute to the therapeutic efficacy of anticonvulsant compounds. Important effects of clinically used compounds have been demonstrated on voltage- and neurotransmitter-activated ionic conductances. The ability of phenytoin (McLean and Macdonald 1983), carbamazepine (McLean and Macdonald 1986), and valproic acid (McLean and Macdonald 1986) to limit high frequency sustained repetitive firing (SRF) of action potentials (APs) by use-dependent block of voltage-sensitive sodium channels (Hondeghem and Katzung 1984) parallels efficacy against generalized tonic-clonic and some partial seizures in man and against tonic hind limb extension in animal seizure models (Macdonald and McLean 1986; Macdonald and Meldrum 1989). Benzodiazepines (McLean and Macdonald 1988) and barbiturates (Macdonald and McLean 1986) share this effect at high concentrations achieved in the treatment of status epilepticus. Blockade of low threshold calcium current involved in paroxysmal bursting by ethosuximide may contribute to control of generalized absence seizures (Coulter et al. 1989; Prince et al. this volume). Pentylenetetrazol injection into animals serves as a pharmacological model of this clinical seizure type. Enhancement of chloride-dependent GABA-mediated inhibition by benzodiazepines and barbiturates (Macdonald and McLean 1986; Macdonald and Meldrum 1989) and blockade of N-methyl-D-aspartate (NMDA)-preferring glutamate receptor-activated cation conductances by MK-801 (Wong et al. 1986) are examples of anticonvulsant effects on neurotransmitter-gated conductances.

We will summarize here ongoing studies of the effects of clinically used (phenytoin, PT; valproic acid, VPA; ethosuximide, ES) and experimental (MK-801; nimodipine, NIM; flunarizine, FLU; memantine, MEM) anticonvulsant compounds on SRF and neurotransmitter responses of mouse spinal cord neurons using electrophysiological recording techniques.

Methods

Neurons were superfused with modified Dulbecco's phosphate-buffered saline (DPBS) containing 10 mM Mg^{2+} at 37°C to suppress spontaneous activity. Anticonvulsants, NMDA, and acetylcholine (Ach) were prepared as concentrated stocks in distilled water or DMSO and then diluted into DPBS for pressure application onto neurons from a blunt-tipped micropipette. Long (400 ms) depolarizing pulses applied through the recording microelectrode elicited SRF (Fig. 1 top, PRE). Action potential rate of rise (dV/dt), an indirect measure of inward sodium current, declined slightly to a steady level with continued firing of APs at the level of depolarization during the pulse. Pressure application for 15 - 30 s of PT and VPA, but not ES, at concentrations equivalent to therapeutic free (unbound to protein) serum levels led to reversible limitation of SRF accompanied by progressive reduction of dV/dt until firing ceased.

Experimental Brain Research Series 20
© Springer-Verlag Berlin · Heidelberg 1991

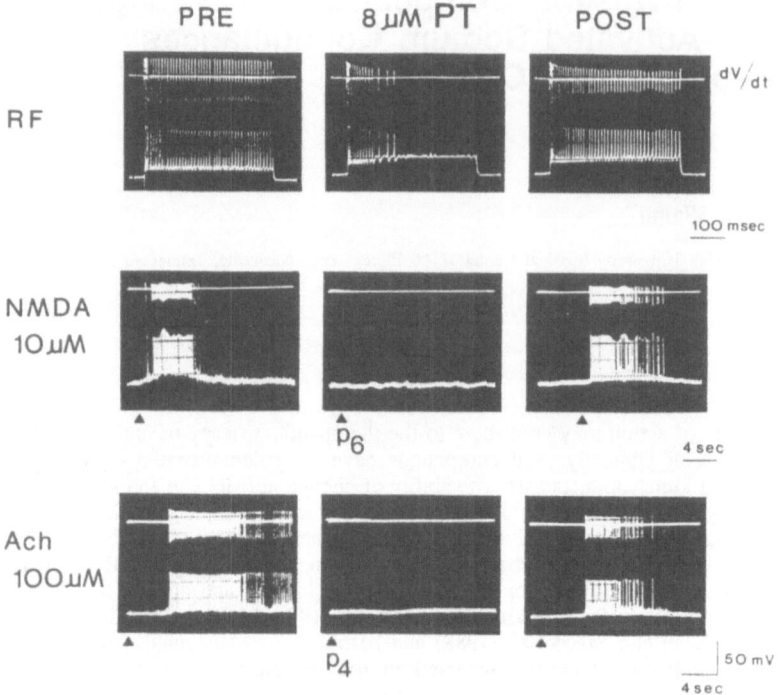

Fig. 1. Effects of phenytoin (*PT*) on repetitive firing (*RF*) and responses to NMDA and Ach by mouse spinal cord neurons in cell culture

Results

The effect of 8 μM PT is shown in Fig. 1 (top , center). Recovery occurred after washout (Fig. 1 top, POST). All four experimental compounds limited SRF. Brief (3 s) pressure applications of DPBS containing NMDA or acetylcholine (Ach) resulted in responses consisting of depolarization and firing of APs (Fig. 1, PRE). Coapplication of an anticonvulsant compound either reduced or had no effect on the response to NMDA or Ach at the concentrations tested. In general, higher concentrations of anticonvulsant compounds were needed to block completely responses to single applications of NMDA or Ach than responses to multiple applications. That is, the block was use-dependent. Figure 1 shows use-dependent blockade of responses to both 10^{-5} M NMDA (by the sixth application, P_6) and 10^{-4} M Ach (by P_4) by 8 μM PT. Anticonvulsants did not effect dV/dt during 3 s applications. The concentration-dependent effects of all compounds tested against SRF and single applications of 10^{-5} M NMDA or 10^{-5} M Ach are shown in Table 1, "yes" indicating blockade, "no" indicating inefficacy. Concentrations tabulated are those required for complete blockade or highest ineffective concentration tested. Pressure application of solution containing 10^{-6} M tetrodotoxin or either choline chloride or sucrose isosmotically substituted for NaCl abolished APs (see Nistri et al. 1985) and responses to NMDA and Ach reversibly.

Discussion

These results demonstrated that a variety of anticonvulsant compounds differentially blocked voltage- and neurotransmitter-gated activities dependent on the influx of sodium. Do these cellular effects predict the range of clinical therapeutic efficacy of the compounds tested? Compounds

Table 1. Effects of anticonvulsant compounds on sustained repetitive firing (SRF) and responses to one application of 10^{-5} M NMDA and 10^{-5} M Ach by mouse spinal cord neurons in monolayer dissociated cell culture. Concentrations shown are those leading to limitation of SRF, complete blockade of neurotransmitter responses, or highest ineffective concentration tested

	SRF	NMDA	Ach
PT	yes (8×10^{-6} M)	yes (8×10^{-5} M)	yes (8×10^{-6} M)
VPA	yes (6×10^{-5} M)	no (3×10^{-4} M)	yes (6×10^{-5} M)
ES	no (10^{-3} M)	no (10^{-3} M)	yes (5×10^{-4} M)
MK-801	yes (3×10^{-7} M)	yes (3×10^{-7} M)	yes (5×10^{-8} M)
NIM	yes (10^{-6} M)	yes (10^{-6} M)	yes (10^{-7} M)
FLU	yes (10^{-6} M)	yes (10^{-6} M)	yes (10^{-8} M)
MEM	yes (5×10^{-4} M)	yes (10^{-3} M)	yes (10^{-4} M)

effective acutely against generalized tonic-clonic and some partial seizures, namely PT and VPA, limited SRF at therapeutically relevant concentrations while ES, effective only against generalized absence seizures, did not. This was shown previously during exposures for up to 1 h (McLean and Macdonald 1983; McLean and Macdonald 1986). All three clinically used compounds blocked responses to Ach-induced excitation. This effect could contribute to the control of diverse seizure types by reducing net excitation, but does not seem to predict efficacy against any particular type.

Blockade of Ach could contribute also to toxicity, e.g., encephalopathy. Notably, PT, which may worsen generalized absence seizures, blocked NMDA-induced excitation. Compounds effective against generalized absence, namely ES and VPA, did not. This suggests that glutamatergic (NMDA subtype) excitation may be important in controlling generalized absence.

The experimental compounds MK-801 (Clineschmidt et al. 1982), NIM (Meyer et al. 1986), FLU, (Desmedt et al. 1975; Overweg et al. 1984) and MEM (McLean 1987; Meldrum et al. 1986) have been shown to be effective against electroshock seizures in animals. MK-801, FLU, and MEM were relatively ineffective against pentylenetetrazol-induced seizures. All four compounds blocked SRF. These findings predict clinical efficacy against generalized tonic-clonic and partial seizures, but not generalized absence. In limited clinical trials, FLU has been shown to be useful as an adjunctive agent in the treatment of refractory complex partial seizures of some patients (Overweg et al. 1984). Clinical trials with MK-801 have been suspended because of poor bioavailability; however, preliminary reports of data from the study suggested some efficacy against complex partial seizures (Troupin et al. 1986). Clinical trials with NIM and MEM have not yet been performed.

Occurrence in vitro at therapeutically relevant concentrations is the most compelling evidence that the observed cellular actions of clinically used compounds may contribute to anticonvulsant efficacy. Of greater mechanistic importance, however, is the finding that repeated applications of PT, MK-801, NIM, FLU, and MEM resulted in use-dependent block of responses to NMDA. Use-dependent block of NMDA responses was shown previously with patch clamp techniques (Huettner and Bean 1988). All compounds tested here blocked Ach responses in a use-dependent manner. This suggests that use-dependent block of various neurotransmitter-gated ion channels may be a common mechanism of action of many anticonvulsant compounds. It also suggests that dynamic electrophysiological assays performed under physiological conditions (e.g., with respect to ionic composition of solutions and temperature) may reveal cellular mechanisms of action not detectable by binding studies. Effects of anticonvulsants on multiple transmitter-gated processes could be explained by blockade of heterogeneous sodium-conducting channels gated by multiple receptors, i.e. non-competitive antagonism.

The cellular effects summarized here have not been demonstrated directly in patients or animals during seizures. Also, it is likely that other actions relevant to therapeutic efficacy remain to be identified. Despite these limitations, continued investigation of effects of anticonvulsant compounds on voltage- and neurotransmitter-gated processes controlling the epileptogenic focus may lead to a mechanistic understanding upon which to base drug selection for the treatment of specific seizure manifestations.

Acknowledgements. The authors thank Mr. Ronald Thomas for technical assistance and Ms. Rhonda Gibson for help in preparing the manuscript. MJM was a Mallinckrodt Scholar supported by grants NS 00817 and BRSG RR05424 from the National Institutes of Health and by a grant from Merz and Co., Frankfurt, West Germany. AWW was supported by a fellowship from the Cissy Patterson Trust.

References

Clineschmidt BV, Martin GE, Bunting PR (1982) Anticonvulsant activity of (+)-5-methyl-10,11-dihydro-5H-dibenzo(a,d)cyclohepten-5,10-imine (MK-801), a substance with potent anticonvulsant, central sympathomimetic, and apparent anxiolytic properties. Drug Dev Res 2: 123-134.

Coulter DA, Huguenard JR, Prince DA (1989) Characterization of ethosuximide reduction of low-threshold calcium current in thalamic neurons. Ann Neur 25: 582-593.

Desmedt LKC, Niemegeers CJE, Janssen PAJ (1975) Anticonvulsant action of cinnarizine and flunarizine in rats and mice. Arzneimittelforschung 25: 1408-1413.

Hondeghem LM, Katzung BG (1984) Antiarrhythmic agents: The modulated receptor mechanism of action of sodium and calcium channel-blocking drugs. Ann Rev Pharmacol Toxicol 24: 387-423.

Huettner JE, Bean BP (1988) Block of N-methyl-D-aspartate-activated current by the anticonvulsant MK-801: Selective binding to open channels. Proc Natl Acad Sci 85: 1307-1311.

Macdonald RL, McLean MJ (1986) Anticonvulsant drugs: Mechanisms of action. In: Advances in Neurology. Ed. Delgado-Escueta AV, Ward AA, Woodbury DM, Porter RJ. Raven Press, New York 44: 713-736.

Macdonald RL, Meldrum BS (1989) General principles: Principles of antiepileptic drug action. In: Antiepileptic Drugs. Ed. Dreifuss FE, Mattson RH, Meldrum BS, Penry JK. Raven Press, New York pp. 59-83.

McLean MJ (1987) In vitro electrophysiological evidence predicting anticonvulsant efficacy of memantine and flunarizine. Pol J Pharm Pharmacol 39: 513-525.

McLean MJ, Macdonald RL (1983) Multiple actions of phenytoin on mouse spinal cord neurons in cell culture. J Pharmacol Exp Ther 227: 779-784.

McLean MJ, Macdonald RL (1986) Sodium valproate, but not ethosuximide, produces use- and voltage-dependent limitation of high frequency repetitive firing of action potentials of mouse central neurons in cell culture. J Pharmacol Exp Ther 237: 1001-1011.

McLean MJ, Macdonald RL (1986) Carbamazepine and 10,11-epoxycarbamazepine produce use- and voltage-dependent limitation of rapidly firing action potentials of mouse central neurons in cell culture. J Pharmacol Exp Ther 238: 727-738.

McLean MJ, Macdonald RL (1988) Benzodiazepines, but not beta carbolines, limit high frequency repetitive firing of action potentials of spinal cord neurons in cell culture. J Pharmacol Exp Ther 244: 789-795.

Meldrum B, Turski L, Schwarz M, Czuczwar SJ, Sontag K-H (1986) Anticonvulsant action of 1,3-dimethyl-5-aminoadamantane. Pharmacological studies in rodents and baboon, Papio, Papio. Naunyn Schmiedeberg's Arch Pharmacol 332: 93-97.

Meyer FB, Tally PW, Anderson RE, Sundt TM, Yaksh TL, Sharbrough FW (1986) Inhibition of electrically induced seizures by a dihydropyridine calcium channel blocker. Brain Res 384: 180-183.

Nistri A, Arensen MS, King A (1985) Excitatory amino acid induced responses of frog motoneurons bathed in low Na^+ media: An intracellular study. Neuroscience 14: 921-929.

Overweg J, Binnie CD, Meijer JWA, Meinardi H, Nuijten STM, Schmaltz S, Wauquier A (1984) Double-blind placebo-controlled trial of flunarizine as add-on therapy in epilepsy. Epilepsia 25: 217-222.

Troupin AS, Mendius JR, Cheng F, Risinger MW (1986) MK-801. In: New anticonvulsant drugs. Ed. by Meldrum BS, Porter RJ. John Libbey, London, pp. 191-201.

Wong EHW, Kemp JA, Priestley T, Knight AR, Woodruf GN, Iversen LL (1986) The anticonvulsant MK-801 is a potent N-methyl-D-aspartate antagonist. Proc Natl Acad Sci (USA) 83: 7104-7108.

Memantine: Suppression of High Frequency Action Potentials Is Caused by Prolongation of the Sodium Current Refractory Period[+]

R. Netzer[1], T. Binscheck[1], R. Koch[2] and H. Bigalke[1]

[1] Medical School of Hannover, Department of Pharmacology and Toxicology, 3000 Hannover 61, FRG
[2] Merz and Co, 6000 Frankfurt-M. 1, FRG

Introduction

Intracellular recordings from cultured spinal cord neurons show a typical pattern of spontaneous synaptic activity. These cells randomly generate inhibitory and excitatory postsynaptic potentials, the latter often triggering action potentials (APs). Cultured spinal cord neurons have been widely used to investigate the mode of action of convulsants and anticonvulsants (Macdonald 1984, 1988). Strychnine, which blocks postsynaptic glycine receptors, alters spontaneous synaptic activity. The neurons develop a rhythmic pattern of bursts; the membrane potential depolarizes by 5 - 30 mV, accompanied by a rapid firing of APs with a frequency of 30 - 120 Hz. Memantine (3,5-dimethyladamantanamin), used in the treatment of different kinds of spasticity, has been shown to suppress seizures in animal models (Meldrum et al. 1986). Furthermore it reduces in various in vitro models different functions such as neurotransmitter responses in cell cultures (Reiser et al. 1988; Bormann 1989), membrane currents in snail neurons (Klee 1982), and muscle contractions (Masuo et al. 1986; Tsai et al. 1989). In spinal cord nerve cell cultures memantine reduces convulsant-induced hyperactivity (Netzer et al. 1988). To elucidate the underlying mechanisms of memantine action, voltage- and time-dependent membrane currents were analyzed by voltage clamp experiments. In this paper we describe blocking effects of memantine on sodium inward currents of spinal cord neurons, which may be the mechanism by which the agent reduces hyperactivity in vitro and in vivo.

Methods

Spinal cord neuron were obtained from 13 day old fetal mice and cultivated on collagen-coated plastic dishes using methods, adapted from Ransom et al. (1977). The experiments were performed on the stage of an inverted phase microscope at room temperature (23°C - 25°C). For intracellular recordings in current clamp mode, high resistance glass microelectrodes filled with 4 M potassium acetate were used (80 - 100 MΩ). Intracellular recordings in voltage clamp mode were performed using low-resistance glass micropipettes filled with a high potassium and low calcium solution (3 - 6 MΩ) (Fenwick et al. 1982). For investigation of the sodium current, other currents were blocked with intracellular cesium chloride (120 mM), tetraethylammonium (TEA, 6mM) and extracellular TEA (6 mM), cobalt chloride (1 mM) and 4-aminopyridine (50 µM). Potassium currents were analyzed when inward sodium and calcium currents were blocked by tetrodotoxin (300 nM) and cobalt chloride (2 mM). Signals were amplified by a standard patch clamp amplifier, low-pass filtered at 10 kHz, and digitally sampled at 25 kHz.

Experimental Brain Research Series 20
© Springer-Verlag Berlin · Heidelberg 1991

218

Results

Memantine Decreased the Duration of Strychnine-Induced Bursts and the Frequency of APs within a Burst

Effects of memantine were examined on cells exposed to strychnine (10 μM), which caused a rhythmic pattern of bursts within 10 min after application. The cultures were then superfused with minimal essential medium containing memantine at different concentrations. Memantine (10 - 100 μM) decreased the duration of bursts evoked by strychnine by 45% - 70%. The frequency of the bursts was unaffected (Fig. 1a). Furthermore, the mean frequency (number of APs within a burst divided by the burst duration) and peak frequency (mean frequency of APs during the first 100 ms of the burst) were decreased by memantine (10 - 100 μM, Fig. 1b). The decrease of the peak frequency, however, which was in the range of 100 Hz, was much more pronounced than the decrease of the mean frequency, which was in the range of 50 Hz.

Memantine Reversibly Decreased the Fraction of Open Sodium Channels (n_0) in a Concentration Dependent Manner

In order to obtain further insight into the blocking mechanism of memantine, the sodium current (I_{Na}) at different concentrations of memantine was recorded voltage-dependently. The fraction of open sodium channels (n_0) was calculated and plotted logarithmically against the applied voltage steps (Fig.2).

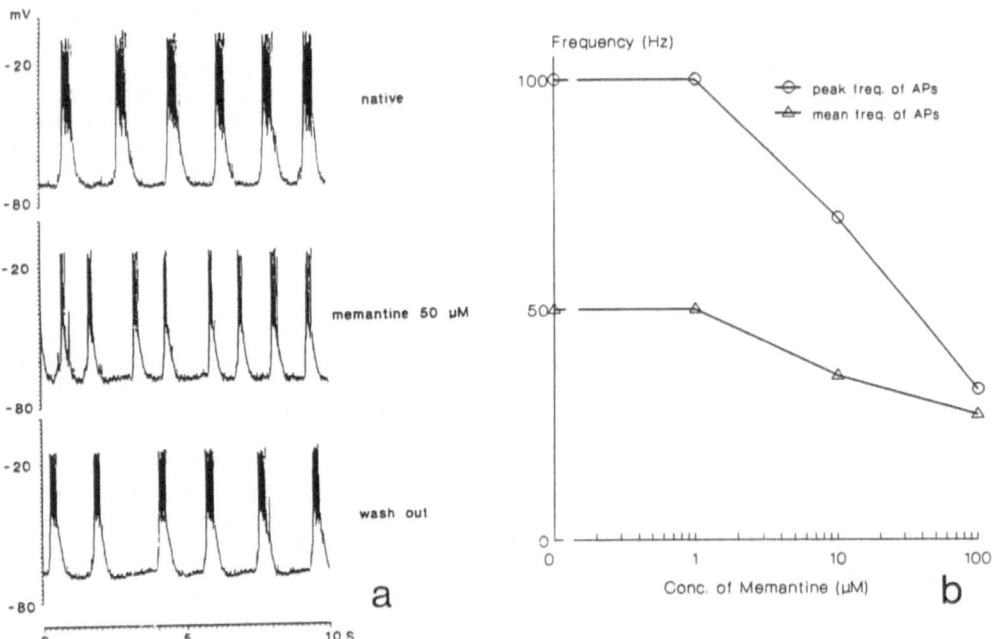

Fig. 1a,b. Alterations of paroxysmal depolarizing events by memantine. **a** Spontaneous activity of a neuron treated with strychnine (10 μM) and continuously superfused with minimal essential medium (*top trace*). The *center* and the *bottom trace* show the spontaneous synaptic activity of the same neuron during perfusion with memantine (50 μM) and after withdrawal of the drug. The resting membrane potential was -68 mV throughout the experiment. **b** Concentration response curves of memantine. *Triangles* represent frequencies of action potentials during strychnine-induced paroxysmal depolarizing events (mean frequency). *Open circles* represent frequencies during the first 100 ms of a paroxysmal depolarizing event (peak frequency)

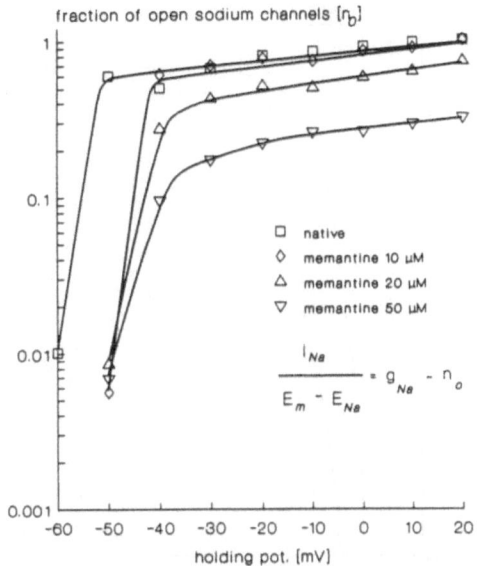

Fig. 2. Reduction of the number of open sodium channels by memantine. Neurons were voltage-clamped at command potentials from -60 to 20 mV for 50 ms. The membrane conductance for sodium ions (g_{Na}) was calculated using the formula indicated. I_{Na} represents the inward current evoked by the potential E_m. The equilibrium potential, E_{Na}, was 59 mV. g_{Na} at 20 mV was set as 1 (all channels open). g_{Na}, at the other potentials, was expressed as a fraction of open sodium channels (n_o). Cultures were continuously superfused within minimal essential medium containing memantine at the indicated concentrations

The steepness of the curve at threshold potentials (-60 mV to -50 mV) represents the limiting voltage sensitivity of the membrane conductance for sodium ions (g_{Na}), or n_0. Due to the logarithmic scaling of n_0, the limiting voltage sensitivity was expressed according to Hodgkin and Huxley (1952) as mV per e-fold, where e means the base of the natural logarithm. The slope within threshold potentials shows limiting voltage sensitivity of 0.3 mV per e-fold increase of g_{Na}. The steady state was reached at depolarizations between -50 and 20 mV. At 20 mV, n_0 was normalized to 1.0. Memantine decreased limiting voltage sensitivity in a concentration-dependent manner. At 50 μM the slope (between -60 and -50 mV) shows a limiting voltage sensitivity of 0.8 mV per e-fold increase of g_{Na}. Furthermore, at steady-state conditions (>-40 mV), n_0 was decreased over the whole examined range of depolarizing steps. The potassium current, (I_K), recorded voltage-dependently, was not affected by memantine (data not shown).

Memantine Prolonged the Relative Refractory Period of the Sodium Inward Current

To investigate the effects of memantine on the relative refractory period of I_{Na}, double pulses, each of 5 ms duration and 60 mV amplitude, were applied every 2 s. Between pulses the cells were clamped at -60 mV. Variations of the interval between the two pulses (2 - 28 ms) yielded frequencies ranging from 30 Hz to 143 Hz. The current elicited by the second pulse decreased in a frequency-dependent fashion (Fig. 3a). Memantine (10 μM) enhanced the decrease mainly at the higher stimulation frequencies (>50 Hz), whereas the first current was hardly affected.

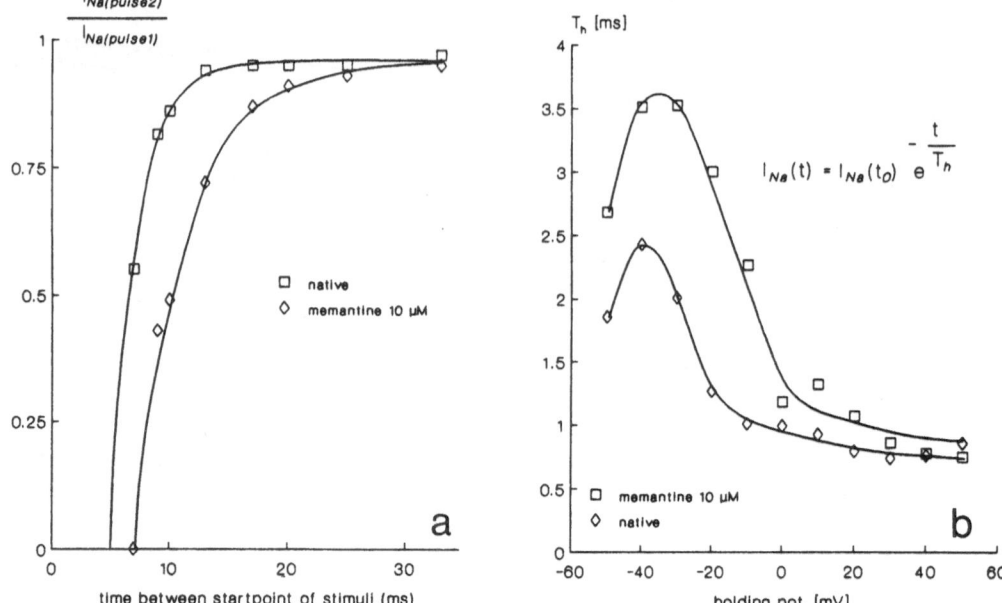

Fig. 3a,b. Prolongation of the relative refractory period of the inward current by memantine. **a** Two depolarizing pulses (5 ms each), separated by intervals between 2 and 28 ms (*abscissa*), were applied to the neuron. The quotient of the inward currents of pulse 2 and pulse 1 (*ordinate*) was calculated before and after superfusion of the culture with memantine. **b** Neurons were voltage-clamped at command potentials from -50 to 50 mV. The time constant, τ_h, was calculated for each E_m using the formula indicated. $I_{Na}(t_0)$ represents the peak sodium inward current and $I_{Na}(t)$, the sodium current at any time during its decay. The cultures were continuously superfused with minimal essential medium. In the figure T_h has been used for τ_h

Memantine Increased the Time Constant τ_h of Sodium Channel Closing/Inactivation

To obtain the time constant of the inactivation of sodium channels (τ_h), the decay of sodium inward current during 50 ms depolarizing pulses from -50 to 50 mV was calculated by a single exponential equation. τ_h was graphed as a function of the membrane potential (Fig. 3b). Memantine increased the time constant, τ_h, by approximately 50% reflecting a delay in the closing and/or inactivating of sodium channels. This effect was voltage-dependent. At positive membrane potentials the delay of closing/inactivating evoked by memantine was considerably reduced, indicating a voltage-dependent action of memantine on inactivation.

Discussion

Memantine, an antispastic agent, suppresses seizures in animals (Meldrum et al. 1986) and reduces repetitive firing evoked by strychnine (Netzer et al. 1988) or by current pulses (McLean 1987) in neuronal cell cultures. The paroxysmal depolarizing events (PDEs) are closely related to paroxysmal depolarizing shifts (PDSs) produced by convulsant-treated neurons in vivo (Frince 1968) which are considered to play an important role in the development of focal epilepsy. The reduction of the PDEs by memantine appears to be caused by a block of the sodium channels. The efficacy of memantine in suppressing APs improves as frequencies increase. The frequency-dependent block of APs could best be explained by a prolongation of the refractory period of I_{Na}. This prolongation is probably caused by a voltage-dependent increase in the time constant, τ_h. The sodium channel has been discussed as a target for many anticonvulsants, especially phenytoin and carbamazepine, which show effects comparable to memantine on current-evoked repetitive firing (Macdonald 1988; McLean 1987) and picrotoxin-induced PDEs (Netzer et al. 1989). It is concluded that memantine belongs to the same group of drugs that express their activity by interfering with the voltage-gated sodium channel (Schwarz and Grigat 1989). The clinical relevance of this finding is difficult to interpret at this stage, because memantine is not used in antiepileptic therapy. However, our results underline the importance of using neuronal cell culture systems not only for the testing of anticonvulsant drugs, but also for investigating their underlying mechanism of action.

Acknowledgements. We thank Drs G. Erdmann and J. Bormann for reading the manuscript. Work was supported by the "Deutsche Forschungsgemeinschaft" (Bi 274/3-1).

⁺This works is part of the PhD thesis of R.N.

References

Bormann J (1989) Memantine is a potent blocker of N-methyl-D-aspartate (NMDA) receptor channels. European J Pharmacol 166: 591

Fenwick E, Marty A, Neher E (1982) Sodium and calcium channels in bovine chromaffin cells. J Physiol 331: 599

Hodgkin AL, Huxley AF (1952) Currents carried by sodium and potassium ions through the membrane of giant axon of Loligo. J PHysiol 116: 449-472

Klee MR (1982) The effects of memantine on membrane properties and postsynaptic potentials in neurons of Aplysia californica. Drug Res 32: 1259

Macdonald RL (1984) Anticonvulsant and convulsant drug action on vertebrate neurons in primary dissociated cell cultures, electrophysiology of epilepsy. Academic, London p 353

Macdonald RL (1988) Anticonvulsant drug action on neurons in cell culture. J Neural Transm 72: 172-183

Masuo K, Enomoto K, Maeno T (1986) Effects of memantine on the frog neuromuscular junction. Europ J Pharmacol 130: 187

McLean ML (1987) In vitro electrophysiological evidence predicting anticonvulsant efficacy of memantine and flunarizine. Pol J Pharmacol 39: 513

Meldrum BS, Turski L, Schwarz M, Czuczwar SJ, Sontag KH (1986) Anticonvulsant action of 1,3-dimethyl-5-aminoadamantane. Naunyn Schmiedeberg's Arch Pharmacol 332: 93

Netzer R, Koch R, Bigalke H (1988) Memantine: electrophysiological evidence from spinal cord neurons in vitro for an anticonvulsant action. Naunyn Schmiedeberg's Arch Pharmacol [Suppl] 337: 458

Netzer R, Binscheck T, Bigalke H (1989) Phenytoin, baclofen, tizanidine and memantine reduce hyperexcitability of neurons in culture by interfering with different currents. Soc Neurosci Abstr 15:1301

Prince DA (1968) The depolarizing shift in "epileptic" neurons. Exp Neurobiol 21: 467

Ransom BR, Neale E, Henkart M, Bullock PN, Nelson PG (1977) Mouse spinal cord in cell culture. I. Morphology and intrinsic neuronal electrophysiologic properties. J Neurophysiol 40: 1132

Reiser G, Binmöller F J, Koch R (1988) Memantine (1-amino-3,5-dimethyladamantane) blocks the serotonin-induced depolarization response in a neuronal cell line. Brain Res in press

Schwarz JR, Grigat G (1989) Phenytoin and carbamazepine: Potential- and frequency-dependent block of Na currents in mammalian myelinated nerve fibers. Epilepsia 30(3): 286

Tsai MC, Chen ML, Lo SC, Tsai GC (1989) Effects of memantine on the twitch tension of mouse diaphragm. European J Pharmacol 160: 133

Different Dose-Dependent Actions of Memantine on Hippocampal Neurons

L. Wagner, M. R. Klee and M. L. Zeise

Max-Planck-Institut für Hirnforschung, Abteilung Neurophysiologie, 6000 Frankfurt, FRG

Introduction

Memantine (1,3-dimethyl-5-aminoadamantane), a derivative of aminoadamantane, is an antiviral agent against A2 influenza and against parkinsonism. It has been used mainly in the treatment of spastic symptoms, parkinsonism, and coma. Recently memantine has been considered as a new candidate for anticonvulsant therapy, since it raised the threshold in the classical maximal electroshock test for electrically induced seizures (Meldrum et al. 1986) and protected mice against the tonical phase of chemically induced convulsions (Chojnack-Wojcik et al. 1983). In addition, it decreased the amplitude of the sustained high-frequency repetitive firing of sodium-dependent action potentials (McLean 1989). Memantine also decreased the posttetanic potentiation of the amplitude of endplate potential and twitch tension of isolated mouse diaphragm (Tsai et al. 1989). However memantine could not prevent the myoclonical response to intermittent stroboscopic stimulation in the photosensitive baboon. High concentrations of memantine even had a proconvulsant effect and caused spontaneous neck and head myoclonus, vertical nystagmus, eyebrow jerks, and ear and scalp tremors (Meldrum et al. 1986). Thus memantine has various effects and actions on mammalian central neurons.

Methods

In order to determine whether this compound may have potentially useful anticonvulsant properties, we conducted electrophysiological and pharmacological experiments utilizing in vitro mammalian brain slice preparations of the guinea pig hippocampus. We cut transverse slices of 300 - 400 µm thickness from guinea pig hippocampus, which were perfused from the lower side in a recording chamber of the Oslo type by a Krebs-Ringer solution at 34°C with a pH of 7.4. We then made recordings from the pyramidal cells (CA3) and granule cells with microelectrodes filled with 3 M potassium acetate with a resistance of 60 - 100 MΩ using bipolar stainless steel electrodes for the electrical stimulation of the perforant path, and mossy fibers. We recorded the field potentials of the granule and pyramidal cells with low-resistance microcapillaries filled with 2 M NaCl. The drugs used (memantine, Merz, FRG, Bicuculline, Sigma, FRG) were added to the Krebs-Ringer solution.

Intracellular recordings were taken from 18 neurons and extracellular recordings from 20 neurons. Cells with a spike amplitude lower than 55 mV were omitted. The resting membrane potential of the granule cells was at least 65 mV and that of pyramidal cells 55 mV.

Results

The action of memantine was analyzed using concentrations from 10 to 1000 µM. Memantine causes different effects at low concentration (i.e., 10 - 100 µM) at high concentration (200 - 1000 µM) in both types of neurons. The action of memantine also depends on the time of exposure to the drug; it has the reverse effect in concentrations greater than 50 µM after 20 min. Low concentrations of memantine increase rather than decrease the excitability of hippocampal neurons.

Synaptically evoked population spikes, which reflect the number of neurons firing, increases by a maximum of 167% in granule cells and 112% in pyramidal cells. Paired synaptic stimuli with an interval of 20 ms display a facilitation of the second response in pyramidal cells; the second stimulus triggers a larger population spike than the first. Memantine 10 - 100 μM clearly enhances this facilitation of the second population spike in CA3 neurons (maximum 40%).

Using the same memantine concentration, intracellular recording of both types of neurons show simultaneously an increased excitatory postsynaptic potential (EPSP) paralleled by a prolonged duration and an increased slope of the rising phase of the action potential (Fig. 1A). The amplitude of the action potentials and their overshoot slightly increase, possibly caused by a membrane hyperpolarization. A 50 μM concentration of memantine decreased the threshold for stimulation in both types of neurons (Fig. 1B,3). In contrast to granule cells, pyramidal cells frequently develop spontaneous activity in control. Memantine intensifies this phenomenon and even induces episodes of spontaneous activity in granule cells. After 20 min, low doses of memantine enhance the inhibitory postsynaptic potential (IPSP) in pyramidal cells and in some granule cells as demonstrated in Fig. 1C, where three sweeps of the oscilloscope are photographically superimposed. A synaptic stimulus is given during hyper- and depolarizing current pulse and without current injection. The synaptic activation of the neuron evokes an EPSP/IPSP sequence. The shunting of the membrane caused by the IPSP is accentuated under the influence of 10 μM memantine. In contrast to pyramidal cells, a paired shock with an interval of 20 ms induces a suppression of the second response in granule cells. In addition, memantine enhances this depression of the second field response of granule cells (maximum 50%). At this memantine concentration, membrane input resistance either remains unchanged (three CA3 cells), slightly increases (six CA3 cells), or decreases (all granule cells) by maximum 70%.

A high memantine concentration and an exposure time of more than 20 min leads to characteristics opposite to those described above. Instead of being increased the amplitude of the postsynaptic

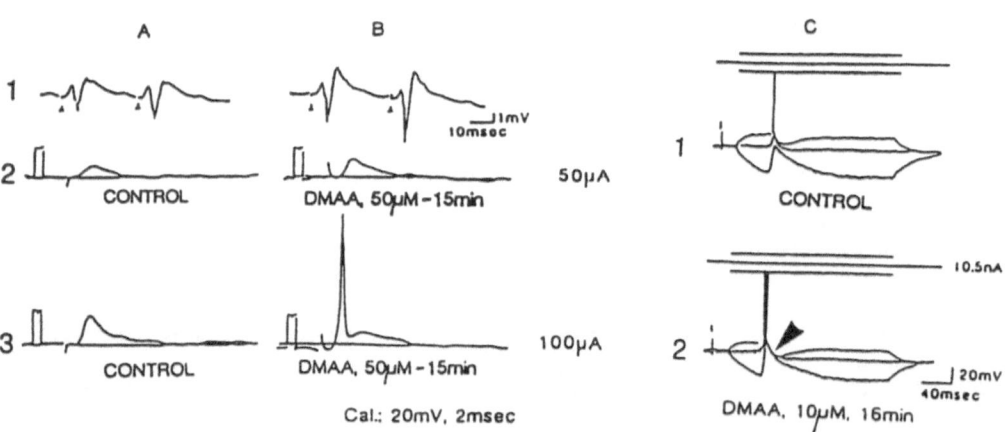

Fig. 1. The influence of 50 μM memantine on the postsynaptic potentials and on the population spike of pyramidal cells. *Row 1 (columns A, B)*, Extracellular recordings of a CA3 neuron. A paired synaptic stimulus of the mossy fibers with an interval of 20 ms leads in CA3 neurons to a facilitation; the second response shows a larger population spike than the first. Memantine 50 μM enhances this facilitation (+37%) and also increases the first population spike (183%).
Rows 2, 3 (columns A, B), intracellular recordings. Memantine 50 μM increases the amplitude (two fold), the duration (130%), and the rising slope of the EPSP. The evoked EPSP reaches spike threshold *(row 3, column B)*.
Column C (row 1, 2) demonstrates the increase in membrane shunting *(arrow in row 2)* by the IPSP caused by 10 μM memantine concentration. *Upper traces* show the duration and amplitude of the current injection (160 ms, 0.5 nA). Three oscilloscope traces are superimposed. Synaptic stimulation is given after the onset of a hyper- and depolarizing current or without a current injection. Memantine induces a decrease in membrane resistance

potentials, i.e. EPSP and IPSP, are reduced to about half and finally abolished (Fig. 2A). Even the strongest stimulus does not evoke any synaptic response. While synaptic transmission is blocked, a direct current injection evokes a new type of cell firing, displaying discharges and afterdischarges with phenomena of inactivation in pyramidal cells as well as in granule cells (not shown). Memantine causes a significant change in the composition of the action potential. As shown in the dV/dt display, the slope is no longer increased but is reduced by at least 200%, action potential is reduced by more than 25%, and the speed of the repolarization is decreased by an average of 40%. Memantine also increases the spike latency by more than 250%. These effects are accompanied by a continuous depolarization of the resting membrane potential by up to 20 mV in granule cells and 18 mV in pyramidal cells.

To test the anticonvulsant activity of memantine we added bicuculline (10 - 50 μM) to the medium. In CA3 neurons bicuculline evokes multiple discharges and afterdischarges as well as increased spontaneous activity. Bicuculline always induces a giant EPSP in granule cells but not necessarily convulsant actions such as multiple discharges or bursts. To counteract bicuculline effects and to reduce or even block epileptogenic activity, the concentration of memantine in the perfusion fluid must be at least 400 μM. In spite of the presence of the convulsant, high memantine concentrations reduce afterdischarge, burst, and spontaneous activity in CA3 neurons.

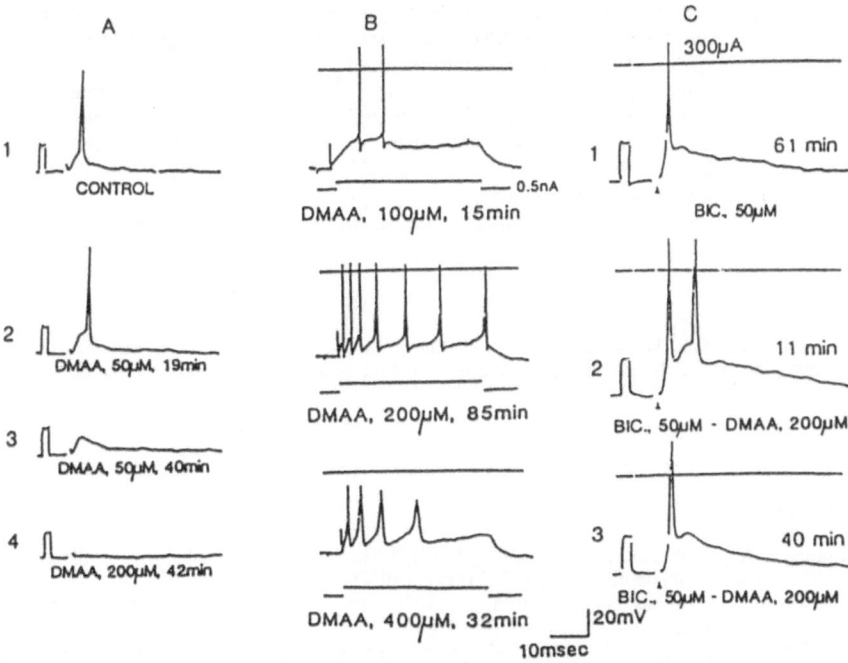

Fig. 2. *Column A*, intracellular recording of granule cells before *(row 1)* and during the addition of 50 μM *(rows 2, 3)* and of 200 μM memantine *(row 4)*. After 19 min memantine causes an increase in the EPSP *(row 2)*, the depolarizing IPSP, and a slight increase of the spike amplitude (+8 mV). The same memantine concentration has an opposite effect after 40 min. The EPSP diminishes and no action potential can be evoked by synaptic stimulation. Memantine 200 μM (40 min) suppresses any synaptic transmission *(row 4)*. *Column B, upper trace*, reference line for the resting membrane potential. *Lowest trace*, monitor of the depolarizing current injection of 100 ms and 0.5 nA. The responsiveness to electrical stimulation of granule cells is first intensified *(row 2)* and the frequency of firing increases. Memantine 400 μM causes a reduction of repetitive firing of the neuron with an increased duration and decreased amplitude of the action potential. *Column C* bicuculline induces a giant EPSP *(row 1)*; memantine 200 μM (11 min) even slightly increases the amplitude and duration of the EPSP, triggering two action potentials *(row 2)*. After 30 min the same concentration diminishes the EPSP, suppresses the second spike response, and reduces spike amplitude *(row 3)*

The EPSP is continuously decreased, the spike amplitude reduced, and the action potentials broadened. Finally, even the highest stimulus intensities evoke neither an EPSP nor an action potential. Lower memantine concentrations cannot prevent the generation of multiple discharges and afterdischarges caused by bicuculline. However, low doses of memantine even enhance the EPSP and induce multiple discharges in those granule cells in which under bicuculline a stimulus evoked only single action potentials (see Fig. 2C).

Discussion

The results show that memantine does not act selectively on hippocampal neurons but induces various effects on postsynaptic potentials and membrane characteristics. In a low range, and only up to 20 min, an augmentation of postsynaptic inhibition can be observed, as expected of a drug which is supposed to act as an anticonvulsant. But the increased postsynaptic inhibition is paralleled by an increased excitatory postsynaptic potential, increased spontaneous activity, and decreased threshold for stimulation; these ultimately result in an increased excitability of the hippocampal neurons. The above results agree with the those described by Sontag et al. (1982), who observed a period of increased spontaneous activity of spinal α-motoneurons of a spastic Han-Wist rat restricted to the first 30 min after intraperitoneal application. Later the same memantine concentration lead to opposite results and reduced the spontaneous hyperactivity of α-motoneurons. Memantine in a low range also increased spontaneous synaptic activity in neurons of *Aplysia californica*. Administration of 10 μM evoked spontaneously and regularly firing EPSPs at the RC-R15 synapse (Klee 1982). The report by Osborne et al. (1982) confirms that memantine neither interferes with the specific uptake mechanisms of various transmitter substances nor interacts with dopamine, opiate, GABA, adrenergic, of benzodiazepine receptors. Osborne et al. showed that a high concentration of memantine (50 μM or higher) released H-monoaminose and led to morphological changes in neurons and glia cells. Our results show that memantine counteracts bicuculline induced convulsant activity when present at a concentration greater than 200 μM, however, at the cost of impaired membrane functions and instability of the hippocampal neurons.

Acknowledgements. We thank Mrs. M. Duesmann for technical assistance, Mrs. H. Thomas and Mrs. M. Ehms-Sommer for the photography, and Mr. N. Steinberg for editing the manuscript

References

Chojnack-Wojcik E, Tatarczynska E, Maj J (1983) The influence of memantine on the anticonvulsant effects of the antiepileptic drugs. Pol J Pharmacol Pharm 35: 511-515

Grossmann A, Grossmann W, Jurna J (1976) The effect of dimethylaminoadamantane on neuronal membranes. Eur J Pharmacol 35: 379

Grossmann W, Schütz W (1982) Memantin und neurogene Blasenstörungen im Rahmen spastischer Zustandsbilder. Arzneimittelforschung 32:1273-1276

Klee MR (1982) Die Wirkung von Memantin auf Membraneigenschaften und postsynaptische Potentiale von Nervenzellen der *Aplysia californica*. Arzneimittelforschung 32:1236-1273

Klee MR, Misgeld U, Zeise ML (1980) Pharmacological differences between CA3 and dentate granule cells in hippocampal slices. Budapest, Adv Physiol Sci 36:145-154

McLean HJ (1989) In vitro electrophysiological evidence predicting anticonvulsant efficacy of memantine and flunarizine. Pol J Pharmacol Pharm (in press)

Meldrum BS, Turski L, Schwarz M, Czuczwar SS, Sontag KH (1986) Anticonvulsant action of 1,3-dimethyl-5-aminoadamantane. Naunyn Schmiedeberg's Arch Pharmacol 332:93-97

Menon MK, Clark WG (1979) GABA-ergic drugs block the locomotor stimulant effects of 1,3-dimethyl-5-aminoadamantane (D-145). Neuropharmacol 18: 223-225

Osborne NN, Beal R, Golombiowska-Nikitin K, Sontag KH (1982) The effect of memantine on various neurological processes. Arzneimittelforschung 32: 1236-1273

Preisendörfer U, Zeise ML, Klee MR (1987) Valproate enhances inhibitory postsynaptic potentials in hippocampal neurons in vitro. Brain Res 435: 213-219

Reiser G, Binmöller TJ, Koch R (1988) Memantine blocks the serotonin induced depolarization response in a neuronal cell line. Brain Res 443: 338-344

Sontag KH, Wand P, Schwarz M, Wesemann W, Osborne NN (1982) Die Wirkung von 1,3-Dimethyl-5-Aminoadamantan auf spinale Motoneurone und Gehalt an Dopamine (DA), Noradrenalin (NA) und Serotonin (5-HT) des Striatums und lumbalen Rückenmarks. Arzneimittelforschung 32: 1236-1273

Tsai MC, Chen ML, Lo SC, Tsai GC (1989) Effects of memantine on the twitch tension of mouse diaphragm. Eur J Pharmacol 160: 133-140

Suppression of Partial and Generalized Seizure Activity by Intracerebroventricular Perfusion of the Organic Calcium Antagonist Verapamil in the Rat

E.-J. Speckmann[1,2] and J. Walden[1]

[1] Institut für Physiologie, Universität Münster, Robert-Koch-Straße 27a, D-4400 Münster, FRG
[2] Institut für Experimentelle Epilepsieforschung, Universität Münster, Hüfferstraße 68, D-4400 Münster, FRG

Introduction

Many observations in the literature have demonstrated an essential role for calcium ions in epileptogenesis (Heinemann et al. 1986; Speckmann et al. 1987). In this context the following findings are of special interest:

(a)   The extracellular calcium ion concentration in cortical tissues is decreased during focal and generalized tonic-clonic seizure activity (Heinemann et al. 1977; Lehmenkühler et al. 1986).

(b)   The calcium ion concentration of the outer surface of single neurons is diminished during paroxysmal depolarization shifts (Lücke et al. 1989).

(c)   Paroxysmal depolarization shifts of single neocortical and archicortical neurons are suppressed by intra- and extracellular application of organic calcium antagonists (Bingmann et al. 1988; Bingmann and Speckmann 1989; Straub et al. 1989; Witte et al. 1987) and enhanced by organic calcium agonists (Walden et al. 1985).

The present chapter describes the antiepileptic action of the calcium antagonist verapamil on focal and generalized tonic-clonic seizures in neuronal populations (neocortex, rat, in vivo).

Material and Methods

The experiments were performed on the motor and somatosensory cortex of anesthetized and artificially ventilated rats. Bioelectrical activity was monitored by recording field potentials (EEG, evoked potentials, DC potential). Focal epileptic activity, elicited by local application of penicillin onto the cortical surface, and generalized tonic-clonic seizures, evoked by repeated intraperitoneal injections of pentylenetetrazol (PTZ), served as epileptic models. The calcium channel blocker verapamil (1 mM), dissolved in artificial CSF, was applied systemically by perfusion of a lateral cerebral ventricle (push-pull technique).

Results and Discussion

Focal interictal epileptiform discharges in the EEG, induced by local application of penicillin, were reduced in amplitude and in frequency of occurrence and, eventually, were most often abolished during an intracerebroventricular perfusion of the calcium antagonist verapamil (Fig. 1A) (Walden et al. 1985). In some experiments the depression of seizure discharges was sometimes preceded by a transient enhancement. Tonic-clonic seizure activity, elicited by repeated administrations of PTZ, were also depressed by an intracerebroventricular perfusion of verapamil (Fig. 1B-1). Together, the negative shift of the cortical DC potential, evoked by intraperitoneal injections of PTZ, shifted to a positive displacement (Fig. 1B-2; 3).

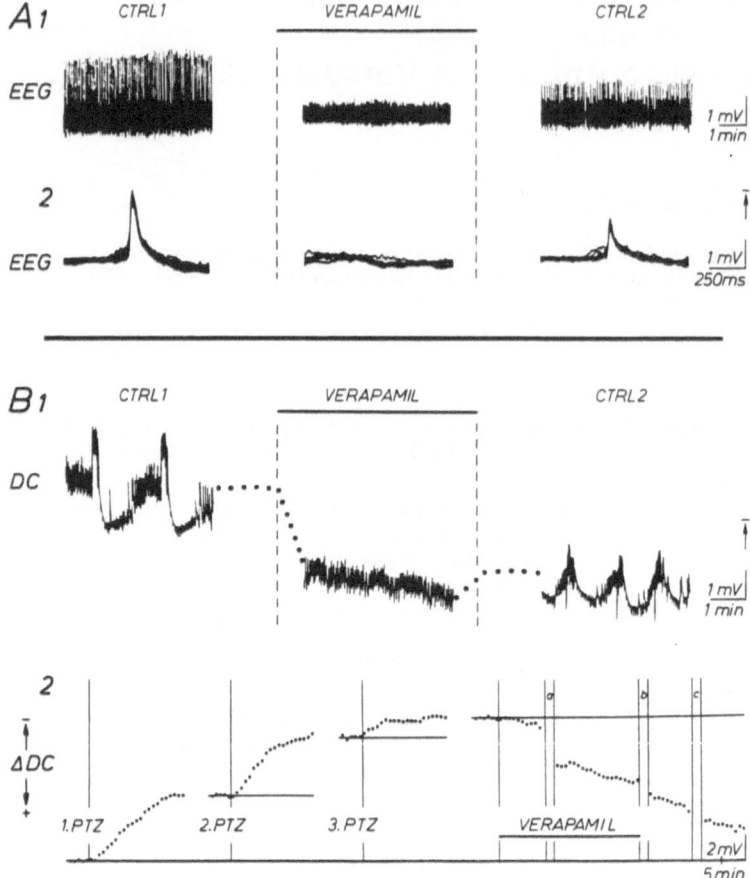

Fig. 1A,B. Effects of verapamil on focal interictal (A) and generalized tonic-clonic (B) epileptic discharges in epicortical EEG and DC potentials. Verapamil was applied into a lateral cerebral ventricle by push-pull perfusion for 30 min (A,B1) and up to 120 min (B2). Recordings were made 27 min (A) and 21 min (B1) after commencement of perfusion. *CTRL1, CTRL2*, recordings before (3 min, A; 10 min, B1) and after (50 min, A; 25 min, B1) verapamil perfusion. A. Focal seizure activity elicited by epicortical application of penicillin; chartrecorder tracing. A2. Single epileptic potentials recorded with an extended time scale; six superimposed sweeps. B1. Generalized tonic-clonic activity induced by repeated intraperitoneal injections of PTZ. B2. Displacement of the epicortical DC potential (*DC*) after the first, second, and third PTZ injections and during intracerebroventricular perfusion of verapamil. Mean value of eight original tracings. Interruption of the curve: up to 45 min (a) and 20 min (b, c).
(A, B1 modified from Specmann and Walden 1987; Walden and Speckmann 1988)

In order to test whether the effects described are specific for epileptic discharges, the action of verapamil on the bioelectrical activity of animals not treated with epileptogenic agents was examined. Typical findings are presented in Fig. 2. It can be seen that, in nonepileptic preparations, verapamil exerted no depressant effects on somatosensory evoked potentials (A) or on the spontaneous waves of the EEG (B) and failed to shift the cortical DC potential to the positive side (C) (Walden et al. 1985; Walden and Speckmann 1988).

In summary, the results demonstrate that organic calcium antagonists are able not only to reduce epileptic activity in single neurons (Bingmann et al., this volume) but also in neuronal populations (see also Meyer et al. 1986; Morocutti et al. 1986; Pockberger et al. 1986).

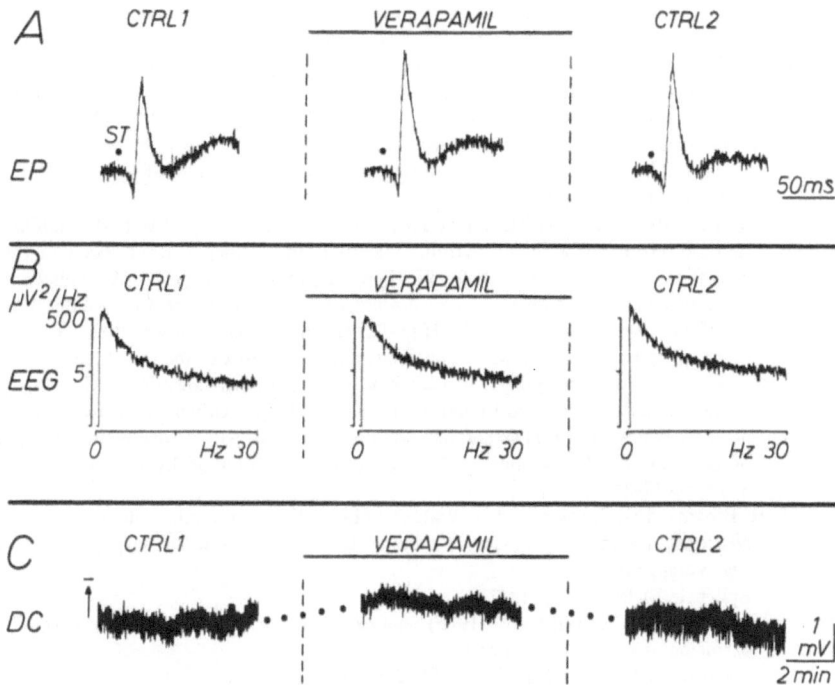

Fig. 2. Somatosensory evoked potentials (*EP*; A), spontaneous waves of the EEG (B) and DC potential (C) at the cortical surface before, during, and after systemic administration of verapamil. Rat somatosensory (A) and motor (B, C) cortex; preparation without application of epileptogenic agents. Verapamil was administered into a lateral cerebral ventricle by push-pull perfusion for 90 min (A, B) and 30 min (C). Recordings were made 60 min (A), 85 min (B), and 22 min (C) after commencement of perfusion. *CTRL1*, *CTRL2*, recordings before (55 min, A; 11 min, B; 5 min, C) and after (50 min, A; 48 min, B; 55 min, C) verapamil perfusion. **A.** EP elicited by stimulation (**ST**) of the sciatic nerve; average potentials; n = 100. **B.** Power spectral analysis of the spontaneous EEG. Epochs of analysis, 2 min. **C.** Chartrecorder tracing of DC potential (Speckmann and Walden 1987)

References

Bingmann D, Speckmann E-J, Baker RE, Ruijter J, de Jong BM (1988) Differential antiepileptic effects of the organic calcium antagonist verapamil and flunarizine in neurons of organotypic neocortical explants of newborn rats.
Exp Brain Res 72: 439–442

Bingmann D, Speckmann E-J (1989) Specific suppression of pentylenetetrazol-induced epileptiform discharges in CA3 neurons (hippocampal slices, guinea pig) by the organic calcium antagonists flunarizine and verapamil.
Exp Brain Res 74: 239–248

Caspers H, Speckmann E-J, Lehmenkühler A (1987) DC potentials of the cerebral cortex: Seizure activity and changes in gas pressure.
Rev Physiol Biochem Pharmacol 106: 127–178

Heinemann U, Lux HD, Gutnick M (1977) Extracellular free calcium and potassium during paroxysmal activity in the cerebral cortex of the cat.
Exp Brain Res 27: 237–243

Heinemann U, Klee M, Neher E, Singer W (eds) (1986) Calcium electrogenesis and neuronal functioning. Springer, Berlin Heidelberg New York (Experimental brain research series, vol 14)

Lehmenkühler A, Kersting U, Richter A, Boerrigter P (1986) Relations of bioelectrical activity, extracellular ion concentrations and extracellular volume in cortical epileptic foci. In: Speckmann E-J, Schulze H, Walden J (eds) Epilepsy and calcium. Urban and Schwarzenberg, Munich, pp 227-245

Lücke A, Speckmann E-J, Lehmenkühler A, Altrup U (1989) Changes in free calcium concentration at the surface of identified snail neurons during pentylenetetrazol induced paroxysmal depolarization shifts. Pflügers Arch 413: R4

Meyer FB, Tally PW, Anderson RE, Sundt TM, Yaksh TL, Sharbrough FW (1986) Inhibition of electrically induced seizures by a dihydropyridine calcium channel blocker. Brain Res 384: 180-183

Morocutti C, Pierelli F, Sanarelli L, Stefano E, Peppe A, Mattioli GL (1986) Antiepileptic effects of a calcium antagonist (nimodipine) on cefasolin-induced epileptogenic foci in rabbits. Epilepsia 27: 498-503

Pockberger H, Rappelsberger P, Petsche H (1986) Calcium antagonists and their effects on generation of interictal spikes: a field potential analysis in the neocortex of the rabbit. In: Speckmann E-J, Schulze H, Walden J (eds) Epilepsy and calcium. Urban and Schwarzenberg, Munich, pp 357-378

Speckmann E-J, Schulze H, Walden J (eds) Epilepsy and calcium. Urban and Schwarzenberg, Munich

Speckmann E-J, Walden J (1987) Generation and inactivation of paroxysmal neuronal activity: involvement of calcium. In: Chalazonitis N, Gola M (eds) Inactivation of hypersensitive neurons. Liss, New York, pp 17-24

Straub H, Bingmann D, Speckmann E-J, Walden J (1989) Bicucullin induzierte epileptische Aktivität in CA3 Neuronen (Hippocampus, Gewebeschnitt): Unterdrückung durch den organischen Calciumantagonisten Verapamil. Epilepsie-Blätter 2 (Suppl): 23

Walden J, Speckmann E-J, Witte OW (1985) Suppression of focal epileptiform discharges by intraventricular perfusion of a calcium antagonist. Electroenceph. Clin Neurophysiol 61: 299-309

Walden J, Pockberger H, Speckmann E-J, Petsche H (1986) Paroxysmal neuronal depolarizations in the rat motorcortex in vivo: intracellular injection of the calcium agonist BAY K 8644. Exp Brain Res 64: 607-609

Walden J, Speckmann E-J (1988) Suppression of recurrent generalized tonic-clonic seizure discharges by intracerebroventricular perfusion of a calcium antagonist. Electroenceph Clin Neurophysiol 69: 353-362

Witte OW, Speckmann E-J, Walden J (1987) Motor cortical epileptic foci in vivo: actions of a calcium channel blocker on paroxysmal neuronal depolarizations. Electroenceph Clin Neurophysiol 66: 43-55

Differential Antiepileptic Effects of Organic Calcium Antagonists in CA3 Neurons of Hippocampal Slices and in Neurons of Organotypic Neuronal Explants

D. Bingmann[1], E.-J.Speckmann[2,3], J. Walden[2], H. Straub[2], R. E. Baker[4] and J. Ruijter[4]

[1] Institut für Physiologie, Universität Essen, D-4300 Essen, FRG
[2] Institut für Physiologie, Universität Münster, D-4400 Münster, FRG
[3] Institut für Experimentelle Epilepsieforschung, Universität Münster, D-4400 Münster, FRG
[4] Netherlands Institute for Brain Research, NL-1105 AZ Amsterdam, The Netherlands

Introduction

Epileptic discharges of neuronal populations in the neocortex (Heinemann et al. 1977) and in the hippocampus (Benninger et al. 1980) are accompanied by a decline in the extracellular calcium concentration. These findings led us to assume that transmembranous calcium currents contribute markedly to the generation of paroxysmal depolarization shifts (PDS) (see Lücke et al. 1989). This is in line with observations that: (a) epileptic neuronal discharges, elicited by different epileptogenic drugs in various in vitro and in vivo preparations of the CNS of rats and guinea pigs, were blocked by organic calcium antagonists (Bingmann and Speckmann 1989; Meyer et al. 1986; Morocutti et al. 1986; Walden et al. 1985; Walden and Speckmann 1988; Witte et al. 1987), and (b) the calcium agonist BAY K 8644 intensified epileptic discharges (Walden et al. 1985). In further investigations the antiepileptic effects of organic calcium antago-nists of different groups (the papaverine derivative, verapamil; the piperizine derivative, flunarizine; and the dihydropyridine derivative, nifedipine) were tested on pentylenetetrazol (PTZ)-induced epileptic activity in neocortical and archicortical neurons. In order to obtain controlled administrations of the calcium antagonists in comparable tissue concentrations, the present experiments were carried out in vitro using hippocampal slices (guinea pigs) and organotypic neocortical explants (newborn rats).

Methods

Hippocampal slices (ca. 400 μm thick) were prepared from adult guinea pigs. Slices were preincubated for 2 h in 28°C warm saline equilibrated with 5% CO_2 in O_2. The saline contained (in mM): NaCl 124, KCl 5, $CaCl_2$ 0.75, $MgSO_4$ 1.4, KH_2PO_4 1.25, $NaHCO_3$ 26, glucose 10. Organotypic neocortical explants were prepared from 6 day old rats. The excised tissue was placed on a plastic grid and superfused by a chemically defined CO_2/bicarbonate buffered nutrient medium (Romijn et al. 1988) for 2-3 weeks.
For intracellular recordings the aforementioned preparations were positioned in a recording chamber mounted over an inverted microscope. The chamber was perfused by a 32°C warm saline bath in which the calcium concentration was raised to 2 mM (control saline). Epileptic activity was elicited by repeated systemic administration of pentylenetetrazol (3-10 mM). The organic calcium antagonists verapamil, flunarizine (dissolved in beta-cyclodextrin) and nifedipine (dissolved in dimethyl sulfoxide) were added to the saline to give final concentrations of between 10-100 μM.

Results and Discussion

Hippocampal Slices.
In all CA3 neurons PDS induced by repeated exposures to PTZ were reduced in amplitude and/or duration and in frequency of occurrence after adding verapamil (n=9; Fig. 1A), flunarizine (n=22; Fig. 1B) or nifedipine (n=4) to the bath saline. PDS were totally blocked after latencies of 20-60 min when the concentrations of the calcium antagonists ranged above 20 μM (Bingmann and

234

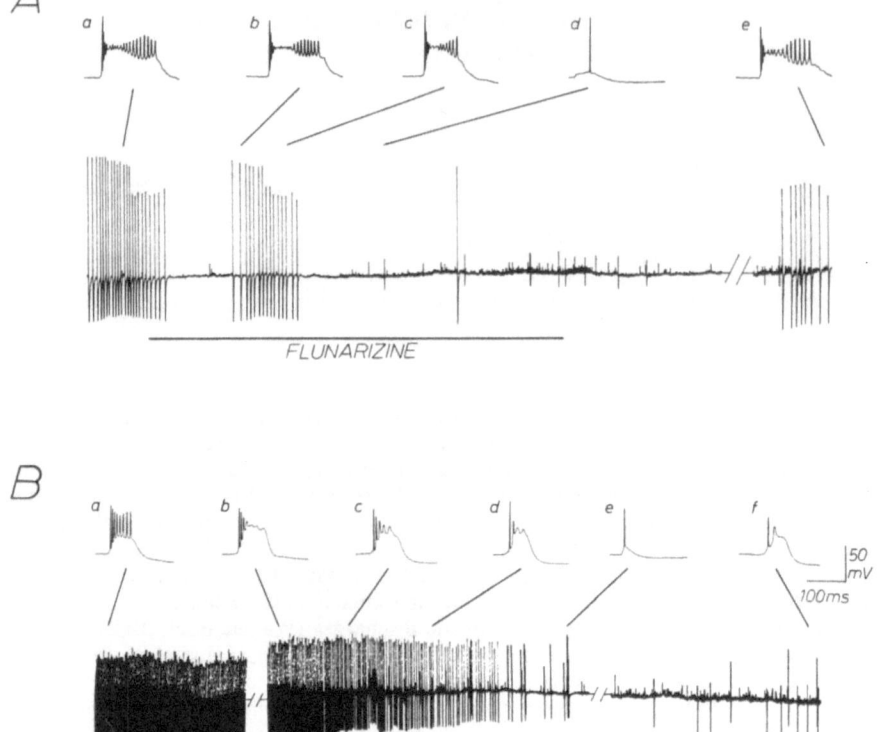

Fig. 1A,B. Effect of A flunarizine (50 μM) and B verapamil (50 μM) on pentylenetetrazol-induced paroxysmal depolarization shifts in CA3 neurons of hippocampal slices (guinea pig). Membrane potentials were recorded by an oscilloscope and a chart recorder, respectively, and are related to each other by lines. Interruptions in A 80 min, in B 12 and 15 min. Same scales in A and B. (From Bingmann and Speckmann 1989 with permission)

Speckmann 1989). In general, all tested organic calcium antagonists exhibited antiepileptic effects in hippocampal neurons.

Organotypic Neocortical Explants.
PDS were elicited in all neocortical neurons by exposure to PTZ. The addition of verapamil (40-80 μM; n=11; Fig. 2A) to the PTZ containing bath blocked PDS reversibly . This suppression occurred within 20-50 min when the verapamil concentration was 40 μM. Raising the verapamil concentration to 80 μM decreased the latency, and the time of reappearance of PDS during washing was prolonged. During exposure to verapamil PDS were shortened in duration in the majority of cases. Flunarizine (40 μM; n=5; Fig. 2B; Bingmann et al. 1988) did not change the rate of occurrence, shape and amplitude during test periods of 40-60 min. Nifedipine (100 μM; n=7) suppressed PDS in 5 neocortical neurons (Fig. 2C) with similar latencies as observed with verapamil. In two neurons the rate of occurrence of PDS was not affected by nifedipine (Fig. 2D). Only with high concentrations of nifedipine were the amplitudes of sodium-dependent action potentials re-

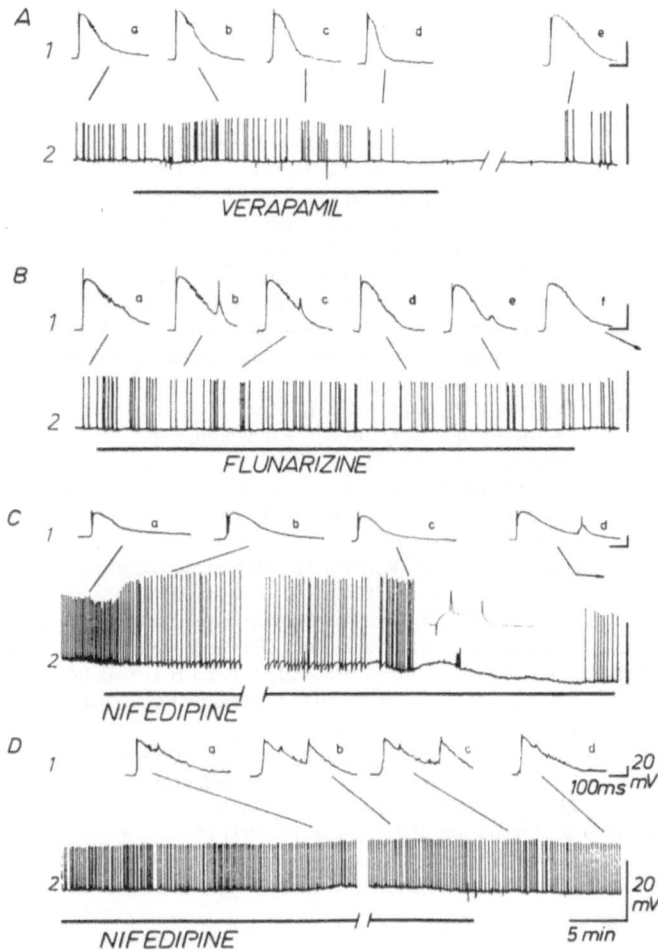

Fig. 2A-D. Effect of **A** verapamil (40 μM), **B** flunarizine (40 μM), and **C,D** nifedipine (80 μM and 100μM, respectively) in neocortical neurons of organotypic cultures exposed to pentylenetetrazol (5 mM). Membrane potentials *1* and *2* were recorded by an oscilloscope and a chart recorder, respectively, and are related to each other by lines. Same scales in **A** as in **B-D**. Interruption in **A** 5 min, in **C** 26 min, in **D** 20 min.

duced. Thus, even a significant reduction in the amplitudes of action potentials due to an unspecific depressant effect of this drug does not necessarily lead to a blockade of PDS sequences.

The presented findings indicate that organic calcium antagonists display differential antiepileptic, and hence highly specific, effects in neocortical and archicortical neurons.

References

Benninger C, Kadis J, Prince D A (1980) Extracellular calcium and potassium changes in hippocampal slices. Brain Res 197: 165-182

Bingmann D, Speckmann E-J, Baker R E, Ruijter J, de Jong B M (1988) Differential antiepileptic effects of the organic calcium antagonists verapamil and flunarizine in neurons of organotypic neocortical explants from newborn rats. Brain Res 72: 439-442

Bingmann D, Speckmann E-J (1989) Specific supression of pentylenetetrazol-induced epileptiform discharges in CA3 neurons (hippocampal slice, guinea pig) by the organic calcium antagonists flunarizine and verapamil. Exp Brain Res 74: 239-248

Heinemann U, Lux H D, Gutnick M (1977) Extracellular free calcium and potassium during paroxysmal activity in the cerebral cortex of the cat. Exp Brain Res 27: 237-243

Lücke A, Speckmann E-J, Lehmenkühler A, Altrup U (1989) Changes in free calcium concentration at the surface of identified snail neurons during pentylenetetrazol-induced paroxysmal depolarization shifts. Pflügers Arch S 413 R 4

Meyer F B, Tally P W, Anderson R E, Sundt T M, Jr., Yaksh T L, Sharbrough F W (1986) Inhibition of electrically induced seizures by a dihydropyridine calcium channel blocker. Brain Res 384: 180-183

Morocutti C, Pierelli F, Sanarelli L, Stefano E, Peppe A, Mattioli G L (1986) Antiepileptic effects of a calcium antagonist (nimodipine) on cefazolin-induced epileptogenic foci in rabbits. Epilepsia 27: 498-503

Romijn H J, de Jong B M, Ruijter J M (1988) A procedure for culturing rat neocortex explants in a serum-free nutrient medium. J Neurosci Methods 23: 75-83

Walden J, Speckmann E-J, Witte O W (1985) Suppression of focal epileptiform discharges by intraventricular perfusion of a calcium antagonist. Electroencephalogr Clin Neurophysiol 61: 299-309

Walden J, Pockberger H, Speckmann E-J, Petsche H (1986) Paroxysmal neuronal depolarizations in the rat motorcortex in vivo: intracellular injection of the calcium antagonist BAY K 8644. Exp Brain Res 64: 607-609

Walden J, Speckmann E-J (1988) Suppression of recurrent generalized tonic-clonic seizure discharges by intraventricular perfusion of a calcium antagonist. Electroencephalogr Clin Neurophysiol 69: 353-362

Witte O W, Speckmann E-J, Walden J (1987) Motor cortical epileptic foci in vivo: actions of a calcium channel blocker on paroxysmal neuronal depolarizations. Electroencephalogr Clin Neurophysiol 66: 43-55

Phenytoin and Carbamazepine Act by Accumulating Inactivated Sodium Channels

J. R. Schwarz

Physiologisches Institut, Universitätskrankenhaus Eppendorf, 2000 Hamburg 20, FRG

Introduction

Phenytoin (PHT) and carbamazepine (CBZ) prevent the generation of paroxysmal activity in neurons and nerve fibers most likely by blocking voltage-dependent Na channels. Effects on synaptic processes, e.g., the potentiation of inhibitory GABAergic synaptic transmission, seem to be less important (Catterall 1987; Macdonald et al. 1985). There are two different mechanisms by which Na channels can be blocked. First, they can be blocked permanently, as occurs with tetrodotoxin. With increasing drug concentration this type of block induces a general depression of excitability which could explain the toxic side-effects. At therapeutic drug concentrations of 10 - 20 μM, however, normal excitability is hardly affected. In addition to the sustained block, Na channels can be blocked transiently. This second type of block was first observed in local anesthetics and was referred to as use- or frequency-dependent inhibition since it could be modulated by the frequency of depolarizing stimuli (Hille 1987). As is shown below, PHT and CBZ also exert a frequency-dependent Na channel block which may account for their anticonvulsant action.

Results

The node of Ranvier of single myelinated rat nerve fibers was voltage clamped and Na currents were recorded. K channels were blocked by tetraethylammonium chloride (TEA) as described previously (Röper and Schwarz 1989; Schwarz and Grigat 1989). As shown in Fig. 1A, Na currents, representing the transient opening of a large number of Na channels, were recorded upon a depolarizing pulse.

At the normal resting potential of -80 mV about 75% of the Na channels can be opened upon depolarization; in the Hodgkin-Huxley terminology they are said to be activated and then inactivated. The transition back to the resting state takes place after repolarization and normally is complete within about 1 ms. Figure 2A shows a scheme of the transitions between the different conformational states of the Na channel. Hyperpolarization removes Na inactivation, i.e., inactivated Na channels are transferred to the resting state. The voltage-dependent distribution of resting and inactivated Na channels can be measured with a two-pulse protocol (see legend to Fig.1). The membrane potential is changed to various positive and negative values for a sufficient length of time and, with a subsequent depolarizing test pulse, a Na current is recorded whose amplitude is proportional to the number of resting, activatable Na channels. From the relation of the peak Na current amplitudes superimposed in the upper panel of Fig. 1A (Ri control), the relative amount of Na channels which can be activated at the resting potential can be determined. The lower pair of Na currents in Fig. 1A was recorded in the presence of 100 μM CBZ which reduced the Na current to small values; however, when the test pulse was preceded by a large negative pulse of long duration, the CBZ-induced Na channel block was totally antagonized. This voltage-dependent drug binding and unbinding induced a shift of the h_∞ curve (Fig. 1B) to more negative membrane potentials. A similar shift has been reported for local anesthetic (Hille 1977) and the anticonvulsant PHT (Schwarz and Grigat 1989; Schwarz and Vogel 1977; Willow et al. 1985) and phenobarbital (Schwarz 1979; Schwarz et al. 1980).

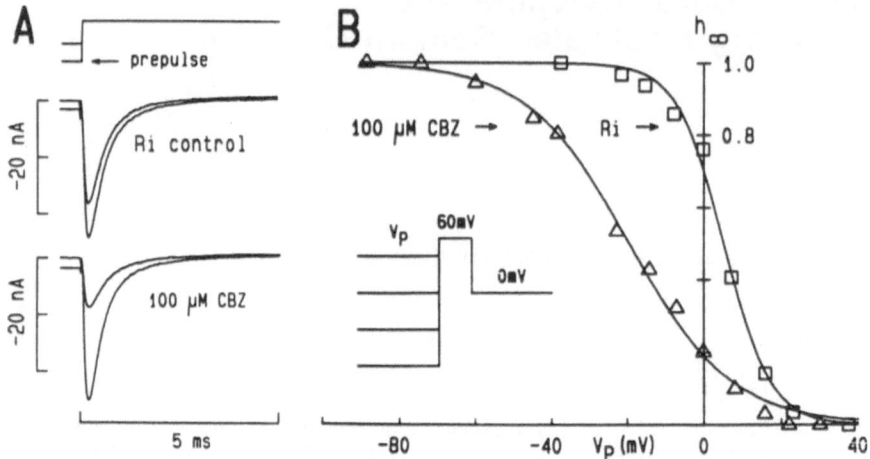

Fig. 1A,B. CBZ induces a potential-dependent block of Na channels. **A** A single myelinated rat nerve fiber was isolated and voltage clamped. Na currents were elicited with a constant test pulse either with or without a negative prepulse (see inset in Fig. 1B). 100 μM CBZ decreased the Na current when recorded without a prepulse. When the test pulse was preceded by a 500 ms hyperpolarization of -80 mV, the drug induced a shift of the Na inactivation curve to more negative membrane potentials. Na currents were elicited with a 60 mV test pulse preceded by negative and positive pulses of various amplitudes, and 500 ms duration. The peak Na inward currents recorded in Ringer's control and in the presence of 100 μM CBZ were normalized and plotted versus prepulse potential. Smooth curves represent fits of a Boltzmann equation (h_∞ = $1/\{1 + \exp[(V-V_h)/k]\}$) to the data points with the following values: Ri: V_h = 5.1 mV, k = 6.0 mV; CBZ: V_h = -19.7 mV, k = 13.1 mV). Data from Schwarz and Grigat 1989

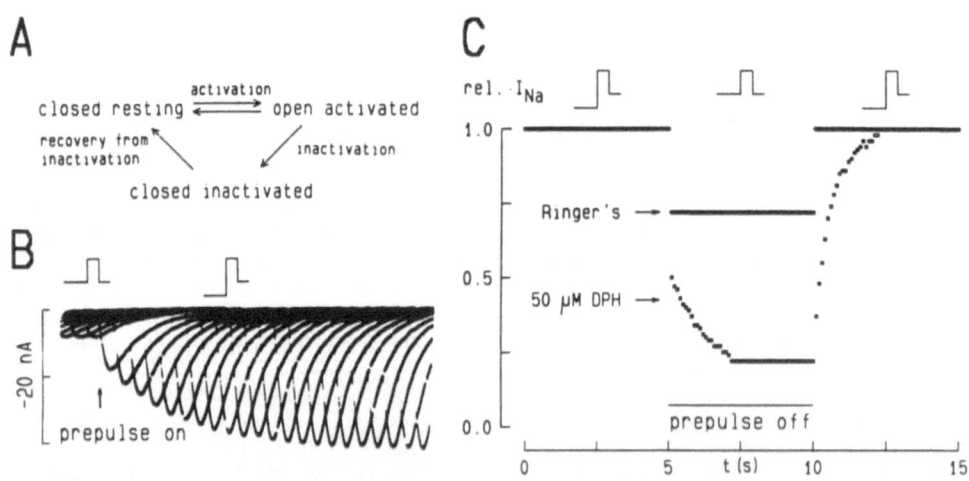

Fig. 2. Time-dependent Na channel block. **A** The Na channel resides in one of three states. The underlying transitions are governed by voltage-dependent rate constants. **B** Time-dependent removal of Na inactivation in the presence of 50 μM PHT. Test pulses of 60 mV and 5 ms duration were applied with intervals of 100 ms and recorded on moving film. Upon addition of a hyperpolarizing prepulse of -50 mV and 50 ms duration the new steady state value of Na inactivation was approached within about 2 s. **C** Time-dependent Na channel block and unblock; evaluation of experiments performed with similar pulse protocols as shown in B. The peak values of Na currents in the beginning of each run were normalized. Data from Schwarz and Grigat 1989

The voltage-dependent receptor-binding and unbinding of CBZ and PHT at the Na channel is slower than the normal gating kinetics of the Na channel. Slow unbinding of the drug from its receptor occurs when the membrane is hyperpolarized. The time constants for the dissociation of PHT and CBZ from their receptors are 90 and 40 ms, respectively (Schwarz and Grigat 1989). This is much slower than normal recovery from Na inactivation, which has a time constant of about 1 ms. The time-dependent transition from the inactivated state to the activatable state is demonstrated in Fig. 2B. In the presence of 50 µM PHT short depolarizing pulses, applied at an interval of 100 ms, elicited small amplitude Na inward currents. Addition of a hyperpolarizing prepulse, which in the control Ringer solution totally removed Na inactivation within the period of one prepulse (not shown), failed to do so in the presence of PHT. It took about 2 s to reach a new steady state of Na inactivation. When the prepulse was removed again, a reduction of the Na currents occurred exhibiting the same time dependence as removal of the Na current block (Fig. 2C). PHT induced a slower inactivation than CBZ. Therefore, the frequency-dependent effect of CBZ was less pronounced than that of PHT (Schwarz and Grigat 1989). The same observation was reported for neuroblastoma cells (Willow et al. 1985).

Discussion

PHT and CBZ induce a potential- and frequency-dependent block of Na channels. The drug-receptor reaction is characterized by a state-dependent block and slow reaction kinetics. Both drugs bind more strongly to inactivated channels than to activated open or resting Na channels. Once bound, dissociation of the drug from its receptor occurs upon a hyperpolarization. It is important to note that this dissociation is slower than the normal recovery from Na inactivation. At therapeutic concentrations the sustained drug-induced Na channel block is small, therefore normal excitability is only slightly affected. At the advent of abnormal cell activity, e.g., sustained bursts of spikes or paroxysmal depolarization shifts, Na channels are more frequently transferred to the activated state in which they are more effectively bound by PHT or CBZ. Accumulation of drug-bound Na channels occurs when the stimulus interval is shorter than recovery from Na inactivation of drug-bound Na channels. Accumulation of inactivated Na channels depresses excitability and blocks the generation of action potentials. Frequency-dependent block of Na channels also provides an explanation for the depression of posttetanic potentiation and limitation of sustained, high-frequency bursts of action potentials observed in the presence of PHT and CBZ.

References

Catterall WA (1987) Common modes of drug action on Na⁺ channels: local anesthetics, antiarrhythmics and anticonvulsants. Trends Pharmacol Sci 8: 57-65
Hille B (1977) Local anesthetics: hydrophilic and hydrophobic pathways for the drug-receptor reaction. J Gen Physiol 69: 497-515
Macdonald RL, MecLean MJ, Skerrit JH (1985) Anticonvulsant drug mechanisms of action. Fed Proc 10: 2634-2639
Röper J, Schwarz JR (1989) Heterogeneous distribution of fast and slow potassium channels in myelinated rat nerve fibres. J Physiol (Lond) 416: 93-110
Schwarz JR (1979) The mode of action of phenobarbital on the excitable membrane of the node of Ranvier. Europ J Pharmacol 56: 51-60
Schwarz JR, Grigat G (1989) Phenytoin and carbamazepine: potential- and frequency-dependent block of Na currents in mammalian myelinated nerve fibers. Epilepsia 30: 286-294
Schwarz JR, Ochs G, Bromm B (1980) Phenobarbital induces slow recovery from sodium inactivation at the nodal membrane. Biochim Biophys Acta 597: 384-390
Schwarz JR, Vogel W (1977) Diphenylhydantoin: excitability reducing action in single myelinated nerve fibres. Europ J Pharmacol 44: 241-249
Willow M, Gonoi T, Catterall WA (1985) Voltage clamp analysis of the inhibitory actions of diphenylhydantoin and carbamazepine on voltage-sensitive sodium channels in neuroblastoma cells. Mol Pharmacol 27: 549-558.

Flurazepam Reduces Axonal Conduction Velocity and Blocks Propagation of High-Frequency Discharges

J. Hoyer

Institut für Neurophysiologie, Universität Wien, 1090 Vienna, Austria

Introduction

It is generally accepted that benzodiazepines (BDAs) exert their anticonvulsive action by enhancing the inhibitory GABA mechanisms at a BDA-receptor GABA-receptor Cl⁻-channel iontophore at synaptic sites (Moehler and Okada 1977; Squires and Braestrup 1977). However, this theory cannot apply to all phenomena observed with the administration of BDAs. In experimental animals BDAs are effective against pentylenetetrazole-induced convulsions in low doses, but high doses are necessary to protect against maximal electroshock-induced seizures. In humans, doses of up to 1 mg/kg bodyweight are necessary to interrupt a status epilepticus. High-affinity BDA receptors work, however, within a nanomolar range. All receptors are already occupied using small doses of BDAs. The different effects at high and low doses, together with the wide spectrum of clinical actions of BDAs, raises the question of whether there are not additional sites of action for the BDAs besides their receptor binding site. BDA receptors appear late in evolution (Nielsen et al. 1978) and thus do not exist in invertebrates. Therefore, the invertebrate provides a preparation without BDA receptors which seems ideal for addressing such questions.

In *Aplysia* neurons, BDAs reduce the amplitude and increase the duration of action potentials (APs) recorded from the soma. Both the reduction of overshoot and the afterhyperpolarization contribute to the reduction in total amplitude of APs. These changes in AP composition are accompanied by a reduction in the maximum rate of rise of the AP (dV/dt) and by an even stronger reduction in the maximum (dV/dt) of the falling phase (Hoyer et al. 1978). The aim of the present investigation was to study the influence of these changes in AP on their conduction along the axon.

Methods

Simultaneous intracellular recordings of the soma and the axon of the neuron R-2 of the visceral ganglion of *Aplysia californica* were used. The penetration of both the soma and the axon with electrodes was carried out without physical (dissection) or enzymatic treatment of the connective tissue. Flurazepam HCl (FLUR), a water soluble BDA, was used at concentrations of 100 and 500 μM and added to artificial sea water to serve as bathing fluid for the ganglion, which was isolated together with the connective nerves. Single and repetitive APs at different frequencies were elicited at the soma using an intracellular electrode (in addition to two recording electrodes). Conduction velocity was calculated as the ratio between the distance of the recording electrodes and the time delay between the soma AP and the axon AP since it is known that the site of origin of the APs is in an axonal region near the soma, the latter being excited only by way of antidromic propagation (Tauc 1960). The distance between the electrode positioned in the soma and the one in the axon varied between 15 and 45 mm. The calculation, using distance and time intervals, gives a range of conduction velocities under control conditions between 0.75 and 1.1 m/s. This value fits well with those of other investigators (Gola and Ducreux 1982). Simultaneous recordings of soma and axon APs demonstrate that the effect of FLUR on the AP is much more pronounced in the axon than in the soma (Fig. 1C versus 1B).

FLUR causes a decrease in the axonal conduction velocity. This decrease is dose dependent and reversible. FLUR 100 µM results in a decrease of up to 5%, and 500 µM in one of up to 25%. The recordings in Fig. 1 are from a preparation in which the two recording electrodes were separated by a distance of 18 mm. The time delay in control (Fig. 1A) is 19.4 ms; the calculated conduction velocity is 0.93 m/s. FLUR 500 µM increases the time delay between the two APs to 24 ms, which is equivalent to a conduction velocity of 0.75 m/s or a reduction of 20% with respect to control. The effect is reversible after a washout of several hours (3 h in Fig. 3A).

Using repetitive stimulation in the range from 0.5 to 10 Hz, FLUR causes an additional reduction in conduction velocity. In a train of APs, each spike has a lower propagation velocity than the one preceding it. This phenomenon can be observed at stimulus rates as low as 0.5 Hz. The use-dependent block becomes more pronounced with increasing frequencies. The conduction velocity within a train of 1.0 Hz diminishes from the first to the tenth APs at 1.0 Hz by only approximately 10% whereas it diminishes by 60% in a train at 7.7 Hz. In the graph of Fig. 2, the conduction velocity from the first to the tenth APs is reduced by 500 µM FLUR from 0.75 m/s to 0.65 m/s at 1.0 Hz and from 0.75 m/s to 0.3 m/s at 7.7 Hz. In trains of APs of a rate of 10 Hz and above, the propagation becomes intermittent; some APs within the train fail to propagate. The AP which follows the pause after a failing AP is conducted faster than the preceding AP, thus demonstrating a rapid recovery of the use-dependent block. The phenomenon of APs failing to propagate can also be observed with spontaneously occurring, drug-induced double discharges, i.e., repetitive discharges with an interval of less than 50 ms. Such doublets occur under the influence of FLUR at both the soma and the axon of R-2, with only a single accompanying AP in the other part of the neuron (Gola and Ducreux 1982).

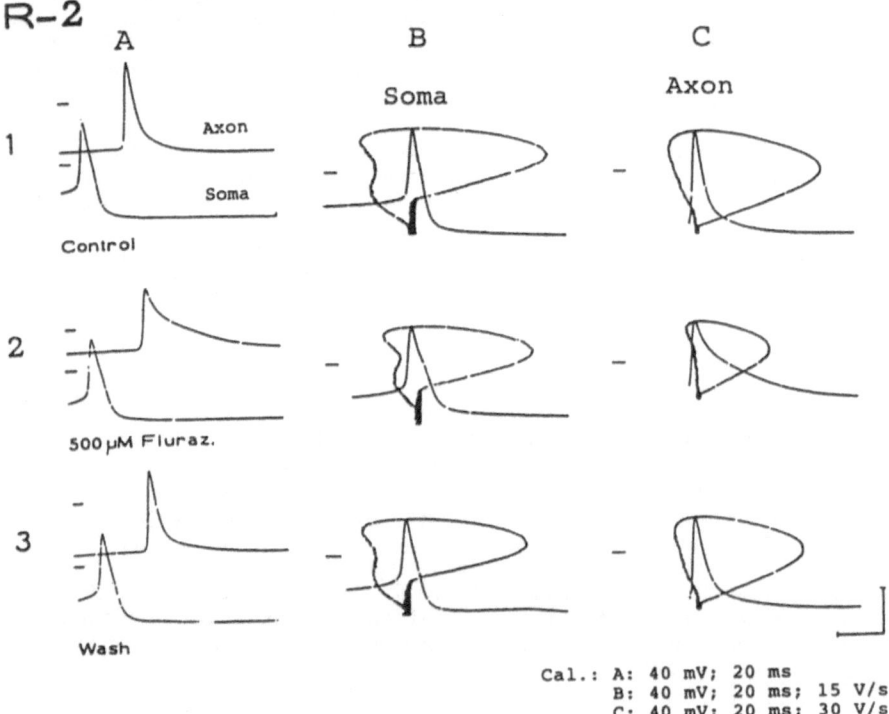

Cal.: A: 40 mV; 20 ms
B: 40 mV; 20 ms; 15 V/s
C: 40 mV; 20 ms; 30 V/s

Fig. 1. A Simultaneous recordings of soma and axon APs. B Soma AP together with its registration in the phase plane mode, i.e. voltage versus dV/dt, the latter positive to the right, negative to the left; the phase plane is drawn counterclockwise, with the rising phase of the AP to the right, and the falling phase to the left. C Axon AP together with phase plane recordings. *Short bar* to the left of each registration indicates 0 mV. Calibration of dV/dt differs in B and C; the slopes of the APs of the axon are much faster than those of the soma (30 versus 15 V/s)

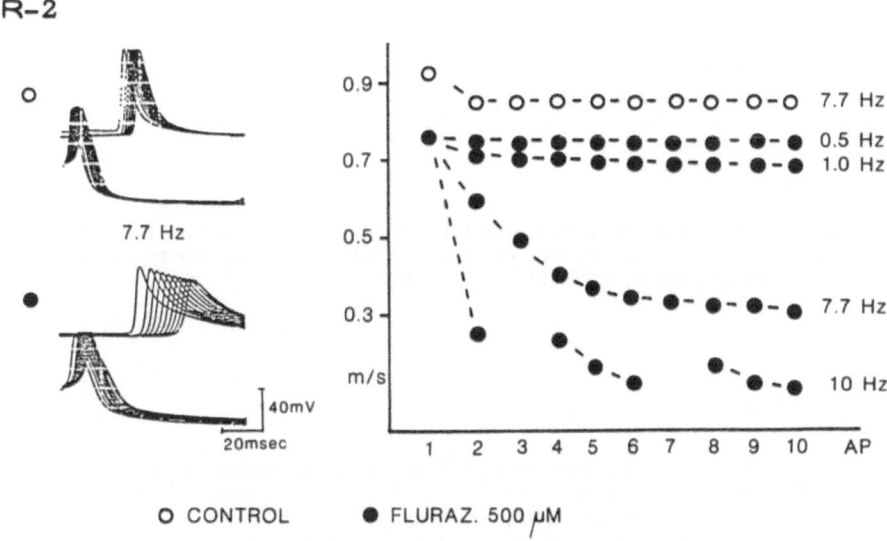

Fig. 2. *Left*, simultaneous recordings of soma and axon APs during repetitive stimulation of the soma at a rate of 7.7 Hz. In control conditions the second to tenth APs of the axon starts from a constant but more hyperpolarized level than the first one. The peak-to-peak interval between soma and axon AP is unchanged for the second to tenth APs. Under the influence of FLUR all axon APs start from the same level, but the peak-to-peak interval increases with each AP. *Right*, graph shows the increase in the peak-to-peak interval expressed as conduction velocity (m/s) for the first to tenth APs during repetitive stimulations at different frequencies

The use-dependent block induced by FLUR differs basically from a reduction of propagation velocity observed under control conditions with repetitive stimulation. As seen in Fig. 2 under control conditions, the first AP of a train is conducted faster than the second to last one. This phenomenon is due to the afterhyperpolarization following the first AP, which is not further increased by the following spikes. The degree of this reduction depends on the interval between the APs, that is the frequency, and occurs when this interval overlaps with the afterhyperpolarization. In R-2, the afterhyperpolarization lasts up to 1 s, so the reduction can be observed at stimulus rates of 1 Hz and above. Up to a frequency of 16.6 Hz, which is equivalent to an interval of 60 ms, and which was the highest frequency tested in this series of experiments, the reduction in propagation velocity increased but was always constant for the second-to-last AP of a train. This phenomenon, observed under control conditions, corresponds to the subnormal phase after an AP, as described by Swadlow and Waxman (1978). In contrast, the BDA-induced reduction of conduction velocity within a train of APs occurs in the absence of afterhyperpolarization and increases for every AP in a train. It occurs already at frequencies of 0.5 Hz and leads to an intermittent propagation at 10 Hz and above.

Discussion

There is a fundamental difference between the two mechanisms which reduce the propagation velocity within a train of APs. The physiological reduction can be explained in terms of changes of the threshold during the afterhyperpolarization, however this interpretation cannot hold for the use-dependent block induced by FLUR due to the lack of an afterhyperpolarization. More likely is that FLUR interacts with ion channels in the activated state, since FLUR is removed from the site of action during an inactive interval. This phenomenon does not seem to occur in *Aplysia* neurons only. A use-dependent block induced by FLUR in guinea pig papillary muscles was described by Liebeswar (1972). Several BDAs limit high-frequency, repetitive firing of action potentials in mouse spinal cord neurons in cell culture (McLean and Macdonald 1988). Since no high-affinity BDA-receptors exist in *Aplysia* neurons, the use-dependent block induced by BDAs might result from a BDA receptor-mediated mechanism of action.

Acknowledgement. I thank Dr. M.R. Klee for discussing the manuscript, Mrs. Duesmann for the layout of the figures, and Mr. N. Steinberg for editing the manuscript. This work was supported by Österreichischer Fonds zur Förderung der wissenschaftlichen Forschung.

References

Gola M, Ducreux C (1982) A reexitation mechanism producing multiple discharges in convulsant-treated molluscan neurons. In: Klee MR, Lux HD, Speckmann EJ (eds) Physiology and pharmacology of epileptogenic phenomena. Raven, New York, pp 289-297
Hoyer J, Park MJ, Klee MR (1978) Changes in ionic currents associated with Flurazepam-induced abnormal discharges in Aplysia neurons In: Chalazonitis N, Boisson M (eds) Abnormal neuronal discharges. Raven, New York, pp 301-310
Liebeswar G (1972) The depressant action of flurazepam on the maximum rate of rise of action potentials recorded from guinea-pig papillary muscles. Naunyn-Schmiedebergers Arch Pharmacol 275: 445-456
McLean MJ, Macdonald RL (1988) Benzodiazepines, but not beta carbolines, limit high frequency repetitive firing of action potentials of spinal cord neurons in cell culture. J Pharmacol Exp Ther 244: 789-795
Moehler H, Okada T (1977) Benzodiazepine receptor: demonstration in the central nervous system. Science 198: 849-851
Nielsen M, Braestrup C, Squires RF (1978) Evidence for a late evolutionary appearance of brain-specific benzodiazepine receptors: an investigation of 18 vertebrate and 5 invertebrate species. Brain Res 141: 342-346
Squires RF, Braestrup C (1977) Benzodiazepine receptors in rat brain. Nature 266: 732-734
Swadlow HA, Waxman SG (1978) Activity-dependent variations in the conduction properties of central axons. In: Waxman SG (ed) Physiology and pathobiology of axons. Raven, New York, pp 191-202
Tauc L (1960) The site of origin of the efferent action potentials in the giant nerve cell of Aplysia. J Physiol 152: 36-37

Effects of Isoflurane and Enflurane on Focal Epileptic Paroxysms

H. Pockberger, K. Lindner and H. Petsche

Institute of Neurophysiology, Univ. Vienna and Brain Research Institute of the ÖAW, Währingerstr. 17, 1090 Vienna, Austria

Introduction

Isoflurane and enflurane are new volatile anesthetics commonly used during surgery. They are chemical isomers which differ only by the position of chlorine and fluorine, respectively. In spite of this close chemical relationship, clinical reports indicate diverse effects with respect to their possible epileptogenic or antiepileptogenic effects.

Isoflurane has been reported to successfully suppress seizures in patients with status epilepticus (Ropper et al. 1986; Kofke et al. 1985). Enflurane, by contrast, might have epileptogenic properties, since myoclonisms and even generalized tonic-clonic convulsions have been observed at various times after enflurane anesthesia (Yamashiro et al. 1985; Grant 1986; Nicoll 1986; Yazji and Seed 1984; Jenkins and Milne 1984).

In experimental studies Komatsu and Ogli (1987) reported the effects of isoflurane, enflurane and halothane anesthesia in mice. All three substances induced severe opisthotonus with EEG spiking activity over the frontal areas; isoflurane was significantly more potent than enflurane or halothane. Oshima et al. (1985), however, tested enflurane in three different epilepsy models in cats and found anticonvulsant properties of this drug rather than an enhancement of seizure activity.

In light of these partly contradictory observations, we decided to test the action of isoflurane and enflurane on penicillin (PNC)-induced paroxysms in the neocortex. This model was chosen since its properties and time course have been studied thoroughly in our laboratory.

Methods

Experiments were performed in 8 rabbits which were initially anesthetized with nembutal (30 mg/kg). After a tracheotomy the animals were placed in a stereotactic frame, relaxed with Alloferin (0.5 mg/h), and artificially ventilated. Anesthesia was continued by either isoflurane or enflurane added by a vaporizer to 100% oxygen.

Field potentials were recorded simultaneously from different lamina of the neocortex with 16-fold microelectrodes (Otto-Sensors), the contacts of which are 150 μm apart. Since each contact was embedded within a small chamber filled with artificial CSF, stable recordings of DC potentials were possible throughout an experiment.

Focal paroxysms were induced by epicortical application of PNC (1000 I.U.) either in the precentral or parietal area of the neocortex.

Results

In all experiments, the concentration of either isoflurane or enflurane was held at 0.4% during the development of interictal spikes (IIS). After epicortical PNC application, IIS undergo a characteristic development, the time course of which has been described elsewhere (Pockberger et al. 1984a). To ensure stable conditions the development of IIS was observed for 45 minutes. Thereafter the concentration of the anesthetic was gradually increased up to 4.5% (measured as end-tidal concentration).

Although the effects of both anesthetics varied considerably between animals, several parameters turned out to be consistent. In general, there was no significant difference between the action of isoflurane and enflurane, inasmuch as we obtained similar results with regard to the shift of DC potential, decrease of IIS frequency, effects on the heart rate, and changes observed with current-source-density (CSD) analysis.

Figure 1 outlines the changes in the DC potential (measured in layer III), in IIS frequency, and in heart rate during increasing concentration of anesthetics in different experiments. Up to a

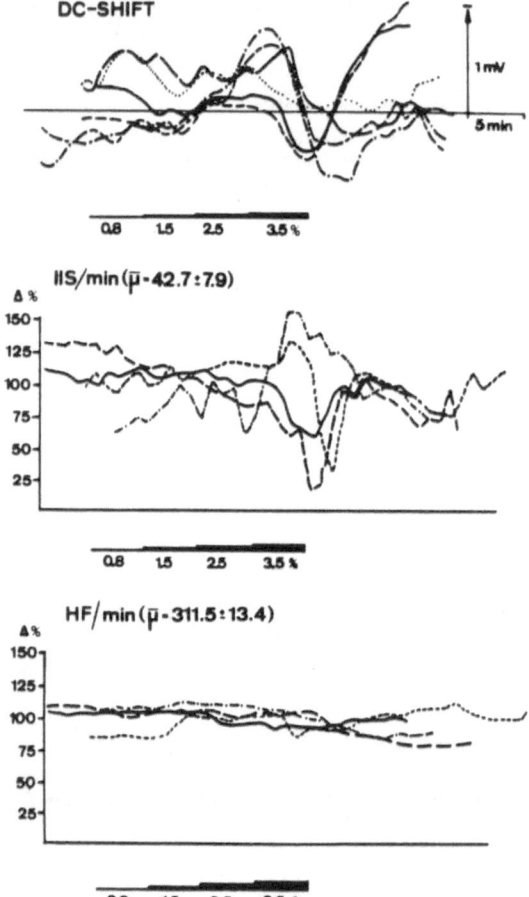

Fig. 1. DC shifts and IIS frequency changes associated with increasing concentration of enflurane and isoflurane (results represent several different experiments). Isoflurane as well as enflurane induced a biphasic DC shift - a negative one in the low concentration range, a positive one with concentrations higher than 2.5%. IIS frequency significantly decreased only with concentrations higher than 2.5%. No significant changes in hear rate were found in the concentration range tested

concentration of 2.5% the DC potentials shifted more or less towards negative. During this phase of the experiment IIS frequency did not change significantly. With concentrations higher than 2.5% the DC potential shifted in the positive directions. This positive DC shift was dependent on the preceding DC level, i.e., starting from a more negative DC level resulted in a greater positive DC shift.

As the concentration was increased to 3.5% IIS frequency initially increased in some experiments and decreased thereafter considerably less than the mean frequency. In other experiments, however, IIS frequency slowly decreased at a concentration of 3.5%. No significant heart rate changes were found with increasing concentration of the anesthetics.

As the concentration of the anesthetic was changed from 3.5% to 0.4%, a prominent negative DC potential shift occurred in all experiments. This negative DC shift was accompanied by an increase of IIS frequency to normal values; however, these was no obvious relation between the amplitude of the DC shift and the increase of IIS frequency.

Current-Source-Density Analysis of IIS

Although the effects of anesthetics on the DC level and the IIS frequency occurred particularly with higher concentrations, CSD analysis revealed significant changes even with low concentrations. As described in previous papers (Pockberger et al. 1984a, b) IIS are generated by a distinct pattern of current source-sink densities. The initial event was described as a sink density in layer IV/Va, which was followed by a more intense and longer lasting, often double-peaked, sink density in layers II/III. With increasing concentration of the anesthetic, sink densities in layer II/III underwent the greatest changes. During the control period the sink density in layer II/III was composed of two components (see Fig. 2). Up to concentration of about 1.5% both components of the sink density in layers II/III decreased in amplitude; the second one decreased on average by more than 50%, however, the first one usually decreased only by 30%. As the concentration was increased to 2.5% the second component was almost completely suppressed, whereas the first increased in amplitude above the level observed during the control situation. A further concentration increase up to 3.5% brought about a decrease of even the first component. At this concentration IIS were significantly smaller in duration than those recorded during lower anesthesia levels. These biphasic changes in sink-source-distribution were observed with both enflurane and isoflurane.

Discussion

Winters (1976) distinguished different anesthetic drugs by their capability inducing different stages of anesthesia. According to this scheme isoflurane belongs to the same group as barbiturates and halothane, whereas enflurane belongs to a different group together with phencyclidine and α-chloralose. Increasing the concentration of isoflurane induces anesthesia stage I, II and IV, whereas enflurane induces only anesthesia stage I and II and, with high concentrations, epileptic seizures.

This general classification scheme which reflects systemic effects of anesthetics is contrasted by studies on the cellular level. A dose-related hyperpolarization and reduced input resistance were found in hippocampal neurons and human cortical neurons with enflurane as well as isoflurane (Nicoll and Madison 1982; Berg-Johnson and Langmoen 1987). The hyperpolarization induced seemed to be due to an increase of a K^+-conductance which was sensitive to intracellular TEA^+ or Ba^{++} (Franks and Lieb 1988). In the hippocampus of rats it was demonstrated that orthodromically evoked potentials were reduced by isoflurane in a dose-dependent manner. This effect was mostly due to a reduction at the postsynaptic side (Berg-Johnson and Langmoen 1986).

In humans Peterson et al. (1986) described a dose-related reduction of amplitude and an increase of latency of somatosensory evoked potentials during surgical anesthesia with halothane, isoflurane and enflurane.

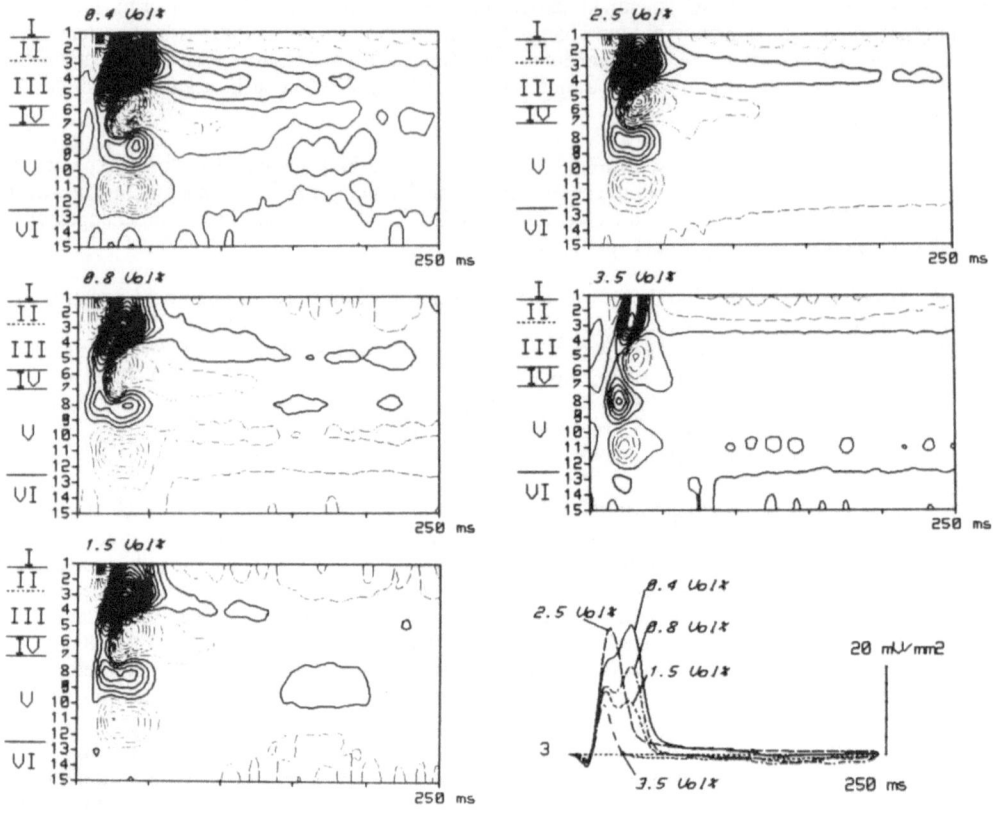

Fig. 2. CSD analysis of averaged IIS recorded under different levels of anesthesia (contour line plots: *dashed lines*, current source densities; *solid lines*, current sink densities; line spacing 1.5 mV/mm². The insert in the lower right corner shows how the different components of the sink densities in layers II/III are effected by the increasing concentrations of anesthetic. Note the increase of the initial component and the simultaneous decrease of the secondary component in the low concentration range. At 3.5% the secondary component is completely suppressed and the initial component is considerably reduced

The DC shifts observed in our studies indicate a biphasic effect of isoflurane and enflurane, in that the DC level became more negative at low concentrations range and turned positive only with higher concentrations. This biphasic shift, however, was not correlated with the excitability of the induced epileptic focus, since IIS frequency remained stable during negative DC shifts and decreased only with positive DC shifts. By contrast we observed considerable changes in the CSD distribution of IIS even with low concentrations of the anesthetic. In the low concentration range (up to 2.5%) we observed a significant reduction of secondary sink-densities in layer II/III, which are thought to be synaptically driven (see Pockberger et al. 1984a, b). Although the secondary components were reduced, the initial components were considerably enhanced. Only in the high concentration range were all components of the IIS were similarly reduced.

In rabbits the minimal alveolar concentration (MAC) was determined as 2.86% for enflurane and 2.15% for isoflurane. This means that drug concentrations lower than 2% induce only stage I or stage II of anesthesia but not surgical anesthesia (stage III).

We therefore conclude that both isoflurane and enflurane reduce mainly synaptically driven epileptic phenomena during stage I and stage II, and thereby reduce the spread of epileptic activity. Under surgical anesthesia (stage III) both drugs seem to interfere with intrinsic membrane mechanisms responsible for the generation of epileptic phenomena.

References

Berg-Johnson J, Langmoen IA (1986) Isoflurane effects in rat hippocampal cortex: a quantitative evaluation of different cellular sites of action. Acta Physiol Scand 128: 613-618

Berg-Johnson J, Langmoen IA (1987) Isoflurane hyperpolarizes neurons in rat and human cerebral cortex. Acta Physiol Scand 130: 679-685

Drummond JC (1985) MAC for halothane, enflurane, and isoflurane in the New Zealand White rabbit: and a test for the validity of MAC determinations. Anesthesiology 62: 336-338

Frank NP, Lieb WR (1988) Volatile general anesthetics activate a novel neuronal K$^+$current. Nature 333: 662-664

Grant IS (1986) Delayed conclusions following enflurane anesthesia. Anesthesia 41: 1024-1025

Jenkins J, Milne AC (1984) Conclusive reaction following enflurane anaesthesia. Anaesthesia 39: 44-45

Kofke WA, Snider MT, Young RS, Ramer JC (1985) Prolonged low flow isoflurane anesthesia for status epilepticus. Anesthesiology 62: 653-656

Komatsu H, Ogli K (1987) Opistotonus during exposure to isoflurane, enflurane, and halothane in mice. Anesthesiology 67: 771-774

Nicoll JM (1986) Status epilepticus following enflurane anesthesia. Anesthesia 41: 927-930

Nicoll RA, Madison DV (1982) General anesthesia hyperpolarize neurons in the vertebrate nervous system. Science 217: 1055-1057

Oshima E, Urabe N, Shingu K, Mori K (1985) Anticonvulsant actions of enflurane on epilepsy models in cats. Anesthesiology 63: 29-40

Peterson DO, Drummond JC, Todd MM (1986) Effects of halothane, enflurane, isoflurane, and nitrous oxide on somatosensory evoked potentials in humans. Anesthesiology 65: 35-40

Pockberger H, Rappelsberger P, Petsche H (1984a) Penicillin-induced epileptic phenomena in the rabbit's neocortex I: the development of interictal spikes after epicortical application of penicillin. Brain Res 309: 247-260

Pockberger H, Rappelsberger P, Petsche H (1984b) Penicillin-induced epileptic phenomena in the rabbit's neocortex II: laminar specific generation of interictal spikes after the application of penicillin to different cortical depths. Brain Res 309: 261-269

Ropper AH, Kofke WA, Bromfield EB, Kennedy SK (1986) Comparison of isoflurane, halothane, and nitrous oxide in status epilepticus. Ann Neurol 19: 98-99

Stevens JE, Fujinaga M, Oshima E, Mori K (1984) The biphasic pattern of the convulsive property of enflurane in cats. Br J Anaesth 56: 395-403

Yamashiro M, Sumimoto M, Furuya H (1985) Paroxysmal electroencephalographic discharges during enflurane anesthesia in patients with a history of cerebral convulsions. Br J Anaesth 57: 1029-1037

Yazji NS, Seed RF (1984) Conclusive reaction following enflurane anesthesia. Anaesthesia 39: 1249

Winters WD (1976) Effects of drugs on the electrical activity of the brain: anesthetics. Ann Rev Pharmacol Toxicol 16: 413-426

Anticonvulsant Actions on Hippocampal Slices

C. Psarropoulou and H. L. Haas

Physiologisches Institut, Johannes Gutenberg - Universität, Saarstrasse 21, 6500 Mainz, FRG

Introduction

As a site of epileptogenesis, the hippocampus has frequently been used to examine the mechanisms of action of anticonvulsant drugs. We have previously investigated their actions on epileptiform activity caused by exposure of hippocampal slices to Penicillin (Oliver et al. 1977; Schwartzkroin and Prince 1978; Rose et al. 1986) and to low Ca^{2+} high Mg^{2+} media (Jefferys and Haas 1982; Taylor and Dudek 1982; Hood et al. 1983; Haas and Jefferys 1984; Olpe et al. 1985; Rose et al. 1986). We have now studied the action of the same prototype antiepileptics (midazolam, carbamazepine, phenytoin, phenobarbital, ethosuximide and valproate) on the multiple discharge which develops upon washing out Mg^{2+} from the bath of hippocampal slices; this discharge is mediated by NMDA-receptor activation (Herron et al. 1985; Psarropoulou and Haas 1989). We report here a lack of specific interaction of the classical antiepileptics indicating that NMDA-receptor antagonism or modulation is not part of their mechanism of action.

Methods

The drugs were tested on 30 hippocampal slices (from 18 Wistar rats) which were kept completely submerged in a recording chamber. Stimulation was delivered to stratum radiatum, and extracellular recordings were made from the CA1 pyramidal layer. The amplitude, latency and number of population spikes in the epileptiform discharges were measured from averaged response (eight sweeps/average). The following drugs were added to the perfusion fluid: D-2-amino-5-phosphonovalerate (APV), phenytoin, phencyclidine (PCP), carbamazepine, phenobarbital, midazolam, ethosuximide, and Na-valproate.

Results

An epileptiform discharge of between five and seven population spikes of declining amplitude developed 15 min after washing out the Mg^{2+} ions. This multiple discharge was suppressed by the NMDA antagonists, PCP (20 µM, Fig. 1B) and APV (10 µM). All anticonvulsants tested diminished this epileptiform discharge in a dose-dependent manner. Characteristic traces representing control (left) and depressed (middle) epileptiform discharge are shown in Fig. 1C.

The EC_{50}s (defined as the effective drug concentration for 50% reduction of the third spike) are used as an estimate of the relative potency of the drugs in this model of epileptic activity [midazolam (0.7 M) > phenytoin (80 µM), carbamazepine (100 µM), phenobarbital (100 µM) > ethosuximide (800 µM), valproate (2 mM)].

These EC_{50}s were compared with the EC_{50}s derived from experiments using penicillin-containing and low Ca^{2+} media and reported in a previous paper (Rose et al. 1986) as well as the therapeutically effective plasma concentrations. The data are shown in Table 1.

252

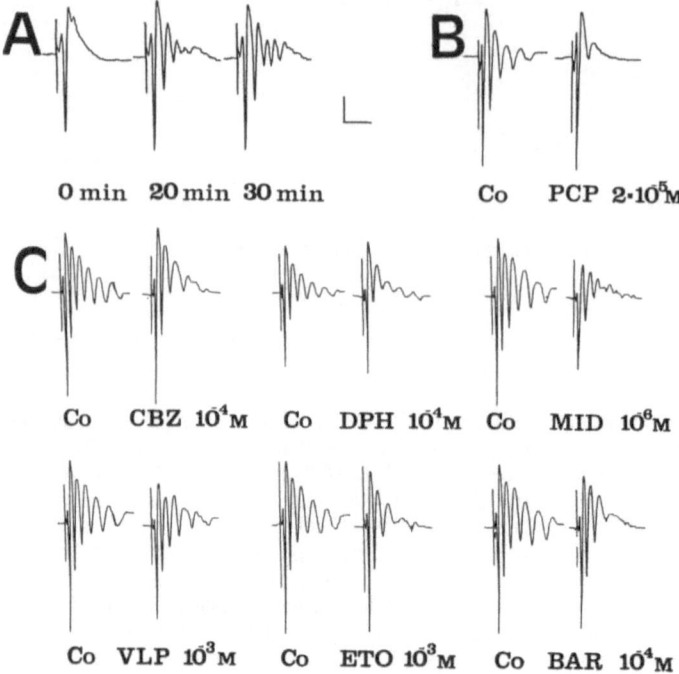

Fig. 1. A Development of epileptiform discharge in CA1 in response to stratum radiatum stimulation upon removal of Mg^{2+} ions. *Left trace* is control in normal medium, *medial* and *right traces* are 20 and 30 min after washing out Mg^{2+} ions. **B** Antagonism by phencyclidine (PCP) 20 μM; *left trace* is control in Mg^{2+} free medium. **C** Effect of anticonvulsants; *left trace* of each set is control in Mg^{2+} free medium. *CBZ*, carbamazepine; *DPH*, diphenylhydantoin; *MID*, midazolam; *VLP*, valproate; *ETO*, ethosuximide; *BAR*, phenobarbital. *Calibration*, 2 mV, 10 ms

Table 1. EC_{50}s of prototype anticonvulsant agents estimated in three different models of epileptiform discharge and effective plasma levels (From Rall and Schleifer 1980, see also Rose et al. 1986)

Drug	EC_{50} (μM) Mg^{2+}-free	Penicillin	Low Ca^{2+}	Effective plasma level
Midazolam	0.7	0.75	100	0.5 - 2
Phenytoin	80	25	25	40 - 75
Carbamazepine	100	15	15	10 - 35
Phenobarbital	100	15	150	40 - 100
Ethosuximide	800	1000	1000	300 - 700
Valproate	>1000	1000	1000	300 - 600

Nonparametric statistics (Wilcoxon signed ranks test for two related samples) revealed a high degree of significance for the differential depressive effect towards the first and the later spikes when all perfusions, at four to five different concentrations, were included (midazolam $p < 0.01$, $n = 7$; phenytoin $p < 0.01$, $n = 17$; carbamazepine $p < 0.01$, $n = 22$; ethosuximide $p < 0.01$, $n = 11$; phenobarbital $p < 0.01$, $n = 14$; Na-valproate $p < 0.05$, $n = 13$).

Discussion

Midazolam, phenytoin, carbamazepine, phenobarbital, ethosuximide, and valproate were effective in depressing the multiple discharge developed in Mg^{2+}-free medium. The late spikes were never completely suppressed even when higher concentrations were employed. A comparison of the $EC_{50}s$ of the drug obtained from different models of epileptiform activity (Rose et al. 1986) suggest that, in Mg^{2+}-free medium, in which excitation is increased, midazolam is as potent as in penicillin containing medium, in which GABAergic inhibition is reduced; it is much less effective in low Ca^{2+} high Mg^{2+} medium in the absence of synaptic transmission. The other agents displayed a decreased (phenytoin, carbamazepine, phenobarbital) or equal (ethosuximide, valproate, phenobarbital) potency compared with the previous experimental paradigm (Table 1).

Phenytoin, carbamazepine, and phenobarbital interact with the adaptation of firing (at higher concentrations) presumably through an effect on sodium channels (Hood et al. 1983; Macdonald et al. 1985) that can also explain their effect on the NMDA-dependent epileptiform discharge. The anticonvulsant action of benzodiazepines and barbiturates is attributed to potentiation of GABAergic transmission (Haefely et al. 1983). This mechanism is also likely to contribute to the preferential reduction of the later (third) spikes, as a strengthened inhibitory postsynaptic potential reduces the excitatory potential.

Hippocampal slices are very useful for studying the mechanism of action of antiepileptic agents but not for the primary screening of new drugs.

References

Haas HL, Jefferys JGR (1984) Low calcium field burst discharges of CA1 pyramidal neurones in rat hippocampal slices. J Physiol 354: 185-201
Haefely W, Polc P, Pieri L, Schaffner R, Laurent JP (1983) Neuropharmacology of benzodiazepines: synaptic mechanisms and neural basis of action. In Costa E (ed) The benzodiazepines: from molecular biology to clinical practice. Raven Press, New York pp 21-66
Herron CE, Williamson R, Collingridge GL (1985) A selective N-methyl-D-aspartate antagonist depresses epileptiform activity in rat hippocampal slices. Neurosci Lett 61: 255-260
Hood TW, Siegfried J, Haas HL (1983) Analysis of carbamazepine actions in hippocampal slices of the rat. Cell Mol Neurobiol 3: 213-220
Jefferys JGR, Haa HL (1982) Synchronized bursting of CA1 hippocampal cells in the absence of synaptic transmission. Nature 300: 448-450
Macdonald RL, McLean MJ, Skerrit JH (1985) Anticonvulsant drug mechanisms of action. Fed Proc 44: 2634-2639
Oliver AP, Hoffer BJ, Wyatt RJ (1977) The hippocampal slice: a system for studying the pharmacology of seizures and for screening anticonvulsant drugs. Epilepsia 18: 543-548
Olpe HR, Baudry M, Jones RSG (1985) Electrophysiological; and neurochemical investigations on the action of carbamazepine on the rat hippocampus. Europ J Pharmacol 110: 71-80
Psarropoulou C, Haas HL (1989) Action of anticonvulsants on hippocampal slices in Mg^{2+} free medium. Naunyn Schmiedebergs Arch Pharmacol 339: 613-616
Rall TV, Schleifer LS (1980) Drugs effective in the therapy of the epilepsies. In: Goodman Gilman A, Goodman LS, Gilman A (eds) The pharmacological basis of therapeutics. MacMillan, New York, pp 448-474
Rose GM, Olpe HR, Haas HL (1986) Testing of prototype antiepileptics in hippocampal slices. Naunyn Schmiedebergs Arch Pharmacol 332: 89-92
Schwarzkroin PA, Prince DA (1978) Cellular and field potential properties of epileptogenic hippocampal slices. Brain Res 147: 117-130
Taylor CP, Dudek FE (1982) Synchronous neural afterdischarges in rat hippocampal slices without active chemical synapses. Science 218: 810-812

Is There a Glycinergic Component in the Mechanisms of Action of Common Antiepileptic Drugs?

S. J. Czuczwar, D. Wlodarczyk, M. Gasior and Z. Kleinrok

Department of Pharmacology, Medical School, 20-090 Lublin, Jaczewskiego 8, Poland

Introduction

A substantial body of evidence exists on the involvement of GABA-mediated inhibition in the mechanism of action of various antiepileptic drugs, such as diazepam, diphenylhydantoin, phenobarbital, and valproate (Haefely 1980; Löscher 1981; Meldrum 1984); however this may not be the case for valproate (Keane et al. 1985). In contrast to the numerous data on GABA-mediated inhibition, the participation of glycine-mediated events in the mechanism of action of antiepileptic drugs has not been extensively studied. There is also much less evidence for the protective activity of glycine and its esters against convulsive events. Nevertheless, glycine (or its esters) were found effective against audiogenic-, 3-mercaptopropionic acid-, or thiosemicarbazide-induced seizures (Laird and Huxtable 1976; Sarhan et al. 1984; Toth et al. 1983).

Glycine fulfills the criteria for an inhibitory transmitter, causing postsynaptic hyperpolarization (Curtis and Watkins 1960; Wermann ct al. 1968) related to increased chloride ion conductance (Alger 1985; Barker 1985). Thus, it was of interest to study the effect of the glycine antagonist, strychnine, on the protective efficacy of common antiepileptic drugs against maximal electroshock-induced convulsion in mice. Any interaction of strychnine with the anticonvulsive potential of the antiepileptics would suggest the involvement of glycine receptor-mediated events in their mechanism of action.

Material and Methods

Experiments were performed on male Swiss mice weighing 22 - 30 g. The animals were kept in colony cages under standard laboratory conditions with free access to food (chow pellets) and tap water. The alternating current (50 Hz) was delivered through corneal electrodes and the stimulus duration was 0.2 s. The convulsive threshold was evaluated as CS_{50} which is the current strength (in mA) necessary to produce tonic hind limb extension in 50% of the animals tested. The efficacy of antiepileptic drugs was reflected by their ED_{50} values (in mg/kg) against maximal electroshock (50 mA). The convulsive test was carried out between 10.00 and 13.00 h.

The following antiepileptic drugs were used: carbamazepine (Amizepin), diphenylhydantoin (Phenytoinum), phenobarbital sodium (Luminalum); all three drugs from Polfa, Warsaw, Poland, and valproate magnesium (Dipromal; Polfa; Rzeszow, Poland). Valproate and phenobarbital were solubilized in saline; the remaining antiepileptics were suspended in a 3% solution of Tween 81. Strychnine nitrate (Sigma) was dissolved in saline. The appropriate doses and treatment times are shown in Table 1. All drugs were given i.p. in a volume of 5 ml/kg body weight. The doses refer to the free drug forms.

To determine the plasma levels of antiepileptic drugs the animals were with the antiepileptic alone or in combination with strychnine (0.2 mg/kg) and killed subsequently by decapitation at appropriate times. Blood samples of approximately 1 ml were centrifuged in eppendorf tubes for 3 min and plasma samples of 70 µl were transferred to Abbott system cartridges. Plasma levels

Experimental Brain Research Series 20
© Springer-Verlag Berlin · Heidelberg 1991

256

of antiepileptic drugs were estimated with the help of the Abbott TDx analyzer using immunofluorescence. The respective drug controls were placed at the beginning and end of each carousel of experimental samples.

Both CS_{50} and ED_{50} values were calculated according to Litchfield and Wilcoxon (1949). The graphical method was modified in that a computer construction of the dose-effect relationship was performed. Statistical evaluation of the CS_{50} and ED_{50} data was carried out according to Litchfield and Wilcoxon (1949). The plasma levels of antiepileptic drugs were expressed as means (in μg/ml) ± S.D. of at least seven determinations and compared by the Student t test.

Table 1. Effect of strychnine nitrate upon the protective efficacy of carbamazepine (CBZ), diphenylhydantoin (DPH), phenobarbital (PhB), and valproate (VPA) against maximal electroshock-induced seizures in mice

Treatment	Strychnine (mg/kg)			
	0	0.05	0.1	0.2
CBZ	15 (14 - 16)	NT	NT	16 (15 - 17)
DPH	11 (10 - 13)	NT	11 (10 - 13)	15 (13 - 18)[*]
PhB	19 (18 - 20)	20 (18 - 22)	27 (24 - 30)[**]	33 (28 - 39)[***]
VPA	210 (196 - 225)	210 (183 - 242)	280 (257 - 305)[**]	>400 [***]

[*]$p < 0.05$; [**]$p < 0.01$; [***]$p < 0.001$ vs respective control group. NT, not tested

Table data are ED_{50} (in mg/kg) with 95% confidence limits in parentheses. Doses of strychnine, PhB, and VPA refer to the free drug. All drug were given i.p.: PhB and DPH 120 min, CBZ 60 min, VPA 30 min, and strychnine 15 min prior to electroconvulsions. Calculation of ED_{50} values and statistical comparisons were carried out according to Litchfield and Wilcoxon (1949).

Results

Strychnine (0.4 mg/kg) lowered the threshold for electroconvulsions from 7.9 to 6.4 mA (p < 0.05) and was ineffective in the lower dose ranges. No seizure activity after strychnine was observed. Strychnine (0.1 and 0.2 mg/kg) strongly reduced the anticonvulsive potency of valproate and phenobarbital. It should be considered that this glycine antagonist (0.2 mg/kg) increased ED_{50} value for phenobarbital by 83% and the value for valproate was elevated above 400 mg/kg (Table 1). By contrast, strychnine (0.2 mg/kg) moderately impaired the protective potential of diphenylhydantoin and remained completely ineffective against carbamazepine. Generally, strychnine (0.1 - 0.2 mg/kg) seemed to reverse the sedation and myorelaxation induced by all antiepileptic drugs.

In no case did strychnine (0.2 mg/kg) affect the plasma levels of antiepileptic drugs. The control levels were as follows: carbamazepine (15 mg/kg; 60 min before sampling), 4.1 ± 1.7; diphenylhydantoin (18 mg/kg; 120 min), 8.7 ± 1.9; phenobarbital (25 mg/kg; 120 min), 26.9 ± 1.2; and valproate (250 mg/kg; 30 min), 288 ± 16 μg/ml; and after combined treatment with strychnine (0.2 mg/kg; 15 min): 4.4 ± 0.7, 8.0 ± 1.6, 27.2 ± 1.6, and 294 ± 20 g/ml, respectively.

Discussion

The obtained results clearly indicate that the protective activities of valproate and phenobarbital were reduced to a considerable degree by strychnine. The potency of diphenylhydantoin was only moderately (but significantly) impaired, whereas the efficacy of carbamazepine remained unchanged after strychnine administration. It is remarkable that no pharmacokinetics interaction was noted. Up to 0.2 mg/kg strychnine did not induce any convulsive activity per se and the CD_{50} value for the induction of tonic seizures was found to be 0.76 mg/kg (Kleinrok et al. 1980). A possibility that strychnine may reduce the protective potential of some antiepileptic drugs due to its general stimulant effects seem unlikely for two reasons: first, bicuculline, in the relatively high dose of 2 mg/kg was unable to affect the anticonvulsive action of a variety of antiepileptic drugs (Czuczwar et al. 1986) and second, strychnine did not affect the activity of carbamazepine.

An intriguing possibility is that the anticonvulsive effects of valproate, phenobarbital, and, to a lesser degree, diphenylhydantoin might be dependent upon glycine-mediated inhibition. A relatively high density of strychnine-sensitive glycine receptors was found in substantia nigra (Snead 1983), the brain region presumed to play a crucial role in the propagation of seizure activity (Iadarola and Gale 1982). There is another population of glycine receptors which is strychnine insensitive and probably allosterically linked to receptors for N-methyl-D-aspartate (NMDA). Moreover, glycine was shown to potentiate the response to NMDA in cultured mouse brain neurons (Johnson and Ascher 1987). Another important finding is that glycine, in fact, potentiates strychnine-induced convulsions which is subsequently reversed by NMDA antagonist (Larson and Beitz 1988). This may lead to the conclusion that the potentiating response could be mediated by strychnine-insensitive glycine receptors. Thus, it should be taken into consideration that the hypothesized anticonvulsant drug (except carbamazepine)-induced enhancement of glycinergic transmission may, in the presence of strychnine, mediate events related to excitatory strychnine-insensitive glycine receptors.

References

Alger BE (1985) GABA and glycine: postsynaptic actions. In: Roganowski MA, Barker JL (eds) Neurotransmitter actions in the vertebrate nervous system. Plenum, New York, pp 33-69
Barker JL (1985) GABA and glycine: ion channel mechanism. In: Roganowski MA, Barker JL (eds) Neurotransmitter actions in the vertebrate nervous system. Plenum, New York, pp 71-100
Curtis DR, Watkins JC (1960) The excitation and depression of spinal neurons by structurally related amino acids. J Neurochem 6: 117-141
Czuczwar SJ, Ikonomidou C, Kleinrok Z, Turski L, Turski WA (1986) Effect of aminophylline on the protective action of common antiepileptic drugs against electroconvulsions in mice. Epilepsia 27: 204-208
Haefely WE (1980) GABA and the anticonvulsat action of benzodiazepines and barbiturates. Brain Res Bull 5 [Suppl 2]: 873-878
Iadarola MJ, Gale K (1982) Substantia nigra: site of anticonvulsant activity mediated by τ-aminobutyric acid. Science 218: 1237-1240
Johnson JW, Ascher P (1987) Glycine potentiates the NMDA response in cultured mouse brain neurons. Nature 325: 529-531
Keane PE, Morre M (1985) Effect of valproate on brain GABA: comparison with various medium chain fatty acids. Pharm Res Commun 17: 547-555
Kleinrok Z, Czuczwar SJ, Turski L, Zarkowski A (1980) Effect of intracerebroventricular injection of kainic acid on electrically and chemically-induced convulsions in mice. Pol J Pharmacol Pharm 32: 265-269
Laird EH, Huxtable R (1976) Effect of taurine on audiogenic seizure response in rat. In: Huxtable R, Barbeau A (eds) Taurine. Raven, New York, pp 267-274
Larson AA, Beitz AJ (1988) Glycine potentiates strychnine-induced convulsions: role of NMDA receptors. J Neurosci 8: 3822-3826
Litchfield JT, Wilcoxon F (1949) A simplified method of evaluating dose-effect experiments. J Pharmacol Exp Ther 96: 99-113
Löscher W (1981) Valproate-induced changes in GABA metabolism at the subcellular level. Biochem Pharmacol 30: 1364-1366
Meldrum BS (1984) Amino acids neurotransmitters and new approaches to anticonvulsant drug action. Epilepsia 25 [Suppl 2]: S140-S149
Sarhan S, Kolb M, Seiler N (1984) The amplification of the anticonvulsant effect of vinyl GABA (4-aminohexenoic acid) by esters of glycine. Arzneimittelforschung 34: 687-690
Snead OC (1983) On the sacred disease: the neurochemistry of epilepsy. Int Rev Neurobiol 24: 93-180
Toth E, Lajtha A, Sarhan S, Seiler N (1983) Anticonvulsant effects of some inhibitory neurotransmitter amino acids. Neurochem Res 8: 291-302
Werman R, Davidoff RA, Aprison MH (1968) Inhibitory actions of glycine on spinal neurons in the cat. J Neurophysiol 31: 81-95

Developmental Defects in El Mouse Neurons and Their Normalization by a New Antiepileptic Drug, TJ-960

E. Sugaya[1], A. Sugaya[2], K. Kajiwara[1], T. Tsuda[2], H. Takagi[1], K. Yasuda[2] and T. Takagi[1]

[1] Department of Physiology, Kanagawa Dental College, Yokosuka, Japan
[2] Faculty of Pharmaceutical Sciences, Josai University, Saitama, Japan

The El Mouse

The El mouse is an epilepsy animal model which was discovered by Imaizumi and registered internationally in 1964 (Imaizumi and Nakano). When the El mouse is trained by tossing it approximately 10 cm high 20 - 30 times once a week from 4 weeks after birth until 6 weeks of age, convulsions are induced. When small doses of pentylenetetrazol (PTZ, 18 mg/kg), which never induce convulsions in normal mice, were injected into El mice, convulsions occurred without fail (Fig. 1) (Sugaya et al. 1987). These PTZ-induced convulsions allow more precise evaluation of the effects of various anticonvulsant drugs than the tossing-up procedures (Sugaya et al. 1987).

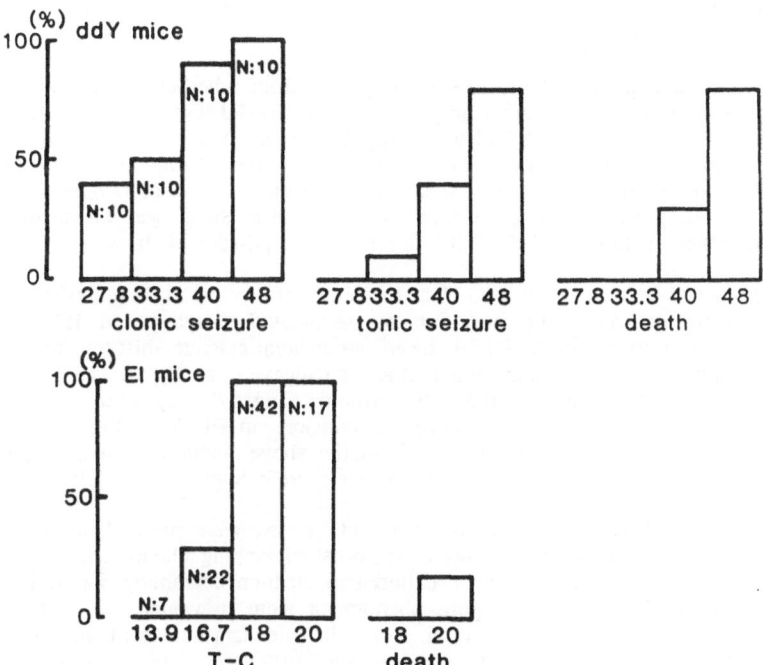

Fig. 1. Dose-dependent seizure manifestation with PTZ in ddY mice *(upper panels)* and El mice *(lower panels)*. *N*, number of experiments. Dose of PTZ are noted below each column in mg/kg (From Sugaya et al. 1986).

Characteristics of El Mouse Neurons and Their Developmental Defects

The characteristics of the El mouse have been investigated mainly from the viewpoint of EEG (Suzuki and Nakamoto 1977) and neurochemistry (Mori et al. 1986), but they have remained obscure. We examined the characteristics of primary cultured cerebral cortical neurons from El mouse embryos. Cultured neurons were prepared by previously described methods (Sugaya et al. 1987). The primary cultured cerebral cortical neurons from 18-day-old ddY mouse embryos underwent cell division in 0.5% of neurons until the third culture day; El mouse neurons never divided. On the seventh culture day, ddY mouse neurons showed long neurite extensions and a well-balanced neuron network, whereas those of El mouse neurons formed bundle-type neurite extensions suggestive of developmental defects in neurite extension (Sugaya et al. 1987).

Gangliosides, especially GD_3 ganglioside, are considered to be important for neurite extension. The total ganglioside content of the El mouse neurons was much lower than that of ddY mouse neurons. When GD_3 gangliosides were immunocytochemically stained with AbR_{24}, a monoclonal antibody, about 40% of the ddY neurons were positive; however, El mouse neurons, GD_3 positive staining was rarely observed in 7-day-old cultured cerebral cortical neurons (Sugaya et al. 1987). Quantitative measurement of gangliosides from 3-day-old cultured cells, which consisted mostly of neurons with a few astrocytes, using high performance thin-layer chromatography showed that the amount of GD_3 in El mouse cultured cells was far less (ca. 1/4) than that in ddY mice (Sugaya et al. 1987).

The above-mentioned results suggest that El mouse neurons have developmental defects, especially in neurite extension, in addition to a high susceptibility to convulsions.

Normalizing Effects on Convulsions and Developmental Defects in El Mouse Neurons by a New Antiepileptic Drug, TJ-960

A new antiepileptic drug, TJ-960, originates from a herbal medicine, Saiko-keishi-to (SK), which is a mixture of nine herbal drugs (Sugaya 1987). TJ-960 is a commercial test formulation of SK which consists of the root of *Bupleuri falcatum*, the bark of *Cinnamomum cassia*, the root of *Panax ginseng*, the root of *Glycyrrhiza uralensis*, the root of *Paeonia lactiflora*, the root of *Scuterallia baicalensis*, the rhizome of *Pinellia ternata*, the rhizome of *Gingiber officinale*, and the fruit of *Zizyphus jujuva*. The original form of this herbal medicine was a decoction of the above-mentioned nine herbal drugs, and TJ-960 is a spray-dried powder of the water extract.

Either TJ-960 or SK showed clear inhibitory effects on PTZ-induced bursting activity in the PTZ-sensitive neurons of the snail, *Euhadra peliomphala* (Sugaya et al. 1985). TJ-960 also showed clear inhibitory effects on the PTZ-induced intracellular calcium shift toward the cell membrane area and PTZ-induced intracellular protein changes (Sugaya et al. 1985). TJ-960 showed complete inhibitory effects on PTZ-induced intracellular protein changes (Sugaya et al. 1985). TJ-960 showed complete inhibitory effects on PTZ-induced convulsions in El mice (Fig. 2) (Sugaya et al. 1988). The inhibitory effects of TJ-960 or SK on the above-mentioned seizure-related phenomena were the same as those of phenytoin (Sugaya et al. 1985; Sugaya et al. 1985).

When cerebral cortical neurons of the ddY mouse were cultured in medium containing 5 µg/ml of cytochalasin B, neurites showed the so-called looping phenomena and growth cones were never observed. When neurons were cultured in medium containing 5 µg/ml of cytochalasin B and 75 µg/ml of TJ-960, the looping phenomena were prevented and normal neurite extension was observed (Fig. 1 of Sugaya et al. 1985). In cytochalasin B containing medium, most neurons died within 10 days, but in medium containing TJ-960 and cytochalasin B, neurite extension continued and neurons survived even on the 10th culture day (Fig. 2 of Sugaya et al. 1985).

Cytochalasin B greatly decreased the amount of total gangliosides and GM_1 ganglioside disappeared; however, with the addition of TJ-960 to the culture medium, GM_1 ganglioside reappeared (Fig. 3 of Sugaya et al. 1987). Thus, TJ-960 showed protective effects on cytochalasin B-induced distortion of neurites and ganglioside disappearance (Sugaya et al. 1987).

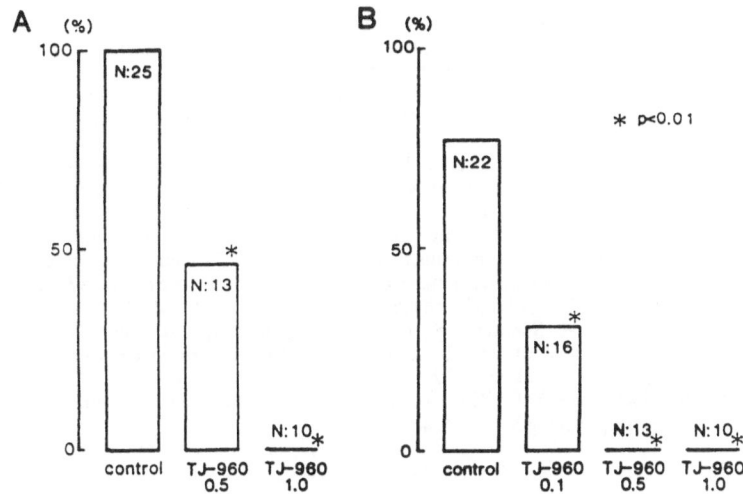

Fig. 2. Anticonvulsant effects of TJ-960 on PTZ-induced convulsions in El mice. **A** 8-week-old El mice. **B** 4-week-old El mice. Doses of TJ-960 are noted below each column in g/kg. Rates of evoked convulsions are expressed by percent values with the numbers of experiments (N) within each column. Asterisks indicate statistically significant values by Student's *t* test (From Sugaya et al. 1988).

El mouse neurons rarely survive more than 10 days in culture even in normal medium. By contrast, with TJ-960 containing medium, El mouse cultured neurons survived and neurite extension continued. Ganglioside content also increased in TJ-960 containing medium.

Type 1 astrocytes from El mice changed into type 2 astrocytes in TJ-960 containing medium within 48 h. In addition, El mouse neurons cultured on TJ-960 treated astrocytes showed enhanced proliferation, about twice that of neurons cultured on the astrocytes without TJ-960 treatment.

Thus, TJ-960 showed protective effects against cytochalasin B-induced neuron damage and promoting effects on El mouse neurite extension, in addition to anticonvulsant action. Such normalizing effects on developmental defects and protective effects on neuron damage are never found with pure, chemically derived, established anticonvulsant agents. From these points of view, TJ-960 is very special and promising drug in the therapy of intractable epilepsy.

References

Imaizumi K, Nakano T (1964) Mutant stocks, Strain El Mouse News Lett 31: 57
Mori A, Hiramatsu M, Kabuto H, Marescau D (1986) Effect of emotional stress on El mouse convulsions and their biochemical background. Neurochem Res 11: 37-45
Sugaya A, Tsuda T, Sugaya E, Takato M, Takamura K (1978) Effect of Chinese herbal medicine 'Saiko-keishi-to' on the abnormal bursting activity of snail neurons. Planta Med 34: 294-298
Sugaya A, Tsuda T, Yasuda K, Sugaya E, Onozuka M (1985) Effect of Chinese herbal medicine 'Saiko-keishi-to' on intracellular calcium and protein behavior during pentylenetetrazole-induced bursting activity in snail neurons. Planta Med 51: 1-6
Sugaya A, Yuzurihara M, Tsuda T, Yasuda K, Sugaya E, Kajiwara K, Takagi T, Komatsubara J (1987) Normalizing effect of Saiko-keishi-to commercial formula on cytochalasin-B distorted neurites using primary cultured neurons of rat cerebral cortex. J Ethnopharmacol 21: 193-199

Sugaya E, Onozuka M, Furuichi H, Kishii K, Imai S, Sugaya A (1985) Effect of phenytoin on intracellular calcium and intracellular protein changes during pentylenetetrazole-induced bursting activity in snail neurons. Brain Res 327: 161-168

Sugaya E, Kishii K, Onozuka M (1985) Inhibitory effect of phenytoin on intracellular cyclic nucleotide and calcium changes during pentylenetetrazole-induced bursting activity in snail neurons. Brain Res 341: 313-319

Sugaya E, Ishige A, Sekiguchi K, Iizuka S, Ito K< Sugimoto A, Aburada M, Hosoya E (1986) Pentylenetetrazol-induced convulsion and effect of anticonvulsant in mutant inbred strain El mice. Epilepsia 27: 354-358

Sugaya E, Asou H, Itoh K, Ishige A, Sekiguchi K, Iizuka S, Sugimoto A, Aburada M, Hosoya E, Takagi T, Kajiwara K, Komatsubara J, Hirano S (1987) Characteristics of primary cultured neurons from embryonic mutant El mouse cerebral cortex. Brain Res 406: 270-274

Sugaya E (1987) SK (TJ-960). Drugs of the Future 12: 360-363

Sugaya E, Ishige A, Sekiguchi K, Iizuka S, Sugimoto A, Yuzurihara M, Hosoya E (1988) Inhibitory effect of a mixture of herbal drugs (TJ-960, SK) on pentylenetetrazol-induced convulsions in El mice. Epilepsy Res 2: 337-339

Suzuki J, Nakamoto Y (1977) Seizure patterns and electro-encephalograms of El mouse. Electroenceph Clin Neurophysiol 43: 299-311

SUBJECT INDEX